ASCE Manuals and Reports on Engineering Practice No. 111

Reliability-Based Design of Utility Pole Structures

Prepared by
Reliability-Based Design Committee of the
Structural Engineering Institute (SEI) of the
American Society of Civil Engineers

Edited by
Habib J. Dagher

ASCE Published by the America.
Society of Civil Engineers

Structural Engineering Institute
of the American Society of Civil Engineers

Library of Congress Cataloging-in-Publication Data

Reliability-based design of utility pole structures : prepared by Reliability-Based Design Committee of the Structural Engineering Institute (SEI) of the American Society of Civil Engineers / edited by Habib J. Dagher.
p. cm. — (ASCE manuals and reports on engineering practice ; no. 111)
ISBN 0-7844-0845-9
1. Structural engineering—Handbooks, manuals, etc. I. Dagher, Habib Joseph. II. Structural Engineering Institute. Reliability-Based Design Committee. III. Series.

TA635.R45 2006
624.1'772—dc22 2005037138

Published by American Society of Civil Engineers
1801 Alexander Bell Drive
Reston, Virginia 20191
www.pubs.asce.org

ISBN 0–7844–0845–9
Manufactured in the United States of America.

MANUALS AND REPORTS ON ENGINEERING PRACTICE

(As developed by the ASCE Technical Procedures Committee, July 1930, and revised March 1935, February 1962, and April 1982.)

A manual or report in this series consists of an orderly presentation of facts on a particular subject, supplemented by an analysis of limitations and applications of these facts. It contains information useful to the average engineer in his everyday work, rather than the findings that may be useful only occasionally or rarely. It is not in any sense a "standard," however; nor is it so elementary or so conclusive as to provide a "rule of thumb" for nonengineers.

Furthermore, material in this series, in distinction from a paper (which expressed only one person's observations or opinions), is the work of a committee or group selected to assemble and express information on a specific topic. As often as practicable the committee is under the direction of one or more of the Technical Divisions and Councils, and the product evolved has been subjected to review by the Executive Committee of the Division or Council. As a step in the process of this review, proposed manuscripts are often brought before the members of the Technical Divisions and Councils for comment, which may serve as the basis for improvement. When published, each work shows the names of the committees by which it was compiled and indicates clearly the several processes through which it has passed in review, in order that its merit may be definitely understood.

In February 1962 (and revised in April 1982) the Board of Direction voted to establish:

A series entitled "Manuals and Reports on Engineering Practice," to include the Manuals published and authorized to date, future Manuals of Professional Practice, and Reports on Engineering Practice. All such Manual or Report material of the Society would have been refereed in a manner approved by the Board Committee on Publications and would be bound, with applicable discussion, in books similar to past Manuals. Numbering would be consecutive and would be a continuation of present Manual numbers. In some cases of reports of joint committees, bypassing of Journal publications may be authorized.

MANUALS AND REPORTS ON ENGINEERING PRACTICE

No.	Title
13	Filtering Materials for Sewage Treatment Plants
14	Accommodation of Utility Plant Within the Rights-of-Way of Urban Streets and Highways
35	A List of Translations of Foreign Literature on Hydraulics
40	Ground Water Management
41	Plastic Design in Steel: A Guide and Commentary
45	Consulting Engineering: A Guide for the Engagement of Engineering Services
46	Pipeline Route Selection for Rural and Cross-Country Pipelines
47	Selected Abstracts on Structural Applications of Plastics
49	Urban Planning Guide
50	Planning and Design Guidelines for Small Craft Harbors
51	Survey of Current Structural Research
52	Guide for the Design of Steel Transmission Towers
53	Criteria for Maintenance of Multilane Highways
54	Sedimentation Engineering
55	Guide to Employment Conditions for Civil Engineers
57	Management, Operation and Maintenance of Irrigation and Drainage Systems
59	Computer Pricing Practices
60	Gravity Sanitary Sewer Design and Construction
62	Existing Sewer Evaluation and Rehabilitation
63	Structural Plastics Design Manual
64	Manual on Engineering Surveying
65	Construction Cost Control
66	Structural Plastics Selection Manual
67	Wind Tunnel Studies of Buildings and Structures
68	Aeration: A Wastewater Treatment Process
69	Sulfide in Wastewater Collection and Treatment Systems
70	Evapotranspiration and Irrigation Water Requirements
71	Agricultural Salinity Assessment and Management
72	Design of Steel Transmission Pole Structures
73	Quality in the Constructed Project: A Guide for Owners, Designers, and Constructors
74	Guidelines for Electrical Transmission Line Structural Loading
76	Design of Municipal Wastewater Treatment Plants
77	Design and Construction of Urban Stormwater Management Systems
78	Structural Fire Protection
79	Steel Penstocks
80	Ship Channel Design
81	Guidelines for Cloud Seeding to Augment Precipitation
82	Odor Control in Wastewater Treatment Plants
83	Environmental Site Investigation
84	Mechanical Connections in Wood Structures
85	Quality of Ground Water
86	Operation and Maintenance of Ground Water Facilities
87	Urban Runoff Quality Manual
88	Management of Water Treatment Plant Residuals
89	Pipeline Crossings
90	Guide to Structural Optimization
91	Design of Guyed Electrical Transmission Structures
92	Manhole Inspection and Rehabilitation
93	Crane Safety on Construction Sites
94	Inland Navigation: Locks, Dams, and Channels
95	Urban Subsurface Drainage
96	Guide to Improved Earthquake Performance of Electric Power Systems
97	Hydraulic Modeling: Concepts and Practice
98	Conveyance of Residuals from Water and Wastewater Treatment
99	Environmental Site Characterization and Remediation Design Guidance
100	Groundwater Contamination by Organic Pollutants: Analysis and Remediation
101	Underwater Investigations
102	Design Guide for FRP Composite Connections
103	Guide to Hiring and Retaining Great Civil Engineers
104	Recommended Practice for Fiber-Reinforced Polymer Products for Overhead Utility Line Structures
105	Animal Waste Containment in Lagoons
106	Horizontal Auger Boring Projects
107	Ship Channel Design (Second Edition)
108	Pipeline Design for Installation by Horizontal Directional Drilling
109	Biologic Nutrient Removal (BNR) Operation in Wastewater Treatment Plants
110	Sedimentation Engineering: Theory, Measurements, Modeling, and Practice
111	Reliability-Based Design of Utility Pole Structures

CONTENTS

ACKNOWLEDGMENTS

The Committee acknowledges the technical support and valuable input received from many individuals including the following: Richard Aichinger, Elias Ghannoum, "Tip" Goodwin, Dave Gromala, Dr. Kathleen Jones, Dr. Leon Kempner, Robert Kluge, Jason Rollins, Jim Rossberg, Dr. Lawrence Slavin, and Dan Snyder.

The Committee also wishes to thank Ms. Doreen Parent at the University of Maine for providing exceptional logistical and clerical support. The Committee also thanks Florida Power and Light, the Salt River Project, and the University of Maine for hosting many of the Committee's meetings. The Committee thanks Osmose, Inc., for hosting many of the Committee's web meetings, and Madelon Wise at the U.S. Department of Agriculture (USDA) Forest Products Laboratory and Roberta Laverty at the University of Maine for their editing assistance.

ASCE/SEI POLE RBD COMMITTEE MEMBERS

Nelson G. Bingel III, Osmose, Inc.
Gary E. Bowles, Electrical Consultants, Inc.
Dr. Habib J. Dagher (Committee Chair), University of Maine
Dr. James W. Davidson, Shakespeare Composites and Electronics
Dr. Fouad Fouad, University of Alabama, Birmingham, Ala.
Dr. Magdi Ishac, Hydro One Network Services
Brian Lacoursiere, International Utility Structures, Inc.
Paul M. Legrand II, Entergy Corp.
Wesley J. Oliphant, Newmark International, Inc.
Ronald E. Randle, EDM International, Inc.
Martin Rollins, H. M. Rollins Company, Inc.
Camille G. Rubeiz, American Iron & Steel Institute
Larry D. Vandergriend, Hughes Brothers, Inc.
Michael Voda, Salt River Project
David West, Duke Energy
Ronald Wolfe, U.S. Forest Products Laboratory
Dr. Jerry Wong, Florida Power & Light
Alec Zolotoochin, BC Hydro

FIGURES

TABLES

Chapter 1
INTRODUCTION

1.1 CURRENT PRACTICE

There is a need to provide a design methodology for distribution and transmission poles that yields consistent structural reliabilities across all material types: wood, steel, concrete, fiberglass, and other new materials. Whereas this document is currently focused on poles, future editions will have an expanded scope to cover reliability-based design (RBD) for all utility structures and components.

Prevailing U.S. practice and most state laws require that transmission and distribution lines be designed to meet the basic safety requirements of whichever edition of the *National Electrical Safety Code (NESC)* is adopted by each state. The current *NESC* rules (IEEE 2002) for the selection of design load and strength factors are largely based on successful experience but they do not have a strong theoretical foundation. Some designers find the rules too restrictive, whereas others adopt more conservative criteria. In fact, individual utilities often develop their own loading criteria to supplement the *NESC* rules.

A desire to achieve more consistent structural reliabilities across materials was the impetus for the development of the American Society of Civil Engineers (ASCE) *Manual 74,*"Guidelines for transmission line structural loading" (ASCE 1991) and the International Electrotechnical Commission (IEC), "Design criteria of overhead transmission lines," *IEC 60826* (IEC 2002). Although *ASCE Manual 74* offers consistent methods to calculate loads, there is still a need to provide consistent methods to calculate strength across pole materials.

The nominal strength (R_n) of a pole is calculated using a variety of approaches and guides, depending on the pole material type. This includes, for example, *ASCE Manual 72* for steel poles (ASCE 1990), *ANSI O5.1-2002* for wood poles (ANSI 2002), and the Precast/Prestressed Concrete Institute (PCI), "Specification guide for prestressed concrete poles" (PCI 1999). *ASCE Manual 104,* "Recommended practice for fiber-reinforced polymer products for overhead utility line structures," has recently been developed for fiber-reinforced polymer (FRP) poles (ASCE 2003).

In addition to these guides, pole manufacturers typically use in-house models for predicting the nominal strength of prestressed concrete poles and FRP poles. These strength guides, standards, and in-house methods have essentially evolved independently so there is little consistency in the definition of nominal strength across pole materials. For some pole materials and design methods, the nominal strength value is close to a mean strength, whereas for other pole materials the nominal strength represents a more conservative value, such as a 5% or 10% lower exclusion limit (LEL). As a result, a wide range of possibilities currently exists concerning what the nominal strength of a pole really represents. This results in inconsistent and unknown reliability levels for different pole materials.

To illustrate the problem, refer to Fig. 1-1. Any strength property of a pole, such as the lateral strength of a Class 1 wood pole in a cantilever test, is a random variable (testing 100 identical poles in the same manner will result in 100 different values of the strength). The many sources of uncertainty in observed pole strength include inherent material property variability, geometric variability, manufacturing effects, and variation in the testing method. The strength of a pole may thus be characterized by a probability density function (PDF) with a mean value (m) and a coefficient of variation (COV), as shown in Fig. 1-1. The COV (the standard deviation divided by the mean) varies with the pole material and pole type. The COV is higher, for example, for wood poles than for steel poles. The higher COV causes the PDF for wood poles to be wider and flatter than for steel poles. The procedure used to calculate the nominal or characteristic strength of the pole (whether per *ASCE Manual 72* [ASCE 1990] or *ANSI O5.1-2002* [ANSI 2002] yields a nominal strength R_n that falls somewhere on the horizontal axis of the PDF in Fig. 1-1.

Figure 1-1 shows three possible positions of the nominal strength: $R_n = R_1$, R_5, and R_{50}, representing a 1%, 5%, and 50% LEL, respectively. For example, a 5% LEL nominal strength (a 5th percentile) is a value not achieved by 5% of the poles. For a normal density function, the 5th percentile is 1.645 standard deviations below the mean:

$$R_5 = m - 1.645\ (m \times \mathrm{COV}). \qquad \text{(Eq. 1-1)}$$

A 50% LEL or 50th percentile nominal strength is a value not reached by 50% of the poles. For a normal density function, R_{50} is also the mean strength. It follows that $R_{50} > R_5 > R_1$ and the distance between these values grows with increasing COVs. Clearly, unless strength design methods for different pole materials yield consistent nominal value definitions (e.g., all materials at the 5% LEL), it will be difficult to achieve consistent reliability among material types.

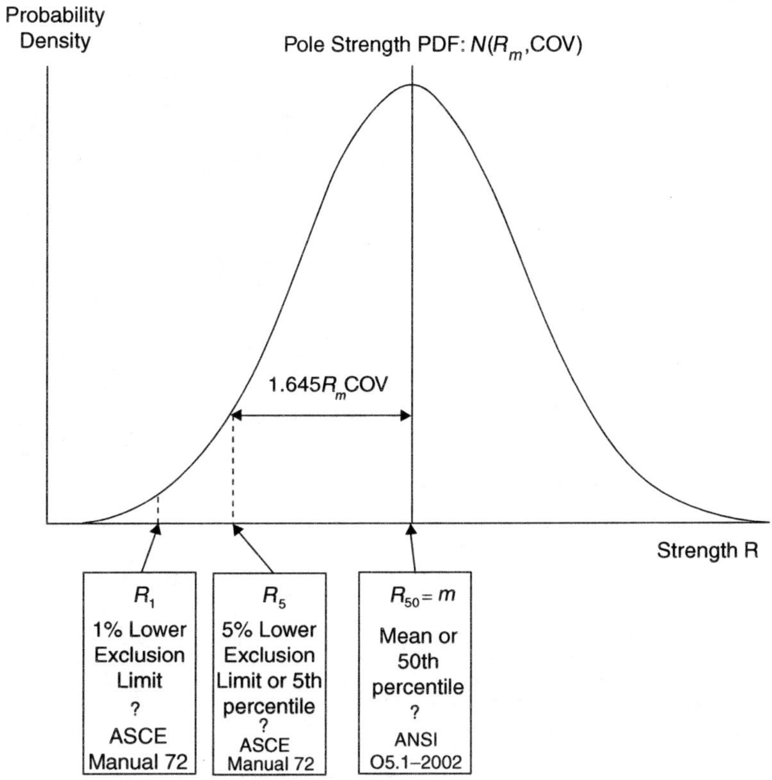

FIGURE 1-1. Probability Density Function (PDF) (Normal Distribution).

1.2 RELIABILITY-BASED DESIGN

Failure of a component or a structure occurs when the load (Q) exceeds the resistance (R). If the PDF of the strength and the load are known, then the probability of failure of a component may be estimated using the overlap region under the two curves (Fig. 1-2). Minimizing the overlap between the two curves is commonly achieved by moving the curves apart by moving the strength PDF to the right (e.g., by applying safety factors or partial safety factors to the mean or nominal design values). The distance between the means of the two curves ($M_R - M_Q$), measured in number of standard deviations of ($R - Q$), is referred to as the reliability index β (Fig. 1-2).

$$\beta = (M_R - M_Q) \;/\; (\sigma_R^2 + \sigma_Q^2)^{0.5}. \qquad \text{(Eq. 1-2)}$$

The calculation of reliability index β in Eq. 1-2 and Fig. 1-2 is valid only when both R and Q are normally distributed and uncorrelated.

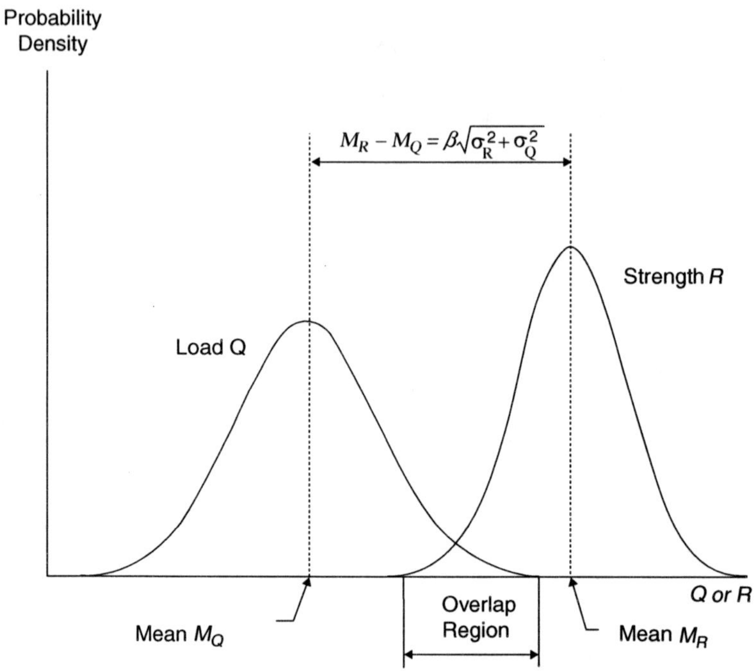

FIGURE 1-2. Failure Occurs in Overlap Region Where the Load Q > Strength R.

In this case, the reliability index β and the probability of failure are related as follows:

$$\text{Probability of Failure} = 1 - \phi(\beta), \qquad \text{(Eq. 1-3)}$$

in which Φ is the cumulative density function (CDF) of the normal distribution. Larger reliability indices β mean more distance between the two curves in Fig. 1-2 and a smaller probability of failure. Components with equivalent reliability indices β have relatively equivalent probabilities of failure.

In RBD, load and strength factors (also called partial safety factors) are selected so that components have relatively equivalent reliability indices β (Ellingwood et al. 1980). Target reliability indices for conventional designs are typically selected by reliability calibration. The RBD methods for transmission lines are described in "Design criteria of overhead transmission lines," *IEC 60826* (IEC 2002), "Reliability-based design of transmission line structures" (Peyrot and Dagher 1984), and "Reliability-based design of foundations for transmission line structures" (EPRI 1995). These documents provide additional references and details for conducting structural reliability evaluations, including reliability index calculations, reliability calibrations of new RBD methods, and load and resistance factor design (LRFD) methodologies.

1.3 OBJECTIVE

The objective of this manual is to outline a methodology to achieve relatively consistent structural reliability across all pole materials. In future editions, this document will be expanded to cover RBD for other types of utility structures, components, and related load considerations.

1.4 SCOPE

The structural reliability of poles designed according to the 2002 *NESC* (IEEE 2002) provisions and existing strength design guides such as *ASCE Manual 72* (ASCE 1990) or *ANSI O5.1-2002* (ANSI 2002) varies depending on the material type. Reasons for the discrepancy relate to the historical development of the deterministic design approach used in the *NESC*, as well as inconsistencies in the nominal strength definitions in the various material-specific strength design guides.

This manual provides a design methodology for achieving relatively consistent structural reliability among poles of different materials. Specifically, this manual provides the following:

1. Target reliability levels for poles of *NESC* Grades B and C construction, applicable to all material types.
2. A consistent definition of nominal strength that is applicable to all pole materials, including the following:
   a. Consistent test methods, sample sizes, and data analysis and selection procedures for establishing the statistical properties of pole strength.
   b. Incentives (reduction in required load/strength factors and pole cost) for manufacturers to reduce variability and better define pole statistical properties.
3. Default statistical strength data (COV, % LEL of nominal strength) for the RBD of poles of all materials.
4. An RBD methodology that includes the following:
   a. General LRFD equations.
   b. Load factors applicable to all material types and material-dependent strength reduction factors.
   c. Characteristic or design load values obtained using the procedures in *ASCE Manual 74* (ASCE 1991).

d. A set of load and strength factors for the LRFD equations that achieve reliability levels corresponding to either *NESC* Grade B or Grade C construction, including the following:
   i. A table to adjust strength factors depending on the LEL of nominal strength and COV of pole strength (see Table 2-2).
   ii. A discussion of how to change reliability levels.

1.5 BENEFITS

This manual provides a design methodology for distribution and transmission pole structures that yields relatively consistent reliabilities across all materials. This manual also addresses the following:

1. Defines minimum reliability levels for both *NESC* Grades B and C construction on the basis of reliability analyses of existing *NESC* pole designs. Therefore, designers can still produce designs to *NESC* Grades B and C with more consistent results.
2. Provides a means for quantifiably adjusting reliability whenever needed or justified. An essential line should be more reliable than a less important line.
3. Achieves relatively uniform structural reliability across all pole materials, thereby allowing utilities to compare the cost of equivalent lines made of different materials.
4. Facilitates innovation by allowing the introduction of new materials into pole design.
5. Encourages manufacturers to continually improve their products by providing design incentives for more reliable poles and better databases for pole strength. A manufacturer that develops better statistical data on pole strength is allowed to adjust the strength factors accordingly.
6. Provides uniform procedures for defining the nominal strength of transmission and distribution structures to be used in conjunction with the available strength guides (such as *ASCE Manual 72* [ASCE 1990] and *ANSI O5.1-2002* [ANSI 2002]). This manual also suggests that all future editions of the strength guides provide nominal or characteristic strength values at or near the 5% LEL. Table 2-2 is provided to be used with existing strength guides to adjust load and strength factors depending on the LEL of nominal strength and COV of pole strength.
7. Complements *ASCE Manual 74* (ASCE 1991). This manual does not provide new methods for calculating loads and load combinations, instead referring to the *ASCE Manual 74* procedures for computing design loads and load factors that are independent of the materials

of the supporting structures. It is consistent with the *ASCE Manual 74* approach that a loading agenda should reflect uncertainties in the loads and the accepted risk that these loads may be exceeded.

8. Brings pole structural design in line with well-established RBD codes such as the American Association of State Highway and Transportation Officials (AASHTO) "Standard specifications for highway bridges" (AASHTO 2002), the American Institute of Steel Construction (AISC) "Load and resistance factor design specification for structural steel buildings" (AISC 1999), and the American Forest and Paper Association "Load and resistance factor design (LRFD) manual for engineered wood construction" (AF&PA 1996).

1.6 RELATION TO *NATIONAL ELECTRICAL SAFETY CODE* AND OTHER ASCE GUIDES

The *NESC* (IEEE 2002) rules contain basic provisions considered necessary for the safety of employees and the public under specified, deterministic load conditions. The *NESC* rules are not intended to be used as a design specification. This manual, which provides a design methodology, should be used in conjunction with any *NESC* safety requirements.

This manual provides reliability-based loads and strength factors that may be used with the *ASCE Manual 74* (ASCE 1991) load conditions.

Chapter 2

RELIABILITY-BASED DESIGN METHODOLOGY

2.1 INTRODUCTION

This section describes a reliability-based design (RBD) methodology for transmission and distribution (T/D) line structures. This RBD approach is calibrated to yield average reliability levels consistent with years of history and practice with existing deterministic design approaches. Therefore, on average, designs will be nearly equivalent to past practice.

This methodology strives to correct the problem of inconsistent reliabilities among different material types. Minimum reliability levels are recommended for different grades of construction. Simply selecting different return periods in designing the T/D line quantifiably varies reliability levels. The method uses one set of load and load factors regardless of the material type. The strength reduction factors are a function of material type; the equations are based on initial strength before material degradation occurs.

2.2 STRUCTURAL SYSTEMS

A T/D line consists of two separate structural systems: the structural support system (towers, poles, and foundations) and the wire system (including insulators and hardware). Each system should be considered separately, even though it is evident they are closely related to each other.

The structural support system has the task of supporting the weight of the wire system and accumulated ice and of resisting the winds acting on both systems. The wire system consists of the conductors and shield wires, which may be severely tensioned by ice loading or extreme wind loading. The wire system includes all components such as dead-end insulators and hardware in series with the wires and all major angle and dead-end structures that are critically affected by the wire tensions. Major loads of a transmission line are generated on or by the wire system on the structures themselves, except for high-intensity-type winds such as tornados. Focus should be on the wire system to understand what happens to transmission or distribution lines, especially when wires are loaded with ice. Whereas the support system can support very heavy vertical loads at relatively low cost, the strongest vertical capacity support system may be overstressed when unusual or unexpected loads, such as wind-blown or falling debris (branches), affect the wire system. A simple break in the wire system may promote a cascading failure that is difficult to control.

2.3 DESIGN OF WIRE SYSTEM

For the reasons discussed in the previous chapter and others, the components of the wire system should be selected conservatively, with loadings limited to damage levels and set well below the rated strengths of the components. The strength levels of the structures and foundations of the support system can be adjusted relative to each other. Generally, it is desirable to ensure that the foundations are more reliable than the supported structures. Some utilities, however, elect to have foundations for directly embedded structures fail (i.e., deflect excessively) prior to structure failure (i.e., buckling or rupture), since a foundation's deflection can be repaired quite easily compared to replacing a complete structure. In some cases the deflection can reduce the loads on the structure. The strength levels of important angle and dead-end structures can also be adjusted upward from those of tangent structures to make them more reliable.

2.4 TYPES OF LOAD-PRODUCING EVENTS AND RETURN PERIOD

When describing loads in a T/D line system, it is convenient to distinguish between the events that produce the loads and the resulting loads in the components of the subsystems. The loads are direct forces on the conductors, ground wires, or structures. The events producing the loads can be classified as weather-related, accidental, and construction and maintenance (C&M) events.

2.4.1 Weather-Related Events

Current design procedures, such as those found in the *National Electrical Safety Code* (*NESC*) criteria (IEEE 2002) for extreme wind, incorporate some probability information about weather-related events. This may be done intuitively through experience or by direct use of the return period (RP) concept, whenever data are available. Design nominal values are normally specified in a code or loading agenda in such a manner that their risk of being exceeded is relatively low. A nominal value often used is that value of the variable that has the probability 1/RP of being exceeded in one year. Such a value is said to have an RP-year return period. A nominal value for a weather-related event or a load with an RP-year return period is indicated herein by the subscript RP. For example, a 50-year RP wind velocity, V_{50}, has a probability of 0.02 (2%) of being exceeded in any one year.

Since a direct relationship exists between weather-related events and corresponding load effects, the load effect from a weather-related event with an RP-year return period also has an RP-year return period. In this manual, design equations that constitute the *reliability* requirement use weather-related events and corresponding loads with a specified RP as the nominal events and loads.

The probability that an event or load with an RP-year return period will be exceeded at least once during the planned lifetime of a line (for example, 50 years) is given in Table 3-1 in Chapter 3. This probability is a useful indicator but it does not correspond to the probability of failure of the line or to that of any of its components. More information on the type of calculations yielding the results of Table 3-1 can be found in the commentary of *ASCE Standard 7-02* (ASCE 2002).

2.4.2 Accidental Events

Some of the events that produce loads in a transmission line system cannot be described statistically because of their nature or lack of data. Accidental events such as breaks of components from defects, wear, fatigue, or impact or failures of entire structures from landslides, tornadoes, sabotage, or any other unforeseen phenomena fall in that category. Design procedures do not control the occurrence of these events but attempt to minimize their consequences. Because of this, the designer must make sure that if a failure is triggered by an accidental or weather-related event, it will not propagate without control. This *security* requirement can be accomplished by designing for special loadings (longitudinal or torsional loads) at all or some structures by load-limiting devices such as mechanical fuses or by inserting stop structures at specified intervals in the line.

2.4.3 Construction and Maintenance Events

Some line components may be subjected to their critical loading during C&M operations. Once the magnitudes of the loads produced by the operations are established, they should be multiplied by a load factor to provide an adequate level of safety. Design for these factored loads, in addition to those specified by national regulations or codes of practice, constitute the *safety* requirement.

2.5 LIMIT STATE DESIGN

A limit state defines acceptable and unacceptable structural behavior evaluated in design. Limit states can be normally classified into the three categories of serviceability, damage, and failure (ultimate strength) limit states. For T/D structures, serviceability limit states include vibration, clearances, deflections, and alignment. Damage and failure limit states include any type of partial or complete failure. Designers have always checked limit states in design by comparing load effects with corresponding limits. Therefore, limit state design does not represent a new design method but, rather, is a collective term to describe all the limits checked in design.

In this document, design equations that deal with ice, wind, and temperature are intended to control the probability of occurrence of damage limit states; the equations are provided to prevent permanent damage of components from weather-related events. The design equations that deal with security-related loads address failure limit states; the concern is not with the damage of one or two structures because the line has already failed but, rather, with the possibility of cascading-type failures.

2.5.1 Loads and Load Effects

Loads on a T/D line, denoted herein as Q, are forces applied on the wires or forces applied directly to the supporting structures. Forces, displacements, and stresses caused by the loads in the various components of the subsystems are called load effects. A load effect herein will be denoted "Effect of [ ]," where the terms inside the brackets are the loads (Table 2-1). Loads are further discussed in Chapter 3.

2.5.2 Component Strength

A component is normally selected in such a way that its strength multiplied by a strength reduction factor exceeds the corresponding design loads multiplied by load factors. The actual strength of a given component is a random variable. In current design applications, the strength of a component is identified by a unique value called its nominal strength. The nominal strength of a component, R_n, is normally calculated with equations described in the appropriate design document (AISC 1999; ANSI 2002; ASCE 1992; PCI 1999). (*ASCE Standard 10-97* [ASCE 2000] was previously American Society of Civil Engineers [ASCE] *Manual 72*, [ASCE 1990].)

The nominal strength may also be provided by a manufacturer in the form of a minimum or guaranteed strength or as a percentage of an

TABLE 2-1. Load Conditions That May Be Considered in Design

Equation	Load Case Description
2-1a and 2-1b Weather Loads	Extreme wind from any direction Extreme ice with reduced wind (combined ice and wind) Unbalanced ice without wind (where applicable) Substantial wind on reduced ice (where applicable)
2-2 Security Loads	Failure containment criterion or loading (for example, broken conductor load)
2-3 Construction and Maintenance Loads	Structure erection loads Stringing loads Worker load (250 lb)
2-4 Legislated Loads	Legislated loads (National Electrical Safety Code)

estimated breaking load. For example, the nominal strength in compression of a given type of steel angle in a lattice tower is given by a compression formula in American National Standards Institute (ANSI)/ASCE *Standard 10-90* (ASCE 1992). A nominal value, R_e, is said to be the e% exclusion limit of strength if it has an e% probability of not being achieved. Chapter 1 described where the value R_e is located relative to the probability density function (PDF) of the strength. Usually, the nominal values are set in a way that most components tested at loads equal to the nominal strength would survive the test. The percentage of components not surviving is equal to the exclusion limit.

Unfortunately, the exclusion limits of R_n for different types of components are not the same and are often unknown. The exclusion limit for the design strengths of many metallic components of lattice or tubular structures are in the range of 1% to 10%, with corresponding coefficients of variation (COV_R) in the range of 5% to 20%. The strength values specified in *ANSI Standard O5.1-2002* (ANSI 2002) for wood poles have exclusion limits that are significantly higher than 5% (Bodig et al. 1986). Design equations for the strength of foundations may have exclusion limits as high as 50%. This manual recommends that, in the future, published nominal values for the strengths of all transmission line components be established at the lower exclusion limit (LEL) of 5%. In any case, if statistical data are available on the strength of a particular component, the strength value's exclusion limit can be determined as shown in Chapter 4.

2.5.3 Load and Resistance Factor Design Format

2.5.3.1 Load and Resistance Factor Design Equations. Load and resistance factor design (LRFD) describes one way of assessing behavior at various limit states. The following set of LRFD design equations is recommended for the design of components in a transmission line:

Weather-Related Loads (Reliability-Based). Equation 2-la (or 2-1b) is the design equation that controls reliability for weather-related events. The limit state considered is damage of a component caused by extreme wind or combinations of ice and wind.

$$\phi R_n > \text{effect of } [1.1 \text{ DL and } \gamma Q_{50}]. \qquad \text{(Eq. 2-1a)}$$

The 1.1 load factor for self-weight is less than the 1.2 factor used in *ASCE 7* (ASCE 2002b), recognizing that self-weight is more predictable in utility structures than it is in building structures.

Or,

$$\phi R_n > \text{effect of } [1.1 \text{ DL and } Q_{RP}], \qquad \text{(Eq. 2-1b)}$$

where

ϕ = strength factor (Table 2-2)
γ = load factor for wind (Table 3-2 in Chapter 3) or ice thickness factor for combined ice and wind (Table 3-3 in Chapter 3)
R_n = nominal strength of the component
DL = dead loads (not including weight of ice)
Q_{50} = wind or combined ice and wind loads based on a 50-year RP
Q_{RP} = loads similar to Q_{50}, based on RP.

If RP >50 years, the structural reliability is increased. For example, Q_{100} wind loads are approximately 15% larger than Q_{50} wind loads. (Refer to Chapter 3, Section 3.4 for more details.)

Security Loads. Equation 2-2 provides for the security of the line. Ideally, the limit state considered in Eq. 2-2 should be an ultimate or failure limit state. The purpose of the equation is not to prevent localized damage but, rather, to prevent failure propagation. Assume, for simplicity, that damage and ultimate limit states are identical. With that conservative assumption, the same R_n can be used in all of the design equations. The conservative use of a damage limit with respect to security loads may impose a cost penalty on the utility.

$$\phi R_n > \text{effect of } [\text{DL and SL}], \qquad \text{(Eq. 2-2)}$$

TABLE 2-2. Strength Factor ϕ[a]

Nominal Strength Lower Exclusion Limit, e (%)	Strength Factor, ϕ for COV_R =		
	0.05	0.10	0.20
0.1	1.00	1.16	1.48
1	0.97	1.07	1.27
2	0.95	1.04	1.21
5	0.93	1.00	1.12
10	0.92	0.96	1.04
20	0.90	0.92	0.95
50	0.86	0.85	0.81
Mean	0.86	0.85	0.79

[a] Assumes log-normal strength property.

where

ϕ = strength factor
R_n = nominal strength of the component
DL = dead loads
SL = security loads.

Construction and Maintenance Loads (Safety). Equation 2-3 considers the damage limit state of a component from C&M loads.

$$\phi R_n > \text{effect of } [\gamma_{CM}(\text{DL and C\&M})], \qquad \text{(Eq. 2-3)}$$

where

ϕ = strength factor
R_n = nominal strength of the component
γ_{CM} = load factor applied to the C&M load
DL = dead loads
C&M = loads produced by construction and maintenance operations.

Legislated Loads. Equation 2-4 is listed to emphasize that requirements from governing codes should always be considered.

$$\varphi_{LL} R_n > \text{effect of } [\text{LL}], \qquad \text{(Eq. 2-4)}$$

where

φ_{LL} = legislated load strength factor
R_n = nominal strength of the component
LL = legislated loads.

The load factor γ in Eq. 2-la (or the RP in Eq. 2-1b) allows a designer to modify the reliability of a line. The absence of load factor in Eq. 2-2 emphasizes that the design loads providing the security level of a line cannot be described statistically. However, simply increasing the nominal loads can increase the security of a line. Suggested minimum security loads (SLs) are presented in Chapter 3; also, refer to the *National Electrical Safety Code* (IEEE 2002) and the working draft of *ASCE Manual 74*.

The strength factor φ in Eqs. 2-1a through 2-3, obtained from Table 2-2, accounts for the nonuniformity of the exclusion limits in published information and formulas for nominal strength R_n for different materials. This factor also accounts for differences in strength coefficients of variation, COV_R. If all strength design guides published nominal strength values at a 5% lower exclusion limit (5% LEL), Table 2-2 could be effectively reduced to one row.

The strength and load factors to use with Eq. 2-la (or strength factor and RP to use with Eq. 2-1b) could be developed by a number of different techniques. They could be chosen by consensus to represent current or projected practice. Preferably, they can be selected to control the reliability (or probability of failure) of the components of the line. That selection process is commonly referred to as reliability-based LRFD. Section 2.6 will describe a method for implementing RBD for T/D line structures.

2.5.3.2 Load Conditions. Equations 2-1a through 2-3 are generic equations to satisfy the basic requirements of reliability, security, and safety. Equation 2-4 represents design requirements for legislated loads. In practice, several load cases are considered in each of the categories of loads covered by the equations. Table 2-1 describes load cases normally considered in design.

Design loads on supporting structures are generally obtained by applying the selected load conditions to assumed maximum vertical span, wind span, and line angle. However, actual spans, line angles, and combinations of loads less than the maximum loads may be critical in some cases. The designer should be aware of conditions where the lesser values result in higher stresses in some components of the structures.

2.6 RELIABILITY-BASED DESIGN

The variability of loads and strengths can be formally considered in design through applications of probability theory. A probability-based design procedure is one that considers the probability of occurrence of a given limit state over a fixed period of time, usually one year or the planned lifetime of the system. The procedure should address two essential points: (1) how the probability of occurrence of the limit state is to be estimated, and (2) how small it should be.

The ultimate goal of RBD is to control the reliability of the line system. Line reliability is directly affected by the reliability of each component in each subsystem. Because accurate control of a line system's reliability is beyond the current state of the art, the approach adopted in this manual is to control the relative failure probability (RFP) of different lines or of different structures within a line. Line RFP is approximately increased or decreased by a given factor by simply increasing or decreasing the RP of the load used in designing the line. For example, traditional *NESC* Grade B or C construction may be achieved by changing the load factor γ, as shown in Table 2-3.

2.6.1 Target Reliability Levels and Corresponding Load Factors

Different load factors are used to achieve relative target probabilities of failure for different types of construction. A larger design load factor results in a more reliable line. Table 2-3 provides the minimum design load factors for *NESC* Grades B and C construction with transverse wind force applied to the ice-covered conductors, as appropriate. Section 3.4 in Chapter 3 provides further information on load factors and their effects. The load factor γ is applied to the wind force and to the ice thickness (Table 2-3).

The RBD method for the load factors in Table 2-3 uses nominal climatic design loads based on a 50-year RP (Q_{50} in Eq. 2-1a). Historical *NESC* Grade C

TABLE 2-3. Minimum Load Factors for Transverse Wind Force and Ice Thickness Corresponding to *NESC* Grades B and C Construction

NESC Grade of Construction	Minimum Load Factors, γ Applied to 50-Year Events (Eqs. 2-1a and 2-1b)	
	Extreme Wind	Combined Ice and Wind
Grade B	Wind Force: 1.0	Wind Force: 1.0 Ice Thickness: 1.0[b]
Grade C	Wind Force: 0.5[a]	Wind Force: 1.0 Ice Thickness: 0.5[b]

[a] If any portion of the structure or its supported facilities exceeds 60 ft above ground or water level, a load factor of 1.0 should be used.

[b] The load factor for the ice thickness is applied to the thickness of ice on the conductor, structure, or component prior to calculating the associated load (such as weight or transverse wind force). The wind force load factor is then applied to the calculated wind load on the ice-covered component. The ice weight load factor, if any, is applied to the resulting weight of ice, including the ice thickness load factor.

construction-factored loads are less than 50-year RP values (i.e., *NESC* Grade C construction is designed for less than a 50-year RP load). This results in RBD load factors less than 1.0 for Grade C construction in Table 2-3.

The load factors in Table 2-3 result from the author's extensive effort to calibrate reliability to historic *NESC* designs. As part of the effort, three types of tangent poles were designed to full utilization at 40 U.S. locations for both *NESC* Grade B and Grade C construction (240 poles). A fully utilized design has a nominal design load effect equivalent to 100% of the nominal design strength. The reliabilities of these tangent poles designed according to *NESC* criteria were evaluated using the best available probabilistic wind and wind-on-ice models at each of these 40 locations. The reliability of *NESC* designs varied with location and pole height, even within the same loading district. Since this calibration study was restricted to tangent structures, line tension effects were not considered at this time. Future editions of this manual will consider angle and dead-end structures.

Using this information, the following target reliability levels for tangent structures were selected to provide equivalence to *NESC* construction in the majority of the United States: reliability index $\beta = 2.0$ for Grade B construction and $\beta = 1.5$ for Grade C construction (see Chapter 1, Section 1.2 for definition of β). These are annual β values obtained through Monte Carlo simulations of poles designed using *NESC* at 40 locations in the United States. The strength was assumed log-normal and the statistical models for wind and for wind-on-ice were the ones used in developing *ASCE 7* (ASCE 2002b).

The target reliability levels selected here will serve to maintain more consistent reliability while minimizing changes in current pole design. Departures from *NESC* traditional designs will result at locations where the wind maps or combined ice and wind maps used in *ASCE Manual 74* (or *ASCE 7*) (ASCE 1991; ASCE draft; ASCE 2002) are significantly different from the *NESC* district map (light, medium, and heavy). Departures from *NESC* designs will also occur for poles close to but less than 60 ft high that have not traditionally been designed for extreme wind.

Techniques for calculating the loads Q_{50}, Q_{RP}, SL, and C&M in Eqs. 2-1a through 2-3 are described in Chapter 3 of this manual as well as in *ASCE Manual 74* (ASCE 1991; ASCE draft). The load factor γ in Eq. 2-la (or the RP in Eq. 2-lb) that corresponds to a given RFP can be obtained from Tables 3-2 and 3-3. The RFP is defined as follows:

$$\text{RFP} = \frac{\begin{array}{c}\text{Probability of Failure of a Component or Structure}\\ \text{Designed to a Load Return Period RP}\end{array}}{\begin{array}{c}\text{Probability of Failure of the Same Component or Structure}\\ \text{Designed Using a 50-Year Return Period Load}\end{array}} \qquad \text{(Eq. 2-5)}$$

The factored load γQ_{50} (with γ from Table 3-2 or 3-3 in Chapter 3) is an approximation of the load Q_{RP}. This is further discussed in Chapter 3.

The commonly designated "15% rule" for wind loadings indicates that every time the wind load factor γ is increased by increments of 0.15 (e.g., 0.85, 1.00, 1.15, 1.30, 1.45 in Table 3-2 in Chapter 3), the return period of the wind load is doubled (e.g., 25, 50, 100, 200, and 400 years). This also results in approximately reducing the probability of failure against wind by a factor of 2.

If statistical data on weather-related events are available, the use of Eq. 2-lb (using Q_{RP}) is recommended over that of Eq. 2-1a (using γQ_{50}). In addition to meeting the minimum values given in Table 2-3, selection of an appropriate RFP should be based on the importance of the line and its location and length. For example, a higher reliability may be selected for a portion of a line located in an urban area.

The reliability of a long line is less than that of a short one, all design parameters being the same. The primary reason for the reduced reliability is that because a long line is exposed to a larger number of severe events, its likelihood of experiencing some kind of failure is greater. Also, weak components are more likely to be exposed in a larger population of components. The line designer may consider decreasing the RFP for long lines in order to increase the line's reliability. Methodology to quantify the reliability of long lines is outside the scope of this manual but is addressed in Dagher et al. (1993) and *IEC 60826* (IEC 2002).

Design for loads with a return period of 50 years (i.e., RFP = 1) is considered the basis for transmission line work. For temporary, noncritical components and for Grade C construction, an RFP >1 may be acceptable; that is, a design with RP <50 years may be used, as shown in Table 2-3, Grade C.

2.6.2 Selection of Strength Factor

The purpose of the strength factor ϕ in Eqs. 2-1a through 2-3 is to account for the nonuniformity of the exclusion limits that currently exist in published formulas for nominal strength R_n and for differences in strength coefficients of variation COV_R. The values of the strength factor ϕ can be obtained from Table 2-2. The factors indicated assume a log-normal distribution of strength, which is considered more realistic than the normal distribution.

In the future, the authors recommend that all T/D line strength design guides publish strength values at the 5% lower exclusion limit (5% LEL), or the estimate of this value—the 5% lower tolerance limit (5% LTL) with a minimum confidence of 50%. Chapter 4 discusses methods for obtaining the 5% LTL as well as the coefficient of variation COV_R, including sample sizes and data reduction techniques.

Larger strength values of the 5% LTL are obtained if more test data are available or product variability is reduced. This gives an incentive for manufacturers to conduct sufficient testing of their products so they can use higher design values and maintain product consistency. For large sample sizes or very small variability (COV_R), the 5% LTL approaches the 5% LEL value.

The use of Table 2-2 requires knowledge of the LEL corresponding to the nominal strength R_n of a component, as well as the COV_R of the component strength. (The 5% LTL, as determined by the method in Chapter 4, may be used for R_e for $e = 5\%$.) Both numbers will vary from one material type to another (steel versus wood pole) and will depend on the strength design guide or method used to calculate nominal strength (e.g., *ANSI O5.1-2002* [ANSI 2002] or *ASCE Manual 72* [ASCE 1990]).

The strength factors of Table 2-2 are to be applied to the design nominal strength used for the pole, using the exclusion limit of the nominal strength and the COV of the pole strength. For example, according to *ANSI O5.1-2002*, Douglas fir poles have a designated fiber stress of 8,000 pounds per square inch (psi) with a COV = 0.20. Since this designated fiber stress value represents the mean groundline fiber strength (i.e., not the 5% LEL strength), this stress level must be multiplied by a strength factor (ϕ) of 0.79 (Table 2-2) for design of the pole:

$$F_{b5\%LEL} = 0.79 \times 8{,}000 \text{ psi} = 6{,}320 \text{ psi}.$$

The factors in Table 2-2 are based on a log-normal strength distribution and have been derived to result in relatively consistent reliability (i.e., approximately the same probability of failure) when structures are subjected to extreme winds, independent of the type of pole product and material (i.e., COV_R). As a result, these factors do not directly correspond to the simple calculation that may otherwise be used to obtain a 5% LEL. (For example, assuming a normal strength distribution without regard to the failure rates for different materials, the simple calculation would have been $[1 - 1.645 \times COV_R]$ applied to a mean strength.) See Chapter 4 and Appendix A for additional discussion.

Information on how to calculate the LEL and COV from existing test data is given in Chapter 4 and the related appendices. Typical values of the LEL and COV_R for different components used in transmission and distribution line are also given in Chapter 4. If no data are available, the following default information may be used:

1. **Steel Components and Prestressed Concrete Poles.** For components of steel towers and steel or prestressed concrete pole structures designed according to the ASCE and Precast/Prestressed Concrete Institute (PCI) publications (ASCE 1990; ASCE 1992; PCI 1997; ASCE 2000; PCI, 1999), assume that R_n (or R_e) has an exclusion limit in the

range of 5% to 10% and COV_R is in the range of 10% to 20%. Therefore, from Table 2-2, ϕ is in the range of 0.96 to 1.12 or, typically, $\phi = 1.0$.

2. **Reinforced Concrete.** For reinforced (non-prestressed) concrete components designed according to the American Concrete Institute (ACI) procedures (ACI 2002), the ACI strength reduction factors can be used in lieu of the ϕ factors given in Table 2-2, since the ACI factors already contain strength derating effects for the various concrete components.
3. **Wood Poles.** For wood pole structures, the statistical data in *ANSI O5.1-2002* (ANSI 2002) can be used to determine a value for R_n at the 5% LEL by using Eq. 4-4 in Chapter 4. The corresponding strength factor can then be obtained from Table 2-2. Alternatively, the COV_R and 5% LTL may be computed from actual data and the corresponding strength factor obtained from Table 2-2, for $e = 5\%$.
4. **Foundations.** Table 2-2 can be used for foundations for which statistical data are available. For foundations for which statistical data are not available, nominal strengths and strength factors based on established practice can be used. In such cases, however, the reliability of the foundation relative to that of the supported structure is unknown.
5. **Conductors and Ground Wires.** The LRFD format described herein is also applicable to the mechanical design of conductors or ground wires. If the nominal strength R_n for a conductor is defined as its rated ultimate strength T_u (AAI 1982), then a strength factor of 0.6 to 0.7 is recommended. Using a strength factor of 0.60 to 0.70 should prevent damage to the conductor and virtually eliminate the possibility of its rupture at that level. These suggestions are not reliability-based but represent current practice.

A summary of the RBD procedure is given in Fig. 2-1.

2.7 MOMENT MAGNIFICATION CONSIDERATION FOR FLEXIBLE POLES

Slender poles may have significant secondary moments that need to be considered in the design. These moments are caused by vertical loads acting through a horizontal displacement near the top of a deflected pole. These magnified moments may be obtained using an iterative calculation, using an amplification factor such as the Gere-Carter method (Appendix A, Example Calculation of P-Δ Effect Using the Gere-Carter Method) or using a computer program that accounts for geometric nonlinearity.

As a result of the reliability calibrations that included designing typical poles at 40 locations around the United States (described at the beginning of this chapter), the recommended load factors in this manual are less than or equal to 1.0 (Table 2-3). The ASCE Pole RBD Committee recommends

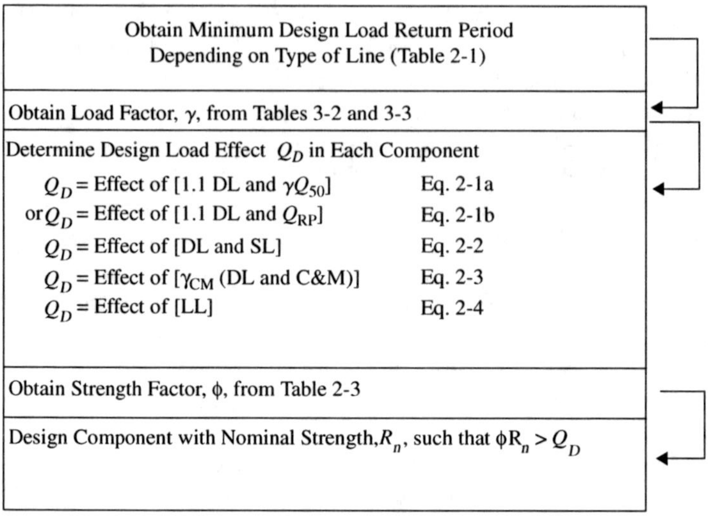

FIGURE 2-1. Summary of Reliability-Based LRFD Design Procedures.

that factored loads should be used in calculating secondary moments (P-Δ effects). The Gere-Carter method is typically conservative and is illustrated in Appendix A. From the reliability calibrations used to evaluate hundreds of poles, researchers determined that the moment magnification factors ranged from 1.0 to over 2.0, depending on the pole geometry, materials, load configuration, and load magnitude.

2.8 COORDINATION OF FAILURE SEQUENCES

There is a natural and well-founded desire to exercise some control over the sequence of failure of the different line components when one item fails. A casual event that cannot be avoided or a weather-related event that exceeds the selected RP load can, if not controlled, lead to unnecessary and excessive damage that will prolong the outage.

2.8.1 Structures versus Foundations

The components of the support system do require some attention regarding their relative strengths. In some cases, it may be desirable to design foundations that are generally more reliable than the structures they support. The rigorous approach to arranging this relative relationship would require knowledge of the dispersion characteristics PDFs and variability (COV_R) of these two major line components. By using their COV_R or using default values such as those suggested here, it is possible to adjust the calculated strength of the foundations by applying a strength factor based on a LEL of 1% and using a factor based on a LEL of 10% for the structures. The failure rate of a LEL of 1% is theoretically equal to 1%, which is not significant in itself. The relative reliabilities of using a LEL of 1% and a LEL of 10% are, however, meaningful numbers that can be used in design. If the goal is to design foundations to be more reliable than the structures, then use a LEL of 1% for the foundations with the structures at a LEL of 10% (or a LEL of 0.1% for foundations and a LEL of 1% for structures). This will provide sufficient probability that the structure will fail before the foundation.

2.8.2 Wire System versus Support System

Structural failure of a component in the wire systems (wires, insulators, hardware) puts critical demands on the failure containment properties of the support system, especially if the threat is from heavy icing and the wires are highly tensioned. Fortunately, many of the wire system components are selected on the basis of their damage limits, usually from 50% to 70% of their rated strengths, so that the remaining strength margins (50% to 30% of their *nominal* strength) should ensure that they will not be the first to fail. For insulators, load duration effects may reduce the allowable design loads to a significantly lower level than their rated strengths, which should be taken into consideration in the design and selection of the insulator system. All of these components should not be part of the critical failure path, since all of the components of the wire systems are governed by their damage limits.

2.8.3 Tangent versus Dead-End Structures

There should also be a hierarchy of strengths of the different classes of structures with the tangent structures considered the base units, whereas heavy-angle and dead-end structures should have an extra margin of strength. Because this calibration study was restricted to tangent structures, line tension effects were not considered at this time. This edition assumes that insulators and hardware are designed so that they are not the weak links in the line. Future editions of this manual will consider angle and dead-end structures, as well as insulators and hardware.

The current edition of this manual addresses the reliability of a single pole subjected to a severe storm. It is outside the scope of this current edition to consider the reliability of an entire line comprising a group of poles subjected to a severe storm, including the reliability of the wires, insulators, hardware, and foundations. The reliability of a line is less than that of an individual pole, and techniques to evaluate line reliability are discussed in Dagher et al. (1993) and *IEC 60826* (IEC 2002).

Chapter 3
LOADS

3.1 INTRODUCTION

Electrical transmission and distribution (T/D) line structures must be capable of withstanding loads generated from weather-related events and construction and maintenance (C&M) events and must provide failure containment to minimize damage from catastrophic events. The basic environmental design loads (also called nominal or characteristic design loads) depend on the structure location, including geography, topography, and elevation. *It is this manual's intention that nominal design loads be calculated following the procedures in ASCE Manual 74* (ASCE draft; please note, the authors are referencing the 2002 draft of the *ASCE Manual 74* throughout this chapter unless otherwise noted). Even so, distribution structures need not be designed to the same criteria as transmission structures. Using *ASCE Manual 74* (ASCE draft), designers may still select *National Electrical Safety Code* (*NESC*) (IEEE 2002) Grade C construction over Grade B for distribution structures, and use a correspondingly reduced load factor as shown in Table 2-3. Grade B construction with the increased load factors in Table 2-3 is typically selected for high-voltage transmission systems. These structures are most likely to have longer spans and higher physical profiles and are more difficult to replace or repair when damaged. They could have a substantial effect when out of service since they cover a large area. The stability of the electrical grid could be disturbed by a single, unplanned failure event of such critical structures.

Structures for lower-voltage distribution systems are typically much shorter (less than 60 ft) and may be designed to Grade C construction (refer to Table 2-3 for the corresponding reduced load factors). These structures usually serve a smaller area and do not need significant manpower or equipment resources to perform replacement or repair tasks if such is needed. In addition, electrical system stability is frequently not dependent upon the availability of one particular structure. Thus, the factored design loads for distribution structures designed to Grade C construction are considerably less than those for transmission structures. As may be seen in Table 2-3, the wind and ice thickness factors γ for Grade C construction are one-half of those used for Grade B construction. In addition, with some exceptions, distribution structures do not typically consider failure containment loads.

With lower-profile distribution-type structures, various uncertainties are difficult to evaluate or quantify. The wind turbulence is more severe and less predictable at lower elevations. Weight from ice-covered, broken tree branches, which may fall onto the structure or supported wires, can be many times higher than ice weight accumulated on the wires alone. Debris from either wind or ice storms occurs more often and could generate higher-impact loads on these structures. Thus, the load predictions are more difficult and the actual structural reliability may depend heavily on the conditions of the surrounding environment. Historically, regulatory bodies provide design guidelines mainly on the basis of past performance. The American Society of Civil Engineers (ASCE) has established a committee to study and recommend appropriate, detailed design procedures for distribution structures.

3.1.1 Weather-Related Load Events

Weather-related events include extreme wind, extreme ice with concurrent wind, and the associated temperature effects. All three variables, taken separately or jointly, are random variables and, as such, can only be described by probability distributions. In certain cases, atmospheric pressure and local topography influence the magnitude of weather-related loads. These influences should be considered when appropriate.

Since extreme values of the variables are considered in design applications, it is customary to associate these extreme values with some predetermined return period (RP), such as 50-year events. This would predict that, on average, the level of the extreme event would be reached or exceeded at least once in that period. Weather-related loads are sometimes referred to as reliability-based loads (IEC 2002).

3.1.1.1 Extreme Wind Loads. T/D structures and all structural components should be designed and constructed to resist extreme wind loads, which are often the controlling condition. In the United States, the design requirements are typically based on a 50-year (approximate) RP, 3-s gust wind event under standard atmosphere (i.e., temperature of 59 °F [15 °C] and sea level pressure of 29.92 in. of mercury [101.325 kPa]). This wind speed is measured at 33 ft (10 m) above ground in flat and open terrain (Exposure Category C, *ASCE Standard 7-02* [ASCE 2002]). Adjustments should be made to reflect the unique topographic conditions for specific structures.

3.1.1.2 Combined Ice and Wind Loads. Ice accretion on a transmission line is often a governing loading criterion in structure design. In addition to imposing substantial vertical loads on the structural system, the ice buildup on the conductors presents a greater projected area exposed to the wind and affects the force coefficient (also called shape or drag coefficient). The weight of the ice on the wire system may also cause significantly higher tension than bare wire conditions. Any resulting unbalanced tensions would correspondingly increase structure loads and should be considered in the design.

3.1.1.3 High-Intensity Wind Loads. High intensity winds (tornadoes and microbursts) are short-lived, randomly occurring, severe storms that cover small areas. They cause severe damage to houses, mobile homes, and automobiles, although engineered structures often survive without damage.

Almost all tornadoes can engulf a house or small structure but very few have a path width as large as other extreme winds that can load the full span of a transmission line. Thus, when designing for high-intensity wind events, special consideration should be included to account for the horizontal profile variation of the high-intensity winds.

3.1.2 Construction and Maintenance Events

Some line components may be subjected to their most critical loading during C&M operations. Thus, these loads must be considered in the design of the structure. Unlike weather-related loads, C&M loads are controllable to a large extent and are directly related to construction methods.

Workers can be seriously injured as a result of structural overstress; therefore, personnel safety should be a paramount factor when establishing criteria for C&M loads. These loads are sometimes referred to as safety loads.

3.1.3 Failure Containment Loads

Some of the events that produce loads in a transmission line system are difficult to describe statistically because of their nature or the lack of data. Accidental events such as failure of components from defects, wear, fatigue, or failures of entire structures from landslides, floods, tornadoes, sabotage, traffic, or any other unforeseen phenomena fall into this category.

Design procedures do not control the occurrence of these events but attempt to minimize their consequences so that failure will not propagate without control. This security requirement can be accomplished by designing some or all structures for special failure containment loads or by load-limiting devices such as mechanical fuses.

3.1.4 Longitudinal Loads

Longitudinal loads on an intact system may be caused by unequal wind or ice on adjacent spans, by unequal wire tension, or from temperature extremes on different span lengths. These intact system loads usually are well-defined and generally do not govern the design of a structure, with the possible exception of longitudinal imbalances resulting from unequal in-cloud ice deposits on adjacent spans.

Longitudinal loads may also be the result of weather-related events, unexpected broken wires, or failure of an adjacent structure. Extreme events caused by breakage of line or structural components can create severe load imbalances in the wire system, which are capable of causing partial or complete structural failure of a line section. The cascading failure risk of a transmission line should be considered to prevent prolonged outage, hazard to public safety, and significant economic losses. Longitudinal loads are sometimes referred to as anticascading or security loads.

3.1.4.1 Other Load Considerations. Although galloping and structure vibration do not generally produce extreme loads on the structures, loads produced by galloping wires can cause damage to cross-arms, cross-arm connections, and hardware. Furthermore, some structure member shapes are particularly susceptible to wind-induced vibration and have failed under fatigue. Consequently, the designer must be aware of the potential problems associated with these phenomena.

3.2 REFERENCES TO APPROPRIATE LOAD DOCUMENTS

This manual recommends the design load criteria as specified by *ASCE Manual 74,* "Guidelines for transmission lines structural loading" (ASCE draft). Appropriate load factors may be selected to account for loads on distribution structures (see Section 3.4).

3.2.1 Weather-Related Loads

3.2.1.1 Extreme Wind Loads. A basic wind-force formula applicable to transmission lines is presented in *ASCE Manual 74* (ASCE draft). This formula accounts for wind characteristics such as wind speed, terrain roughness, and air density, as well as for structure and line characteristics such as force coefficient, gust response factor, and projected area of the structure and overhead line components. The wind forces recommended in *ASCE Manual 74* (ASCE draft) are primarily based on the provisions of *ASCE Standard 7-02* (ASCE 2002). Where the values deviate, sources of the recommended values are indicated in that document.

3.2.1.2 Combined Ice and Wind Loads. *ASCE Manual 74* (ASCE draft) provides a map of 50-year RP ice thickness from freezing precipitation with concurrent 3-s gust wind speeds. This ice load map is contained in *ASCE Standard 7-02* (ASCE 2002) and includes extrapolations in the western states using information from *Storm Data* (NOAA 1959–1995) to cover the contiguous 48 states. The extrapolation was reviewed by state and regional climatologists and the map was revised on the basis of their comments. Ice thickness zones in the eastern half of the country and in the Pacific Northwest have also been revised, based on *Storm Data,* reanalyzing weather data for a longer period of record, and incorporating comments from state climatologists.

The map values in ASCE Manual 74 (ASCE 1991; ASCE draft) do not include in-cloud icing or wet snow accretions, which are caused by meteorological conditions that can produce significantly different loads. Where more detailed icing data have been compiled for a service area, those data should take precedence over the information in ASCE Manual 74 (ASCE 1991; ASCE draft). Electric utilities are urged to develop ice and concurrent wind loading criteria established specifically for their service regions based on historical data.

3.2.1.3 High-Intensity Wind Loads. The probability of a tornado strike at a given point is very small, even in areas where tornadoes are prevalent. However, the probability of a transmission line being crossed somewhere along its length by a tornado is significantly larger (Twisdale 1982). Since most tornadoes have a very narrow width, this provides an opportunity for improving transmission line resistance to tornado damage at a reasonable cost. *ASCE Manual 74* (ASCE draft) suggests that, where justified and required, structures should be designed to withstand an F2 scale tornado (wind speed of 157 mph or less). Historical data indicate that 86% of categorized tornadoes have intensity rated at F2 scale or less.

3.2.2 Construction and Maintenance Loads

ASCE Manual 74 (ASCE 1991; ASCE draft) provides guidelines for loads and load factors related to C&M events. Most of these recommendations are referenced from other national regulations or codes of practice, such as the Institute of Electrical and Electronics Engineers (IEEE) *Standard 524* (IEEE 1992) and *IEEE Standard 1307* (IEEE 1996).

3.2.3 Failure Containment Loads

Three methods are suggested by *ASCE Manual 74* (ASCE 1991; ASCE draft) to prevent cascading failure. A set of theoretically sound, simplified equations is also presented to calculate unbalanced longitudinal load caused by the initial structural failure. In addition, other guidelines and recommendations are provided to address design issues such as structural vibration and galloping. Typically, these security loads are applied directly with no further adjustments (load factor equal to 1.0).

3.3 REGIONAL AND LOCAL WEATHER-RELATED LOADS

3.3.1 Extreme Wind Loads

Basic wind speed can be determined by using regional wind data for a specific location. The *ASCE Standard 7-02* (ASCE 2002) provisions permit use of regional weather-related data, provided the following three criteria are met:

1. Proper extreme-value statistical analysis procedures have been employed in reducing the database.
2. The length of record, sampling error, averaging time, anemometer height, data quality, and terrain exposure of the anemometer have been taken into account.
3. If meteorological data are used to justify a wind speed lower than the 85-mph, 50-year RP, 3-s gust at 33 ft, then an analysis of sampling error is required to demonstrate that the wind record could not occur by chance.

In addition, in hurricane-prone regions, wind speeds derived from simulation techniques based on regional data can only be used when the following conditions exist:

1. Proper simulation or extreme-value statistical analysis procedures are used. The use of regional wind speed data obtained from anemometers to define the hurricane wind speed is not permitted for the U.S. Gulf and Atlantic coasts, the Caribbean, or Hawaii.
2. The design wind speeds resulting from the study shall not be less than the resulting 500-year RP wind speed divided by $\sqrt{1.5}$.

Note that the gust response factors and velocity pressure exposure coefficients in the equations of *ASCE Manual 74* (ASCE draft) are intended for use with the 3-s gust wind speed at 33 ft above ground in open country terrain. Therefore, it is necessary to make the appropriate adjustments when using regional climate data based on different parameters. Advice from a wind engineer or meteorologist may be needed since some of these adjustments are not always straightforward.

In using local data, sampling errors can lead to large uncertainties in specification of the 50-year wind speed. Sampling errors are the errors associated with the limited size of the climatological data samples (years of record of annual extremes). It is possible to have a 20-mph error in wind speed at an individual station with a record length of 30 years. Although local records of limited extent often must be used to define wind speeds in special wind areas, this should be done with care and conservatism.

3.3.2 Combined Ice and Wind Loads

Very little data exists in North America on equivalent uniform ice thickness from natural ice accretions on overhead transmission lines. Therefore, ice load studies often rely on mathematical models based on the physics of the various types of icing and on meteorological data (precipitation amount and type, temperature, and wind speed). Results from an ice accretion analysis typically give calculated ice thickness for past storms in which freezing precipitation has occurred. An extreme-value analysis can then be applied to determine the ice accumulation. Wind speeds during and after periods of freezing precipitation can also be extracted from the meteorological database and analyzed to determine the wind speed to apply concurrently with the ice thickness.

In lieu of using the recommended values in the *ASCE Manual 74* (ASCE 1991; ASCE draft) ice and wind map, the 50-year RP ice thickness and the concurrent wind speed for a structure may be determined from local meteorological data, provided the following are considered:

1. The quality of the precipitation, wind, and present weather data have been taken into account.
2. A robust ice accretion algorithm has been used to estimate uniform ice thickness and concurrent wind speeds from these data.
3. Proper extreme-value statistical analysis procedures have been used in analyzing the ice thickness and concurrent wind speed data.
4. Both lengths of record and sampling errors have been taken into account.

Recommendation: A meteorologist familiar with atmospheric icing should be consulted in areas with high elevations and complex relief and in areas where little information on ice loads is available.

3.4 EFFECTS OF LOAD FACTORS OR LOAD RETURN PERIODS

Practical design procedure requires that single values of the magnitude of the events be used. These nominal design values are normally specified in a code or loading specification so that their risk of being exceeded is relatively low. For example, a 50-year RP extreme wind velocity has a 2% (0.02) probability of being exceeded in any one year.

The probability that an event or load with a given RP will be exceeded at least once during the planned lifetime (reference period) of a line is given in Table 3-1. This probability is a useful indicator but it does not

TABLE 3-1. Probability of Exceeding Design Load during Reference Period

Return Period (year)	Annual Probability	Reference Period, n (years)					
		1	5	10	25	50	100
25	0.040	0.04	0.18	0.34	0.64	0.87	0.98
50	0.020	0.02	0.10	0.18	0.40	0.64	0.87
100	0.010	0.01	0.05	0.10	0.22	0.40	0.64
200	0.005	0.005	0.02	0.05	0.10	0.22	0.39

correspond to the probability of failure of the line or to that of any of its components. More information related to the calculations that yield the results in Table 3-1 can be found in the commentary of *ASCE Standard 7-02* (ASCE 2002).

The methods for estimating loads, especially those for weather-related events, are mostly based on statistical models. These models, although scientifically correct, have limitations on the precision of their estimates. To ensure structural reliability, load factors are introduced to compensate for this uncertainty.

The load factor allows a designer to modify the reliability of a line. Selection of an appropriate load factor should be based on the importance of the line and its location and length. For example, a higher load factor may be selected for a portion of a line located in an urban area.

All given design parameters being the same, the reliability of a long line is lower than that of a short line. The primary reason for the reduced reliability is that a long line is exposed to a larger number of severe weather-related events and, therefore, its likelihood of experiencing some type of failure is greater. Also, weak components are more likely to be exposed in a larger population. Thus, an essential long line may require a higher load factor.

It is possible to convert wind and ice data collected at a point by a suitable multiplier to reflect the spatial aspect of a typical line. This aspect of defining weather-related load criteria was theoretically demonstrated in Dagher et al. (1993) and used in development of actual line loading criteria by Behncke (1998).

Design for load with an RP less than 50 years may be considered for distribution or for temporary transmission structures. The designer should also be aware that electrical grids are distributed systems and may have multiple redundancies. Multiple looped paths may be provided from the generation sources to the point of service. The reliability of the whole system may be generally addressed by means of looping and extra redundancies. Thus, the stability of the electrical grid typically would not rely on one individual structure.

Multipliers to convert 50-year-event extreme wind loads to 25- to 400-year RP values are presented in Table 3-2. The factors in Table 3-2 show that increasing the 50-year RP design wind load by 15% approximately doubles the relative reliability level, and increasing by 30% approximately quadruples it.

Multipliers to convert 50-year-event ice thickness and concurrent wind speeds to 25- to 400-year RP values are presented in Table 3-3. The factors in Table 3-3 indicate that the uniform ice thickness for a 50-year RP is increased by 25% to approximately double the relative reliability level, and by 50% to approximately quadruple it. Using these factors, the concurrent wind load to be applied with the extreme uniform ice thickness need not be adjusted for RP.

TABLE 3-2. Approximate[a] Load Factors to Convert Extreme Wind Loads from a 50-Year Return Period to Another

Return Period (year)	Wind Load Factor
25	0.85
50	1.00
100	1.15
200	1.30
400	1.45

[a] ASCE *Standard 7-02* (ASCE 2002). Exact conversion factors depend on wind storm statistics.

TABLE 3-3. Approximate[a] Load Factors for Combined Ice and Wind Loads

Return Period (year)	Ice Thickness Factor	Concurrent Wind Load Factor
25	0.80	1.00
50	1.00	1.00
100	1.25	1.00
200	1.50	1.00
400	1.85	1.00

[a] Applied to 50-year combined ice and wind loads. Exact conversion factors depend on wind storm statistics, ice storm statistics, and wire diameter.

Chapter 4

STRENGTH OF SINGLE-POLE UTILITY STRUCTURES

4.1 INTRODUCTION

Single-pole utility structures are generally loaded as cantilever beam-columns. Their load capacity varies with geometry, structural material, manufacturing process, and support conditions. These parameters, often characterized with varying degrees of uncertainty, are used with mechanics-based as well as empirical-based models (that also introduce some uncertainty) to obtain load capacity or resistance estimates. For these reasons, the design procedure described in Chapter 2 incorporates resistance factors (ϕ factors in Eq. 2-1a) to account for the uncertainties inherent in the estimates of pole capacity.

Pole resistance is a random variable that may be characterized using a probability density function (PDF). For the typical range of pole resistance and load coefficients of variation (COVs), a relatively consistent reliability can be achieved across material types by setting nominal resistance values that represent a 5% to 10% lower exclusion limit (LEL) equal to the design load effect corresponding to a predetermined return period (RP) (IEC 2002; Peyrot and Dagher 1984). For example, setting the 5th percentile pole strength, $R_{5\%LEL}$, equal to the 50-year RP load effect Q_{50}, $R_{5\%LEL} = Q_{50}$, yields resulting reliabilities that are relatively insensitive to the respective PDFs and COVs of the load and strength parameters.

In this manual, nominal resistance (R_n) is defined as the strength that will be exceeded by 95% of poles in the target population (see Fig. 1-1 in Chapter 1). In statistical terms, this value is often referred to as the 5th percentile (5% of the population lies below it) or the lower 5% LEL. Because it is not practical to require a precise evaluation of the 5% LEL, statistical methods are commonly used to obtain an estimate that has an associated level of confidence. Confidence refers to the probability that a randomly selected sample will have an LEL greater than or equal to the target value. The lower confidence bound on a LEL is called a lower tolerance limit (LTL). In this manual, we refer primarily to a 5% LTL having a 50% or 75% confidence.

4.2 OBJECTIVE

This section presents three basic methods for deriving and documenting R_n as an LTL value along with the coefficient of variation (COV_R) for single-pole structures. These include the following:

1. An empirical analysis based primarily on tests of full-sized poles.
2. A theoretical analysis of mechanics-based models used in conjunction with Monte Carlo simulation.
3. A default assignment of material distribution parameters.

These three approaches are intended to address the range of complexity and experience associated with conventional as well as potential utility pole structural materials. The empirical approach relies heavily on test data to verify modeling assumptions. This is especially applicable for wood that is a nonuniform orthotropic material. The variable nature of wood leaves the designer with the options of using a minimum clear wood strength and designing to the minimum specification (for example, largest knot, minimum dimension, greatest grain angle), or referencing full-scale tests of a sample of poles that represent the quality range expected when ordering to a minimum specification (e.g., *ANSI O5.1-2002* [ANSI 2002]). If the critical stress in a wood pole occurs predominantly at the groundline and does not involve knots, strength can be predicted using distribution parameters published for clear wood (ASTM 1998). Knots, cross-sectional dimensions, and taper may present some questions about variability; in such cases a full-scale pole test database may provide a better estimate of the R_n for full-sized poles.

The uniform, isotropic nature of materials such as steel and concrete make them more amenable to the use of mechanics-based models. These models may be coupled with Monte Carlo simulation of material properties to predict their 5% LTL strength. In other cases, where model parameters exhibit little or no covariance the analytical models may be used to estimate mean behavior, and independent parameter variances may be summed to provide an estimate of the pole strength variance with no need for computer simulation.

Default assignment is used when there are insufficient data to characterize the pole strength PDF empirically and when demonstrably reliable models have not been developed to provide accurate estimates of structural performance for a particular pole material or configuration. The default method provides a conservative approach to assigning parameters for estimating R_n.

Market forces will likely control the evolution of the reliability-based design (RBD) approach and ultimately bring all pole configurations to a relatively uniform level of reliability. It is the responsibility of the pole suppliers to provide the parameters and supporting documentation for their poles' strengths, but it is the responsibility of the design engineer to select the correct pole for a given load condition. It is the responsibility of the purchasing agency to review and verify that the poles selected by the engineer are provided by the pole supplier. If competing suppliers are unsure of values being used by their competition, it is their responsibility to fully understand the competitive products and their evaluation. It is this system of checks and balances resulting from competition and documented system performance that will force evolution of increasing reliability.

4.3 SCOPE

The following discussion is related to the derivation of resistance PDFs for single-pole structures subjected to transverse wind and ice loading. In this application, the primary structural component (the pole) is considered to behave as a cantilever beam. Its resistance is characterized by a bending-strength PDF, although the actual failure mechanism may be a localized buckling or tensile failure.

Procedures outlined in this section rely on the fundamental assumption that all poles meet or exceed established minimum manufacturing and process quality standards for poles (AISC 1999; ANSI 2002; ANSI/ASCE *Standard 10-90* [ASCE 1992], which was previously *ASCE Manual 72* [ASCE 1990]; PCI 1999]). The organizations that set these standards provide for acceptable tolerance variations in the manufacturing or processing of all of the pole types. A pole's nominal strength shall be determined and reliability-based strength factors shall be calculated against these acceptable variations. This document does not address pole strength reliability associated with initial substandard quality poles or of pole deterioration because of damage in service or use in hostile environments. Pole deterioration, whether cumulative or due to a single event, is highly variable and should be handled as part of regular pole inspection and maintenance schedules.

4.4 CHARACTERIZING POLE STRENGTH

This document does not cover all possible design criteria for single-pole structures. It illustrates general methods to assign nominal design properties for common pole configurations. The general procedures outlined here can be adapted to less-common designs.

4.4.1 Loads

As previously noted, loads considered in this characterization of pole strength are limited to wind and ice. While the magnitude and variability of these loads are different within and between geographic regions, their effect is manifested on single-pole structures predominantly as cantilever-bending moments. Potential failure modes that result from bending vary with pole material type, geometry, and support conditions. All potential failure modes must be considered in deriving a PDF to characterize pole strength. For wood poles, extreme fiber-bending stress generally controls the pole's bending load capacity. For reinforced concrete poles, the ultimate strength is often controlled by the compressive strength of the concrete. For tubular steel and fiber-reinforced polymer (FRP) structures, local buckling and bending are generally the governing factors.

4.4.2 Nominal Resistance

To achieve relatively uniform structural reliability across all material types, a uniform definition of the characteristic of nominal strength or resistance R_n is required. Resistance or strength (R_n) defined in terms of a limiting stress should represent a consistent estimate with regard to the strength PDFs for all pole materials. To this end, R_n for single-pole structures is herein defined to represent the 5% LEL with a noted level of confidence (= LTL), regardless of material type. The following discussion describes three approaches to characterizing pole strength and identifying:

1. The nominal resistance R_n corresponding to a designated LTL.
2. The resistance coefficient of variation (COV_R).

Ideally, R_n (defined in terms of a limiting stress) should represent the same point estimate with regard to strength PDF for all pole materials. To this end, R_n for single-pole structures is herein defined to represent the 5% LEL regardless of material type. The strength factors in Table 2-2 can be used when pole resistance values are stated at other than 5% LEL.

4.4.2.1 Method 1: Empirical Basis. Empirical derivation of pole strength generally involves some combination of full-sized pole tests and mechanics-based models. Because it is not economically feasible to test every possible combination of pole size, processing variable, and load configuration, standard tests are used to establish a baseline evaluation of pole capacity. This baseline is then adjusted to account for influences specific to a given application.

For example, the wood pole industry has traditionally endorsed a conservative approach to the selection of wood poles by adopting a standard test procedure, American Society for Testing and Materials (ASTM) *Standard D 1036-99* (ASTM 1999b) that uses either a cantilever or a simply supported beam test to evaluate a maximum groundline moment capacity for a green pole (green wood has a moisture content above the fiber saturation point—roughly 35%). At the present time, the best source for information on empirically derived wood pole design values is *ANSI O5.1-2002*, Annex C (ANSI 2002). This annex includes mean and variance values for strength and modulus of elasticity for commercial pole species as well as adjustments for conditioning, height, and size.

This test imposes boundary conditions that are at least as critical as any imposed on a pole in service. Green values have traditionally been used in the design of heavy timber because drying during service has counteracting effects: wood shrinkage reduces the effective section property while fiber strength and stiffness increase. Pole strength has traditionally been assessed as stress at the maximum moment location (groundline in the standard test), regardless of the actual failure location. Failure above the groundline is normally associated with knots or the reduced section property due to natural taper. A reduction in fiber strength with height may also be an influence for fast-grown poles. Tests conducted to assess the effect of variable material quality along the length of the pole (Bodig et al. 1986) show that when this failure occurred above groundline, the groundline stress was generally within 10% of the stress at the failure location. The pole capacity may therefore still be determined on the basis of groundline stress. The ASTM *Standard D 1036-99* (ASTM 1999b) cantilever test method provides conservative estimates of pole groundline moment (GLM) capacity when the groundline circumference is more than 1½ times the circumference at the centroid of load application.

Similar standard test procedures exist for concrete and FRP poles (e.g., ASTM *Standard C 1089-97* [ASTM 1997] for spun-cast prestressed concrete poles, and ASTM *Standard D 4923-92* [ASTM 1992] and ANSI *Standard C 136.20* [ANSI 1996] for FRP poles). ASCE is planning to publish a standard test procedure for steel poles.

Methods used to assess nominal resistance should take a conservative approach in accounting for processing and common service conditions that might affect pole performance. The design engineer should request information on the assumptions made in the derivation of the nominal values and recommendations for modifying them in certain cases: where a pole will be used under uncommon conditions, such as extremely wet or arid conditions, salt air, high temperature, or alkali soil, or where loads or boundary conditions differ from those assumed in deriving R_n. Adjustments for the derivation of design stress in round timber are presented in ASTM *Standard D 2899-01* (ASTM 2001).

A well-documented database established with strict adherence to standard test procedures can provide long-term benefits for the development and evolution of design standards. Periodic updating provides a record of changing trends in pole production and pole performance sensitivity to influencing variables. Over time, this can lead to refined methods for pole classification, reducing variability, and increasing reliability.

For manufactured poles having uniform strength along their length, it may be advantageous to develop a simple beam-bending test procedure in lieu of the full-scale cantilever pole test used for wood. The point of maximum stress in a uniformly tapered cantilever beam can be derived theoretically. If engineering models are available to accurately predict critical stress location but verification data are required to predict effects of change in geometry with applied stress, the standard test could be conducted to concentrate maximum bending stress at a specific cross section. A simple beam test having a high enough span-to-depth ratio to minimize the significance of shear effects would be less costly than a full-size cantilever test and would yield similar results. Alternative test methods must be universally accepted by users and producers and the appropriate design code authority.

4.4.2.1.1 Probability Density Functions. A critical part of any empirical evaluation is ensuring that the referenced database accurately represents the target population. Test samples must be selected to truly represent pole production and test results should be classified so values are characterized on the basis of probability of occurrence.

For full-sized pole test data to be statistically valid, the test specimens must be representative, in terms of size and quality, of the poles to be used in service applications. This may require samples to be selected over time and to be selected on the basis of population proportions. Methods to ensure the statistical validity of the sampling conducted can be found in *ASTM Standard D 2915-99* (ASTM 1999a) and *Standard D 5457-93* (ASTM 1993).

Using a test sample to project a statistical probability of occurrence of pole strength requires the adoption of a PDF. PDFs are generally classified as either parametric (described by a mathematical function) or nonparametric and are assumed to have a frequency profile rather than a strength profile representative of the parent population of pole strengths.

The nonparametric approach imposes no assumed shape on the PDF; it sorts data by order of magnitude. The sequence number corresponding to a particular datum in the ordered set is referred to as its order statistic; the initial estimate of the probability of getting a value less than or equal to that datum is the order statistic divided by the number of data points in the data set. For example, the lowest value in a sample of 20 ($1/20 = 0.05$) or the second value in a sample of 40 ($2/40 = 0.05$) is assumed to represent the 5th-percentile order statistic. In other words, 5% of a sample will have values less than or equal to this value. This value is often referred to as a point estimate of the 5% lower exclusion limit (LEL) of the parent population.

It is generally accepted practice in the wood industry, as in *ASTM Standard D 2915-99* (ASTM 1999a) to select an order statistic that will provide a level of confidence in an estimate of the 5% LEL. For an infinite population, the value at the 5th-percentile order statistic is the 5% LEL. For small samples, there is roughly a 50% chance that the LEL will exceed 5% of the parent population. Statistical tables are available that provide nonparametric order statistics associated with confidence bounds on the estimates of lower exclusion values (FPL 2005). Table 4-1 lists the order statistics associated with the 5% LEL given the 50% minimal confidence value as well as 75% confidence. The higher the level of confidence, the lower the order statistic.

TABLE 4-1. A 5% LTL with 50% or 75% Confidence for a Nonparametric Estimate and for a Normal Distribution[a]

Sample Size	Nonparametric		Normal	
N	I_{50}	I_{75}	K_{50}	K_{75}
5	NA	NA	1.78	2.464
15	NA	NA	1.68	1.99
20	1	—	1.67	1.93
28	—	1	1.66	1.88
40	2	—	1.66	1.83
53	—	2	1.65	1.81
60	3	—	1.65	1.79
70	—	—	1.65	1.78
78	—	3	1.65	1.77
80	4	—	1.65	1.77
90	—	—	1.65	1.76
102	5.1	4	1.65	1.76
125	6.25	5	1.65	1.75
200	10	9	1.65	1.72
∞	15	13	1.65	1.71

[a] For a nonparametric distribution, the order statistic (I = order of magnitude, or rank in an ordered list) is used to identify the 5% LEL (I_{50}) and the 5% LTL (I_{75} = 75% confidence). For a normal distribution, the 5% LTL is derived with 50% (K_{50}) or 75% (K_{75}) confidence standard deviations below the mean.

As specified in *ASTM Standard D 2915-99* (ASTM 1999a), the non-parametric distribution is preferred for structural wood design value derivations when using a small sample (<100) basis. Unlike parametric distributions, this approach requires a minimum sample of 21 to estimate a 5% LEL. Small samples may provide a reasonable estimate of mean trend but they do not provide sufficient information to accurately assess variability or point estimates at low probability levels.

Parametric PDFs are defined by closed-form equations, which can be used to generate an entire population of values that fit a designated distribution shape. Parametric PDFs therefore provide a means of extrapolating beyond the bounds of a given data set to estimate values with a low probability of occurrence. It is generally considered risky to extrapolate too far beyond the range of the available test data. Sample size (N) should therefore be selected to provide a prescribed level of assurance (ξ) that $100 \times \eta\%$ of a population will be included between the largest and smallest values. Wilkes (1944) derived a function to characterize the relationship (Eq. 4-1) between N, ξ, and η:

$$N\eta^{N-1} - (N-1)\eta^{N} = 1 - \xi. \qquad \text{(Eq. 4-1)}$$

This function indicates that a sample size of $N = 93$ is required to encompass 95% of a population with 95% assurance, and a sample size of $N = 472$ to encompass 99% of a population with that same level of assurance.

Many forms of parametric PDFs have been formulated. The three that are most commonly referenced to model strength are the normal, log-normal, and Weibull distributions. These PDFs demonstrate a range of control of shape and range of predicted values and imply varying degrees of knowledge about the population being represented.

The normal PDF has the following mathematical form:

$$f_X(x) = \frac{1}{\sigma\sqrt{2\pi}} e^{-\frac{1}{2}[(x-m)/\sigma]^2} \qquad \text{(Eq. 4-2)}$$

where

m = the mean value of x or first moment of the area under the PDF curve, where $-\infty < x < +\infty$

σ = the standard deviation. It is the square root of the second moment of the area under the PDF about the mean (variance) $\sigma^2 = \int (x - m)^2 f_x(x)\, dx$. For a discrete data set, this relationship yields:

$$\sigma = \left[\frac{\sum_{i=1}^{N} x_i^2 - Nm^2}{N-1} \right]^{0.5} \qquad \text{(Eq. 4-3)}$$

in which N is the sample size.

If there are sufficient data to warrant its use, the normal PDF will generally provide conservative estimates of low tail values. Using the normal distribution, R_n may be determined by

$$R_n = R_m - K \cdot \sigma_R \qquad \text{(Eq. 4-4)}$$

where

R_m = the mean strength
σ_R = standard deviation of strength
K = the distance from the mean to the point on the PDF that corresponds to the target lower tolerance limit.

The K values given in Table 4-1 were derived using a noncentral t-distribution inverse approach discussed by Guttman (1970). These values are derived to consider either a 50% or 75% confidence in the 5th percentile of a normal distribution for sample sizes ranging from 5 to 300.

The normal distribution has historically been used for characterizing the strength of wood. It is easy to use, is widely recognized, and generally provides conservative estimates of low tail values that are referenced when deriving design values. When the available data show that strengths are not distributed symmetrically about the mean, other PDFs are referenced to provide a more accurate characterization. A weakness often cited for the normal PDF is that it presents the possibility of having strength values less than zero. This is not a problem when the standard deviation is less than 30% of the mean and the PDF is used only to provide an estimate of a value having greater than a 1% probability of occurrence.

If the PDF is known to be right-skewed (i.e., having values much farther above than below the mean), the normal distribution may be considered to be overly conservative in estimating LTL values. Wood utility poles are often selected from a truncated normal distribution, where the lower tail represents poles that do not meet the *ANSI O5.1-2002* (ANSI 2002) minimum specifications. An alternative to the normal PDF that addresses the issue of right-skewness (and negative values) is the log-normal PDF. This PDF is derived assuming that the logarithms of the test data are normally distributed. In general, however, when the standard deviation is in the range of 20% of the mean, the log-normal distribution will give a 5th percentile point estimate only slightly larger than that obtained assuming a normal PDF. The log-normal PDF is defined as follows.

The log-normal PDF is applicable if the natural logarithms of strength data (x) are normally distributed with a mean λ and standard deviation μ. In this case, the PDF is of the same form as Eq. 4-2, substituting $y = 1n(x)$ for x, λ for m, and μ for λ.

$$f_y(y) = \frac{1}{\mu\sqrt{2\pi}} e^{-\frac{1}{2}\left(\frac{y-\lambda}{\mu}\right)^2} \qquad \text{(Eq. 4-5)}$$

where λ is the first moment of the area under the PDF or the mean of the 1n, (x) and μ^2 is the second moment or variance of the 1n, (x).

Because there are closed-form transformation equations to relate normal and log-normal PDF parameters, there is little need to actually work with the logarithms of data in order to make point estimates using the log-normal distribution. The mean strength and standard deviation conversion from normal to logarithm have the following form:

$$\text{Log-normal variance} = \mu^2 = 1n\,(\Omega^2 + 1). \qquad \text{(Eq. 4-6)}$$

$$\text{Log-normal mean} = \quad \lambda = 1n(R_m) - \frac{\mu^2}{2} \qquad \text{(Eq. 4-6)}$$

where

1n ( ) = natural logarithm (base e)
Ω = COV_R = coefficient of variation of the strength test data (σ_R/R_m)
R_m = mean of the strength test data
σ_R = standard deviation of the strength test data.

Equations 4-8 and 4-9 provide a point estimate of the LTL nominal strength R_n for a log-normal distribution using the mean R_m and coefficient of variation Ω of the test data.

$$R_n = k_N x R_m. \qquad \text{(Eq. 4-8)}$$

$$k_N = \frac{1}{\exp(\frac{1}{2}1n(\Omega^2+1) + K_N\sqrt{1n(\Omega^2+1)})} \qquad \text{(Eq. 4-9)}$$

Note that K_N and k_N are different variables. The multiplier k_N in Eq. 4-8 converts the mean test data strength to a LTL with a confidence value dependent on K_N. The K_N is the normal distribution tolerance adjustment corresponding to sample size N as listed in Table 4-1.

The Weibull PDF is a versatile alternative that can also be used to represent a distribution of all-positive values. It can be made to fit a wide range of distribution shapes. The Weibull distribution may be characterized using either two or three parameters. The three-parameter function has the following form:

$$f_x(x)=\frac{\omega}{\theta}\left(\frac{x-x_0}{\theta}\right)^{\omega-1}\exp(-(\frac{x-x_0}{\theta})^{\omega})\quad x\geq x_0\quad \omega,\theta>0 \qquad \text{(Eq. 4-10)}$$

where

ω = slope or shape parameter that reflects the relative scatter in the data; the larger the shape parameter, the lower the spread. A shape parameter of 3.5 is symmetric while a value >3.5 gives a negative or left-skewness, and a value <3.5 provides a positive skewness.

θ = scale parameter. As the scale parameter increases, the mode (location) where most events occur moves toward the upper end of the distribution.

x_0 = location parameter. If the location parameter is set equal to zero, Eq. 4-10 reduces to a two-parameter Weibull PDF.

The added versatility of the Weibull PDF also allows a greater chance for misrepresentation. For poles that have a COV in the range of 20%, the two-parameter Weibull distribution is likely to give more conservative estimates of a lower fractile than will a normal distribution. When the COV of a data set has a value less than 30%, a two-parameter Weibull PDF will generally have a shape parameter greater than 3.5, resulting in a negative skewness. Including the third (location) parameter will shift the distribution, reduce the shape parameter, and change the skewness.

The point estimate for the 5% LEL of a Weibull distribution can be calculated using Eq. 4-11.

$$5\%LRL_{Weibull}=\theta\cdot(-1\text{n}(0.95))^{1/\omega}+x_0$$
$$\text{or } 5\%LRL_{Weibull}=\theta\cdot(0.0513)^{1/\omega}+x_0. \qquad \text{(Eq. 4-11)}$$

4.4.2.1.2 Selecting a Probability Density Function. When selecting a PDF, it is important to consider how representative the data are of the population being modeled and how conclusions to be drawn from the data are to be used. Small data sets are generally assumed to be representative of mean trend but have a low probability of accurately representing variability in a parent population. When the PDF is being used only to select a lower tolerance value and not to characterize the shape of the low tail of the resistance distribution, there is limited value in attaining a precise fit. In such cases a normal PDF is the easiest of the established parametric functions to work with. For products subject to some level of quality control or quality assurance, there is generally some justification for assuming that the parent PDF will be skewed to the high side. Normal and log-normal PDFs give similar results with COVs under 20%. In this case, the normal PDF will give slightly more conservative 5% LTL values.

Documentation of the derivation of R_n should include discussion of the PDF and the process used for its selection. *ASTM Standard D 2915-99*, Section 4.5.7 (ASTM 1999a) suggests comparing a histogram or empirical cumulative distribution function to one or more overlaid parametric distribution functions as a means of justifying the PDF selection. Anderson (1952) discusses goodness-of-fit models and how they vary with distribution type.

Individual pole producers who maintain their own database on pole strength may select any PDF that can be supported by their data as a means of estimating a 5% LTL. The nonparametric PDF assumption is the most conservative and is best used with limited samples. If the sample size is large and pole strengths are supported by simple, conservative models that recognize basic material properties (from small clear tests, coupon tests, and cylinder tests) and permissible defects, the normal or log-normal assumptions are likely to give reasonably conservative estimates of a lower fractile of the PDF, as well.

Any organization interested in using a strictly empirical basis for the derivation of nominal resistance should maintain an up-to-date database for poles representative of those being used. Increasing the size of the database leads to greater confidence in the nominal resistance value. Increasing the sample size over time provides a basis for judging trends in materials and manufacturing that might affect the strength PDF. Larger samples also provide the opportunity for adopting a more rigorous approach to assessing the reliability of a utility line.

4.4.2.1.3 Empirical Analysis. Test data generally require some degree of interpretation. For example, ANSI wood pole dimensions are typically used in design, rather than the measured dimensions of the pole. If empirical strength values are derived using measured pole dimensions and applied using the ANSI size-class minimum dimensions, predicted GLM capacity will be less than the measured value. For this reason, values referred to by *ANSI O5.1-2002* (ANSI 2002), Annex C as "adjusted groundline modulus of rupture" are derived as the average failure moment at groundline, divided by the pole-class minimum groundline section modulus. Here the groundline section modulus was derived using the ANSI 6-ft-from-the-butt value adjusted to groundline using the ANSI-tabulated minimum dimensions to estimate taper. Pole modulus of elasticity estimates are also based on the class minimum dimensions at the butt and tip, assuming a linear taper and constant modulus of elasticity (MOE) value over the length of the pole. These values are therefore intended only for use with the ANSI-tabulated minimum dimensions

4.4.2.1.4 Confidence. A number of factors affect the confidence or assurance that an estimate based on a test sample provides a conservative representation of the target point of the parent distribution. The greater the sample size, the greater the probability that the sample mean and variance will closely approximate the parent population values. For a nonparametric distribution, confidence is characterized in terms of order statistics or the order of magnitude. The smallest value in a sample of 20 is the 5th percentile for that sample but only a 50% probability exists that it will be a conservative estimate of the parent population 5th percentile. The first-order statistic in a sample of 28, on the other hand, has a 75% probability of lying at or below the parent population 5th percentile. For a normal distribution, confidence/tolerance adjustment factors represent the distance from the mean of a sample to the point estimate in terms of the number of standard deviations. Basically, the confidence bound is set to provide some level of assurance that values derived on the basis of a small sample will encompass or provide a conservative estimate of the value for the parent population.

Table 4-1 provides a listing of order statistics used to estimate a lower 5% tolerance limit with a nonparametric distribution and adjustment factors representing the number of standard deviations from the mean to the 5% LTL of a normal PDF.

It is apparent from this discussion that an empirically derived value for R_n will vary, depending on the PDF assumed to represent the data. It is imperative for the pole supplier to provide documentation to support the assumptions made in the selection of a PDF and the derivation of the nominal resistance.

In Appendix B, the Method 1 section provides examples of the application of the empirical method to obtain the 5% LTL R_n.

4.4.2.2 Method 2: Mechanics-Based Models Used in Conjunction with Monte Carlo Simulation. Maintaining a database of full-sized pole tests can be prohibitively expensive. As an alternative, basic material properties can be used in conjunction with mechanics-based models to estimate mean pole strength. Strength variability, however, is a more complex issue. If there is no covariance between any of the independent variables, variance of a strictly linear model can be estimated as the sum of variances of the individual input parameters, eliminating the need for simulation. When using a nonlinear model with no covariance, variance may be influenced by parameter effects on any nonlinear function. Simulation provides a tool for characterizing this effect. However, models that rely on covariant input parameters are more complex because, for example, wood fiber strength and stiffness both vary with density, age, and moisture content, and the variability in weld strength may be larger with thicker steel plate. Application of Monte Carlo simulation in these cases requires establishment of an accurate covariance matrix and interaction equations to ensure realistic combinations of input parameters. Any influence that one input parameter has on another should be recognized in the development of the virtual structures being evaluated.

Computer simulation routines are designed to randomly generate physical and mechanical properties from defined PDFs assumed to represent the properties found in service, and are parameters of theoretical models used to predict performance. The advantage of Monte Carlo simulation is that statistical strength data are obtained using relatively inexpensive material coupon tests (small, clear samples for wood; cylinder tests for concrete) rather than testing a large population of full-size poles. Basically, the simulation routine compiles a large sample of computer-generated pole strength estimates. The resulting samples are then treated similarly to the empirical data, with the added adjustments for modeling error.

A few pitfalls to simulation must be considered. The most obvious is the question of mechanics-model accuracy. It is difficult to develop and verify a model that accounts for all variables that may influence strength and variability of the full-scale structure. When a model is used to predict performance of a complex system, it should be verified over the full range of input parameters for which it will be used.

Although the model being used may be accurate at predicting performance for any known combination of parameters, it may not accurately represent expected behavior in the tails of a distribution. For this reason, verification tests should be conducted to assess the prediction accuracy at the extremes of the influencing variables. A verification test should include accurate measurement of raw material mechanical properties as well as physical properties of the test poles. The more variable the material and the wider the range of structural configurations to be modeled, the larger the verification database should be. The model verification database should be well-documented and included along with simulation results as support for nominal resistance values to be used.

Nonlinear mechanics-based models employ iterative techniques to predict failure. These models account for change in material as well as geometric properties with increased strain levels. Verification tests are conducted to assess the prediction accuracy at the extremes of the influencing variables. Confidence in simulated data varies with the accuracy of the models as well as the input data. Model accuracy should be verified by comparing model predictions to full-scale pole test data using the actual material and geometric properties of the corresponding test specimen. The data used to establish input PDFs for mechanics-based models should be subject to the same assessment of confidence as the full-scale pole test data.

The number of simulations required to get a satisfactory confidence on estimates of distribution parameters will vary with the complexity of parameter interactions and symmetry of their assumed distribution functions. These topics are discussed in greater detail by Law and Kelton (2000), Hammersley and Handscomb (1964), and Balci and Sargenti (1984). It is often preferable to run a number of trials, each consisting of 200 to 500 simulations, to generate a distribution of point estimates rather than one run of 10,000 simulations. This provides a better indication of variability and confidence bounds. The number of simulations conducted needs to be large enough, however, to provide stable predictions of the 5th percentile.

The PDFs used to characterize the raw data input for simulation models should be based on large enough sample sizes to ensure a standard error (SE) no greater than 10% of the estimated 5th percentile. If normality is assumed, the tolerance limit is estimated using Eq. 4-4 (Natrella 1963). The SE of this statistic varies with sample size (*N*) and sample standard deviation(s) of the sample. It can be approximated using the equation:

$$SE = s\sqrt{\frac{1}{N} + \frac{k^2}{2(N-1)}} \qquad \text{(Eq. 4-12)}$$

where

K = confidence level factor (Table 4-1).

In Appendix B, the Method 2 section provides examples on the application of Monte Carlo simulation along with mechanics-based models to obtain the 5% LTL R_n.

4.4.2.3 Method 3: Default Basis. The default basis is used if there are insufficient data to characterize the pole strength PDF empirically or if demonstrably reliable models have not been developed to provide accurate estimates of pole strength. The default method provides a conservative approach to assigning parameters for estimating R_n. The National Institute of Standards and Technology (NIST, formerly the National Bureau of Standards) proposed guidelines (Ellingwood 1980) for estimating strength variability as a function of so-called professional, material, and fabrication influences.

A simple approach is to obtain a best estimate of mean with some degree of confidence and establish a conservative estimate of variability until more data become available. Pole strength variability, expressed here as COV, is influenced by a number of factors that should be considered. These include inherent material variability (COV_M^2), which can be evaluated using standard material property tests. The geometric variability includes inherent or fabrication-related dimensional and thickness tolerances. Fabrication-induced variability (COV_{FA}^2) for steel, concrete, and FRP poles include manufacturing process effects on geometry and on material strength properties. Finally, the accuracy of the predictive model of pole strength is referred to as the professional factor or model accuracy (COV_P^2). In estimating the strength of a full-sized pole on the basis of raw material test data, confidence in the result is dependent on the accuracy of the model being used.

Finally, consider so-called other effects (COV_O^2) such as deterioration, design error, and environmental risk. Poles may be damaged due to mishandling during installation or they may experience deterioration from environmental exposure such as high temperatures, grass fires, ultraviolet radiation, decay, corrosion, cracking, and spalling. These effects are not generally included in a design model and they do not have the same effect on all poles in a line. Poles removed from a line after 30 years of service are likely to have neither the same strength nor the same strength variability they had when they were installed.

Combining these individual effects can provide an estimate of the pole strength COV_R:

$$COV_R^2 = COV_M^2 + COV_{FA}^2 + COV_O^2 + COV_P^2 \qquad \text{(Eq. 4-13)}$$

Further information is given in the American Iron and Steel Institute's (AISI) "Specification for the design of cold-formed steel structural members" (AISI 1996) and "Development of a probability-based load criterion for American National Standard A 58" by Ellingwood et al. (1980). These and other publications support overall default values for COV_R of steel and concrete poles of 0.15, and 0.20 for wood poles. A number of variables with fairly broad ranges affect the strength of FRP poles; therefore, useful default values cannot be established for these poles at this time.

4.5 PROOF LOADING

Proof loading to a design value provides some degree of quality assurance but, in the absence of pole failure, this procedure provides little useful information on the strength distribution. Even when the proof loading does result in occasional failures, such results can only provide a basis for assigning some level of confidence about the relative proximity of the proof load and some fractile of the strength distribution. The drawback of proof loading to a level that results in occasional failure is that it provides some risk of causing undetected damage to the pole.

If backed by research to correlate nondestructive evaluation (NDE) parameters to strength, proof loading methods might be developed to enable estimates of LELs of strength. In general, however, NDE parameters are not used to define strength since NDE parameters are poorly correlated with strength.

Appendix A

DESIGN EXAMPLES

A.1 INTRODUCTION

The following examples are for unguyed tangent transmission and distribution (T/D) poles. They have been included to illustrate some of the concepts presented in this manual. The examples use force coefficients (drag coefficients, shape factors) that are based on the minimum recommendations of the 1991 edition of American Society of Civil Engineers (ASCE) *Manual 74* (ASCE 1991). For wind on poles, the force coefficient values were selected using *ASCE Manual 74*, Table 2-3 (ASCE draft). For wire loads, force coefficients of 1.0 are used for all wires, with or without ice. In the calculation of wind forces on both wires and poles, the selection of appropriate force coefficients is very important. Supplemental information on force coefficients can be found in Appendix H of *ASCE Manual 74* (ASCE draft) as well as in other specifications such as in Appendix B of ASCE 7-02 (ASCE 2002) and International Electrotechnical Commission (IEC) *Standard IEC 60826* (IEC 2002). Information in *ASCE Manual 74* (ASCE draft), Appendix H, for example, suggests that force coefficients greater than 1.0 may be appropriate for small-diameter (<~ ½-in.) wire, and *IEC 60826* recommends force coefficients between 1.0 and 1.4 for ice-covered wires.

The design parameters used for these examples do not represent all possible load conditions, structure types, or components but do provide insight into how to properly apply the reliability-based design (RBD) methodology discussed herein. These examples demonstrate how the loading requirements prescribed in the working draft of *ASCE Manual 74* can be used to determine the size of various pole types for different grades of construction. Examples are given for wood, steel, concrete, and fiber-reinforced polymer (FRP) poles based on pole bending (strength being the only design criterion). These examples do not consider other design criteria such as electrical clearances or seismic effects. In each example the pole size is initially established based on a calculated groundline moment (GLM), and then the pole strength is verified at other locations along the pole. As implemented, this GLM accounts for the deflected shape (P-Δ) effect.

The poles in each of the examples are sized for *National Electrical Safety Code* (*NESC*) (IEEE 2002) Grades B and C construction using the load factors given in Table 2-3 in Chapter 2 of this manual. As illustrated in the examples, weather-related loads on poles are independent of material type. Wind loading on the pole structure depends on the geometry of the pole (including the projected wind area of the pole above groundline), the height of the vertical centroid of the applied wind pressure, and the pole force coefficient (round, polygonal).

A.2 EXAMPLE LOAD REQUIREMENTS

In the following examples, two different pole configurations will be considered, each assumed to be governed by different loading conditions. A transmission pole will be designed for an extreme wind loading and a distribution pole will be designed for a combined ice and wind loading, both in accordance with the criteria set forth in the working draft of *ASCE Manual 74* (ASCE draft). (In practical applications, the controlling condition will often correspond to that of extreme wind loading, for both transmission and distribution poles.) For all examples, both pole configurations assume weight spans that are equal to the wind spans, although this is not often the case in actual practice. Note that the wind force formula used in the working draft of *ASCE Manual 74* (ASCE draft), Eq. 2.1-1, is the same as formula specified in the *National Electric Safety Code* (IEEE 2002) for extreme wind loading. This design process is an iterative one. Most methods require that an assumption be made regarding pole size. This pole size is then analyzed for the forces it must support. Based on this analysis, if a different pole size is required the analysis should be repeated to verify the adequacy of the pole.

Transmission Pole Design (Las Vegas, Nevada)

Consider a 75-ft-long pole (65.5-ft height above ground), of the configuration indicated, and subject to the following conditions and parameters (Fig. A-1):

- ASCE Extreme Wind: 90 mph, Exposure C
- Design for two grades of construction: *NESC* (IEEE 2002) Grade B and Grade C
- Wire Parameters:
    - Conductor: 795 aluminum conductor steel-reinforced (ACSR) (26/7) Dia. = 1.108 in., Wt = 1.091 lb/ft

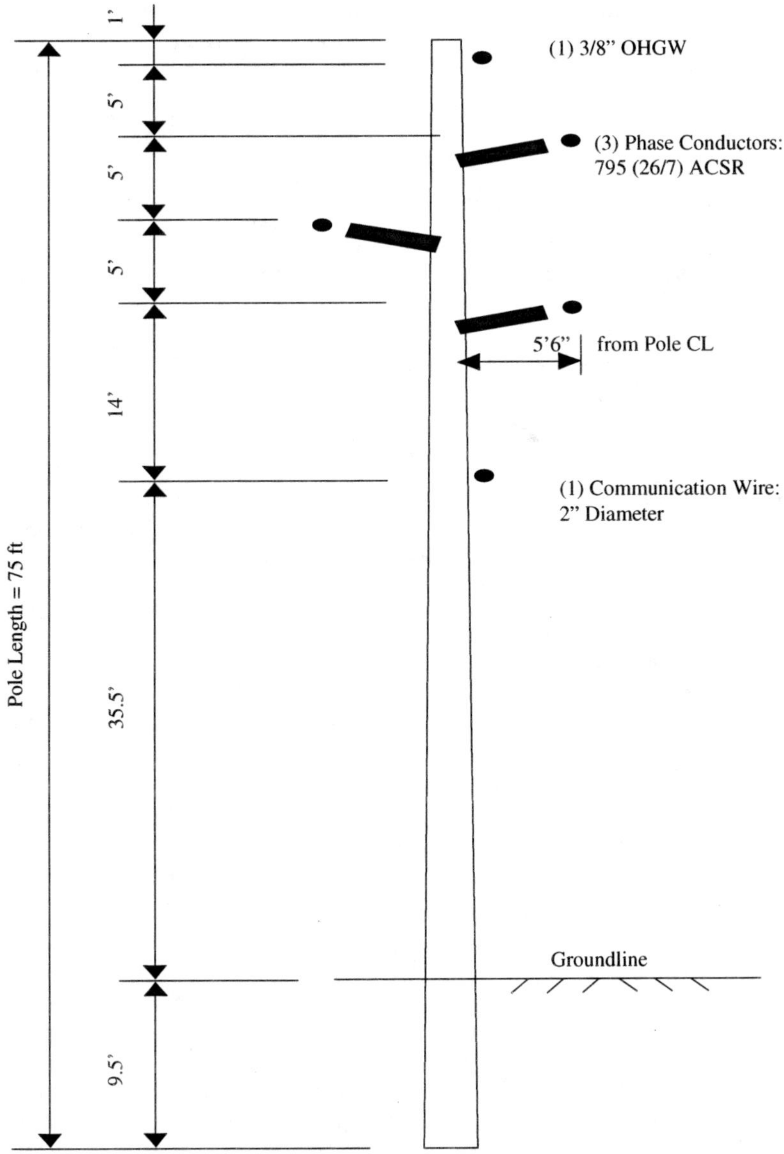

FIGURE A-1. Transmission Pole Design.

- Shield Wire: 3/8-in. high-speed steel (HSS)
  Dia. = 0.36 in., Wt = 0.273 lb/ft
- Communication Wire:
  Dia. = 2.0 in., Wt = 2.25 lb/ft

- Span Parameters:
  - Wind and Weight Spans = 500 ft

In accordance with Eq. 2.1-1 of the working draft of *ASCE Manual 74* (ASCE draft), the wind force, F, in pounds is calculated:

$$F = Q \times K_z \times (V)^2 \times G \times C_f \times A$$

where

Q = air density factor, 0.00256

K_z = velocity pressure exposure coefficient, given by 2.01 $(H/900)^{(2/9.5)}$, where H = effective height (ft), which yields the following values for K_z

1.114 for conductors,
1.154 for shield wire,
1.018 for communication wire, and
1.063 for pole.

V = 3-s gust wind velocity, 90 mph

G = gust response factor, given by $(1 + 2.7 \times E \cdot B^{0.5}) / k_v^2$, as described in the working draft of *ASCE Manual 74* (ASCE draft) or the *National Electric Safety Code* (IEEE 2002) for wires and structures, where

E = exposure factor,

B = response term, and

k_v = 1.43,

which, based upon the respective heights and span lengths, yields the following values for G

0.743 for conductors,
0.737 for shield wire,
0.759 for communication wire, and
0.906 for pole.

C_f = shape factor (force, or drag, coefficient):
1.0 for wires; for poles, see *ASCE Manual 74*, Table 2.6-3 (ASCE 1991).

A = projected wind area (ft^2).

The corresponding loads are provided in Table A-1

These forces must be adjusted by the appropriate load factors (γ) to meet the requirements for the desired grade of construction. For *NESC* Grade B, (IEEE 2002), the corresponding load factors are $\gamma_{wind} = 1.0$ and $\gamma_{dl} = 1.1$, and for Grade C the corresponding load factors are $\gamma_{wind} = 0.5$ for poles that do not extend more than 60 ft above ground; for those that do, $\gamma_{wind} = 1.0$ and $\gamma_{dl} = 1.1$. The GLM for each pole design is calculated by multiplying each of these forces by their respective load factors and their corresponding distances from the groundline. In addition, the eccentricity of the conductor arrangement is accounted for by multiplying the weight of one conductor (616 lb) times its offset from the centerline of the pole (~5.5 ft). The calculations for the wire-related GLMs follow. Since for all

TABLE A-1. Loads for Transmission Pole Design Example

Line Component	Wind Pressure (psf)	Wind Area (ft^2)	Wind Force (lb)
Wind on Conductors (ea)	17.16	46.17	792
Wind on Shield Wire	17.64	15.00	265
Wind on Communication Wire	16.01	83.33	1,334
Wind on Pole	$19.97 \times C_f$	Varies with pole geometry	

transmission pole examples the pole extends 65.5 ft above ground and extreme wind is the applied load case, the load factors ($\gamma_{wind} = 1.0$ and $\gamma_{dl} = 1.1$) and the wire-related loading are the same for both Grade B and Grade C construction. The moment due to wind on the pole is specific to pole geometry and will be calculated for each specific design example. The net result is that for each transmission example, the same-sized poles are required for Grade B and Grade C construction to meet the extreme wind load case (Table A-2).

Distribution Pole Design (Portland, Oregon)

Consider a 45-ft long pole (38.5 ft height above ground), of the configuration indicated (Fig. A-2) and subject to the following conditions and parameters:

- ASCE Combined Ice and Wind: 50-mph wind plus 1¼-in. ice
- Design for two grades of construction: Grade B and Grade C
- Wire Parameters:
  - Conductor: 336.4 ACSR (26/7)
    Dia. = 0.72 in., Wt = 0.462 lb/ft
  - Neutral Wire: 3/0 all-aluminum conductor (AAC)
    Dia. = 0.464 in., Wt = 0.156 lb/ft
  - Communication Cable (+ Messenger):
    Dia. = 1.5 in., Wt = 1.12 lb/ft
  - Span Parameters:
    - Wind and Weight Spans = 275 ft.

In accordance with Eq. 2.1-1 of the working draft of *ASCE Manual 74* (ASCE draft), the wind force, *F*, in pounds is calculated:

TABLE A-2. Groundline Moment (GLM)

			Grade B		Grade C	
Line Component	Force (lb)	Distance (ft)	γ	GLM (ft-lb)	γ	GLM (ft-lb)
Wind on Top Conductor	792	59.5	1.0	47,124	1.0	47,124
Wind on Middle Conductor	792	54.5	1.0	43,164	1.0	43,164
Wind on Bottom Conductor	792	49.5	1.0	39,204	1.0	39,204
Wind on Shield Wire	265	64.5	1.0	17,093	1.0	17,093
Wind on Communication Wire	1334	35.5	1.0	47,357	1.0	47,357
Conductor Eccentricity	546	5.5	1.1	3,300	1.1	3,300
Subtotal (without pole wind force)				197,242		197,242

$$F = Q \times K_z \times (V)^2 \times G \times C_f \times A$$

where

Q = air density factor = 0.00256
K_z = velocity pressure exposure coefficient, given by $2.01 \times (H/900)^{(2/9.5)}$, where H = effective height (ft), which yields the following values for K_z

1.029 for conductors,
1.005 for neutral wire,
0.975 for communication wire, and
0.951 for pole.

V = 3-s gust wind velocity = 50 mph
G = gust response factor, given by $(1 + 2.7 \times E \cdot B^{0.5}) / k_v^2$, as described in the working draft of *ASCE Manual 74* (ASCE draft) or the *National Electric Safety Code* (IEEE 2002) for wires and structures, where
E = exposure factor,

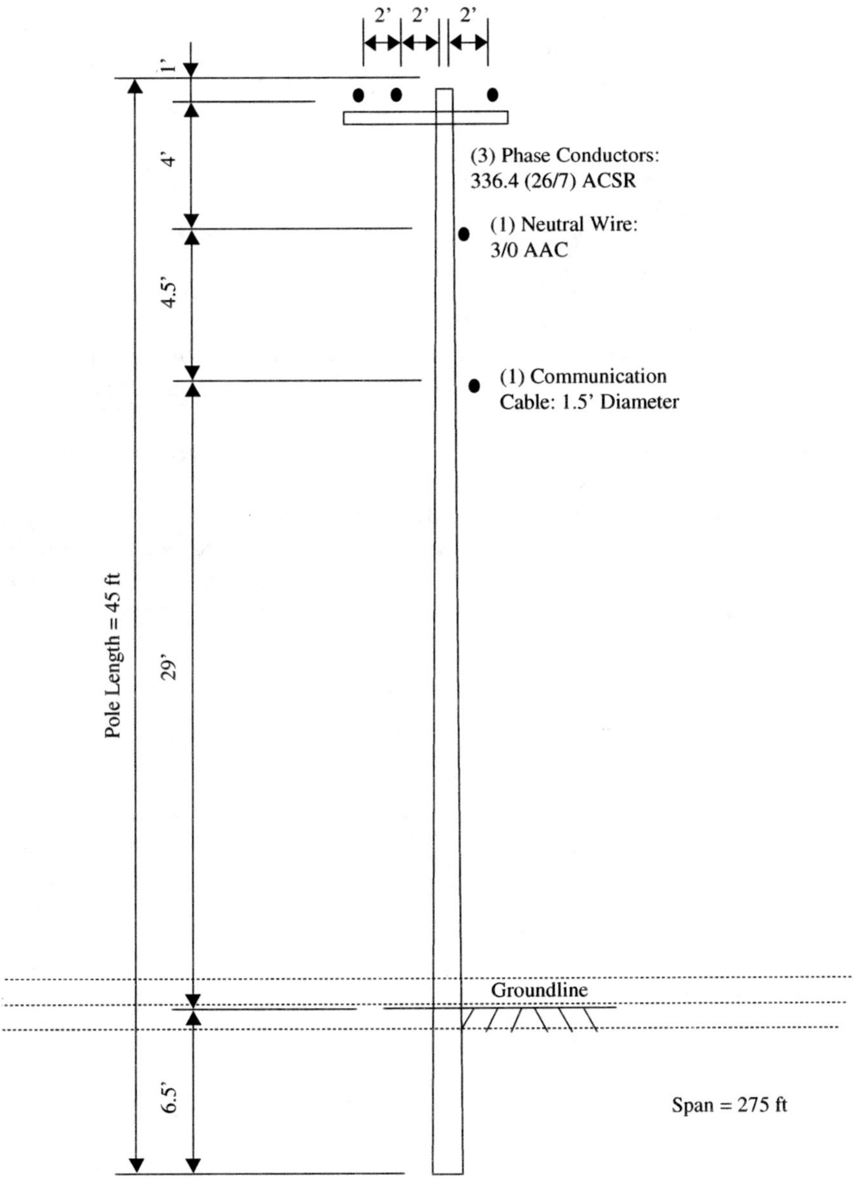

FIGURE A-2. Distribution Pole Example.

B = response term, and
k_v = 1.43,
which, based upon the respective heights and span lengths, yields the following values for G
0.807 for conductors,
0.812 for shield wire,
0.819 for communication wire, and
0.948 for pole.
C_f = shape factor (force, or drag, coefficient):
1.0 for wires, and
for poles, see *ASCE Manual 74* (1991), Table 2.6-3.
A = projected wind area (ft^2).

The corresponding loads are given in Table A-3.

These forces are adjusted by the appropriate load factors (γ) for transverse wind to meet the requirements for the desired grade of construction. In this case, the difference in wind loading between Grade B and Grade C construction is due to the factor applied to the ice thickness for calculating wind area. For Grade B, $\gamma_{ice} = 1.0$ and for Grade C, $\gamma_{ice} = 0.50$, and is reflected in the wind area indicated in Table A-3. The ice weight (eccentricity) is indicated in Table A-4. The GLM for each pole design is calculated by multiplying each of the forces by their respective transverse wind load factors and their corresponding distances from the groundline.

TABLE A-3. Loads for Distribution Pole Example

		Grade B		Grade C	
Line Component	Wind Pressure (psf)	Wind Area (ft^2)	Wind Force (lb)	Wind Area (ft^2)	Wind Force (lb)
Wind on Conductors (ea)	5.31	73.79	392	45.15	240
Wind on Neutral Wire	5.22	67.93	355	39.28	205
Wind on Communication Wire	5.11	91.67	468	63.02	322
Wind on Cross-arm	5.31	0.36	2	0.36	2
Wind on Pole	$5.77 \times C_f$	Varies with pole geometry			

In addition, the eccentricity of the conductor arrangement is accounted for by multiplying the weight of one conductor and insulator (137 lb) times its offset from the centerline of the pole (~4.3 ft). The calculations for the groundline moments are shown in Table A-4. The moment due to wind on the pole is specific to the pole geometry and will be calculated for each specific design example.

A.3 EXAMPLE 1: WOOD TRANSMISSION POLE

In this example, solid, round Douglas fir wood poles will be sized to support the loads on the transmission pole described in Section A.2, for construction Grade B and Grade C. The following wood examples use the strength values indicated in Table 1 of *ANSI O5.1-2002,* (ANSI 2002), including the recommended height adjustment factors and critical section analyses. Pole circumferences have been calculated using the minimum pole class dimensions in Table 8 of *ANSI O5.1-2002* and

TABLE A-4. Groundline Moments for Distribution Pole Example

			Grade B		Grade C	
Line Component	γ	Distance (ft)	Force (lb)	GLM (ft-lb)	Force (lb)	GLM (ft-lb)
Wind on Three Conductors	1.0	37.5	1,176	44,100	720	27,000
Wind on Neutral Wire	1.0	33.5	355	11,892	205	6,868
Wind on Communication Wire	1.0	29	468	13,572	322	9,338
Eccentricity: Conductor/ Insulator	1.1	4.3	137	648	137	648
Eccentricity: Ice on Conductor	1.0	4.3	842	3,621	287	1,234
Wind on Cross-arm	1.0	37.0	2	74	2	74
Subtotal (without pole wind force)				73,097		45,162

assume a linear taper in between. Thus, Douglas fir poles have a designated fiber stress of 8,000 psi with a coefficient of variance (COV) of 0.20. Since this designated fiber stress value represents the mean groundline fiber strength (i.e., not the 5% lower exclusion limit [LEL] strength), this stress level must be multiplied by a strength factor (ϕ) of 0.79 (Table 2-2) for design of the pole:

$$F_{b\,5\%\,\mathrm{LEL}} = 0.79 \times 8{,}000 \text{ psi} = 6{,}320 \text{ psi}.$$

Grade B Construction or Grade C Construction. A Class H2 Douglas fir pole is initially assumed to be sufficient (ANSI 2002). This pole has the following properties:

Pole length:	75 ft (65.5 ft above ground)
Tip diameter:	9.87 in. (circumference = 31.0 in.)
6 ft from butt diameter:	18.78 in. (circumference = 59.0 in.)
Groundline diameter:	18.33 in. (circumference = 57.58 in.)
Pole wind area:	76.95 sq ft
Distance to centroid:	29.47 ft
Pole shape coefficient:	0.9 (Table 2.6-3 in *ASCE Manual 74* [ASCE 1991])
Wind pressure on pole:	19.97 psf × pole shape coefficient.

The moment at the groundline due to applied wire loads equals 197,242 ft-lb, and that due to wind on the pole is calculated as:

$$\begin{aligned}&1.0 \text{ (load factor } \gamma) \times 76.95 \text{ sq ft (wind area)} \times 19.97 \text{ psf (wind pressure)}\\ &\quad \times 0.9\ (C_f) \times 29.47 \text{ ft (distance to centroid)}\\ &\quad = 40{,}758 \text{ ft-lb (GLM)},\end{aligned}$$

yielding a total GLM of 238,000 ft-lb. However, since this example uses a linear analysis technique, the deflected unbalance (P-Δ effect) must be added before the pole can be properly sized. In this example, an amplification factor of 1.112 is calculated using the Gere-Carter method to account for the P-Δ effect. (The Gere-Carter method, described in Section A.9, tends to be conservative—that is, it yields higher amplification factors—in comparison to that of more sophisticated computer-model values. The resulting design moment is 264, 656 ft-lb.)

The section modulus, S, required to support the calculated GLM is determined by dividing the GLM by the design stress value:

$$S = 264{,}656 \text{ ft-lb} \times (12 \text{ in./ft}) \,/\, 6{,}320 \text{ psi} = 502.5 \text{ in.}^3.$$

This corresponds to a pole with a groundline circumference of 54.14 in. A 75-ft Class H2 Douglas fir pole, which has a calculated minimum groundline circumference of 57.58 in., satisfies the groundline requirement. A Class H1 pole with a groundline circumference of 54.16 in. also meets this requirement. However, a recalculation of the *P*-Δ effect for a Class H1 pole results in a larger amplification factor, which increases the GLM to more than what a Class H1 pole can theoretically resist. In addition, because the point of peak stress on tall poles is not necessarily at the groundline, and Section 9 of *ANSI O5.1-2002* (ANSI 2002) specifies that a reduction in stress because of height effect should be applied, these tall poles need to be checked at various points above the groundline to verify that they are not overstressed at these locations. An analysis of stresses at points above the groundline indicates that *the Class H2 pole is sufficient*.

A.4 EXAMPLE 2: WOOD DISTRIBUTION POLE

In this example, Southern pine wood poles will be sized to support the loads on the distribution pole described in Section A.2 for construction Grade B and Grade C. Note that for relatively short poles, such as used for typical distribution applications, the critical stress point is commonly at the groundline, and therefore a GLM check is generally sufficient. Per *ANSI O5.1-2002* (ANSI 2002), Table 1, Southern pine poles have a designated fiber stress of 8,000 psi and a COV of 20%. Since this designated fiber stress value represents the mean groundline fiber strength, this stress level must be multiplied by a strength factor (ϕ) of 0.79 (Table 2-2 In Chapter 2) for design of the pole:

$$F_{b5\%LEL} = 0.79 \times 8{,}000 \text{ psi} = 6{,}320 \text{ psi}.$$

Grade B Construction. A Class 2 Southern pine pole is initially assumed to be sufficient. This pole has the following properties:

Pole length:	45 ft (38.5 ft above ground)
Tip diameter:	7.96 in. (circumference = 25.0 in.)
6 ft from butt diameter:	12.89 in. (circumference = 40.5 in.)
Groundline diameter:	12.83 in. (circumference = 40.31 in.)
Pole wind area:	33.35 sq ft
Distance to centroid:	17.75 ft
Pole shape coefficient:	0.9 (Table 2.6-3 in *ASCE Manual 74* [ASCE 1991])
Wind pressure on pole:	5.77 psf × pole shape coefficient.

The moment at the groundline due to applied wire loads equals 73,907 ft-lb, and that due to wind on the pole is calculated as:

1.0 (load factor γ) $\times$ 33.35 sq ft (wind area) $\times$ 5.77 psf (wind pressure)
$\times$ 0.9 (C_f) $\times$ 17.75 ft (distance to centroid)
= 3,074 ft-lb (GLM),

yielding a total GLM of 76,981 ft-lb. However, since this example uses a linear analysis technique, the deflected unbalance (P-Δ effect) must be added before the pole can be properly sized. In this example, an amplification factor of 1.312 is calculated using the Gere-Carter method to account for the P-Δ effect (this method is described in detail at the end of this Appendix). The resulting design moment is 100,999 ft-lb.

The section modulus, S, required to support the required GLM is then calculated:

$$S = 100{,}999 \text{ ft-lb} \times 12 \text{ in./ft} \ / \ 6{,}320 \text{ psi} = 191.8 \text{ in.}^3.$$

This corresponds to a groundline circumference of 39.27 in. *Thus, a 45-ft Class 2 pole*, which has a calculated minimum groundline circumference of 40.30 in., *is acceptable.*

Grade C Construction. A Class 4 Southern pine pole is initially assumed to be sufficient. This pole has the following properties:

Pole length:	45 ft (38.5 ft above ground)
Tip diameter:	6.68 in. (circumference = 21.0 in.)
6 ft from butt diameter:	11.14 in. (circumference = 35.0 in.)
Groundline diameter:	11.08 in. (circumference = 34.82 in.)
Pole wind area:	28.49 sq ft
Distance to centroid:	17.66 ft
Pole shape coefficient:	0.9 (Table 2.6-3 in *ASCE Manual 74* [ASCE 1991])
Wind pressure on pole:	5.77 psf $\times$ pole shape coefficient.

The moment at the groundline due to applied wire loads equals 45,162 ft-lb, and that due to wind on the pole is calculated as:

1.0 (load factor γ) $\times$ 28.49 sq ft (wind area) $\times$ 5.77 psf (wind pressure)
$\times$ 0.9 (C_f) $\times$ 17.66 ft (distance to centroid)
= 2,613 ft-lb (GLM).

The addition of the previously calculated GLM (subtotal) due to wind on wires of 45,162 ft-lb yields a total GLM of 47,775 ft-lb. However, since this example uses a linear analysis technique, the deflected unbalance (P-Δ effect) must be added before the pole can be properly sized. In this example, an amplification factor of 1.271 is calculated using the Gere-Carter method to account for the P-Δ effect. The resulting design moment is 60,722 ft-lb.

The section modulus, S, required to support the required GLM is then calculated:

$$S = 60{,}722 \text{ ft-lb} \times 12 \text{ in./ft} / 6{,}320 \text{ psi} = 115.3 \text{ in.}^3.$$

This corresponds to a groundline circumference of 33.15 in. *Thus, a 45-ft Class 4 pole,* which has a calculated minimum groundline circumference of 34.83 in., *is acceptable.*

A.5 EXAMPLE 3: STEEL TRANSMISSION POLE

In this example, a 12-sided steel pole will be sized to support the loads on the transmission pole described in Section A.2, for construction Grades B and C.

Grade B Construction or Grade C Construction. A 12-sided, 75-ft steel pole of the following dimensions and characteristics is considered:

Pole length:	75 ft (65.5 ft above ground)
Tip diameter:	8.0 in.
Butt diameter:	20.32 in.
Groundline diameter:	18.76 in.
Wall thickness:	0.1875 in.
Effective steel yield strength:	65 ksi
Taper:	0.164 in./ft
Pole wind area:	73.03 sq ft
Distance to centroid:	28.33 ft
Pole shape coefficient:	1.0 (Table 2.6-3 in *ASCE Manual 740* [ASCE 1991])
Wind pressure on pole:	19.97 psf × pole shape coefficient
Pole bare weight:	2,249 lb
Specified ultimate moment capacity of pole at groundline:	284,000 ft-lb
Lower exclusion limit (% LEL) of specified strength:	5%
COV of pole strength:	0.10
Strength factor, ϕ:	1.00 (Table 2-2, for 5% LEL value and $COV_R = 0.10$).

The moment at the groundline due to applied wire loads equals 197,242 ft-lb, and that due to wind on the pole is calculated as:

$$\begin{aligned}&1.0 \text{ (load factor } \gamma) \times 73.03 \text{ sq ft (wind area)} \times 19.97 \text{ psf (wind pressure)}\\&\quad \times 1.0\ (C_f) \times 28.33 \text{ ft (distance to centroid)}\\&\qquad = 41{,}317 \text{ ft-lb (GLM)},\end{aligned}$$

yielding a total GLM of 238,559 ft-lb. Using a finite element analysis (FEA) modeling program to evaluate the nonlinear P-Δ effect, the resulting deflected unbalance is then determined to be 11,490 ft-lb. Thus, the final required moment capacity at the groundline equals 250,049 ft-lb.

Because the specified moment capacity, or strength, represents the 5th percentile, a strength factor (ϕ) of 1.00 (Table 2-2) is applicable for design of the pole:

$$M_{5\%LEL} = 1.00 \times 284{,}000 \text{ ft-lb} = 284{,}000 \text{ ft-lb}.$$

Since the corresponding capacity of the pole exceeds the required capacity of 250,049 ft-lb, the pole design appears to be sufficient. Note that other points along the length are also checked to verify the local pole strength exceeds the corresponding local moment. *The selected 12-sided steel pole is determined to be acceptable.*

A.6 EXAMPLE 4: STEEL DISTRIBUTION POLE

In this example, a round steel pole will be selected to support the loads on the distribution pole described in Section A.2 for construction Grades B and C. To demonstrate the use of the material strength factor to adjust for a strength value expressed as an LEL less than 5%, this steel distribution pole example assumes a specified strength corresponding to a 1% LEL and a COV = 0.10. It is important to note that the inclusion of this hypothetical example by the ASCE RBD Committee does not suggest an opinion regarding the validity of the 1% LEL value at the specified strength for this type of pole.

Grade B Construction. A round, 45-ft steel pole of the following dimensions and characteristics is considered:

Pole length:	45 ft (38.5 ft above ground)
Tip diameter:	6.0 in.
Butt diameter:	14.33 in.
Groundline diameter:	13.13 in.
Wall thickness:	0.133 in.
Effective steel yield strength:	65 ksi
Taper:	0.185 in./ft
Pole wind area:	30.69 sq ft
Distance to centroid:	16.86 ft
Pole shape coefficient:	0.9 (Table 2.6-3 in *ASCE Manual 74* [ASCE 1991])
Wind pressure on pole:	5.77 psf × pole shape coefficient
Pole bare weight:	667 lb

Specified ultimate moment capacity of pole at groundline:	92,740 ft-1b
Lower exclusion limit (% LEL) of specified strength:	1%
COV of pole strength:	0.10
Strength factor, ϕ:	1.07 (Table 2-2, for 1% LEL value and $COV_R = 0.10$).

The moment at the groundline due to applied wire loads equals 73,907 ft-lb, and that due to wind on the pole is calculated as:

$$1.0 \text{ (load factor } \gamma) \times 30.69 \text{ sq ft (wind area)} \times 5.77 \text{ psf (wind pressure)} \times 0.9\ (C_f) \times 16.86 \text{ ft (distance to centroid)} = 2{,}687 \text{ ft-lb,}$$

yielding a total GLM of 76,594 ft-lb. Using a computer modeling program to evaluate the nonlinear P-Δ effect, the resulting deflected unbalance is then determined to be 14,770 ft-lb. Thus, the final required moment capacity at the groundline equals 91,364 ft-lb.

Because the specified moment capacity, or strength, represents the 1st percentile, a strength factor (ϕ) of 1.07 (Table 2-2) is applicable for design of the pole:

$$M_{5\%LEL} = 1.07 \times 92{,}740 \text{ ft-lb} = 99{,}232 \text{ ft-lb.}$$

Since the corresponding capacity of the pole exceeds the required capacity of 91,364 ft-lb, the pole design appears to be sufficient. Note, however, that the above discussion ignores the combined effect of axial load and bending stresses. While axial stresses in poles such as this one are normally small (<1% to 2%), they can sometimes be sufficient to cause slight overstressing. For this reason, the use of a computer program that checks this combined effect over the entire length of the pole (instead of only checking the moment at its base) is the preferred method of analysis. Such a procedure was used for this example and it was determined that *the selected round steel pole is acceptable.*

Grade C Construction. A round, 45-ft steel pole of the following dimensions and characteristics is considered:

Pole length:	45 ft (38.5 ft above ground)
Tip diameter:	4.5 in.
Butt diameter:	11.89 in.
Groundline diameter:	10.82 in.
Wall thickness:	0.120 in.
Effective steel yield strength:	65 ksi

Taper:	0.164 in./ft
Pole wind area:	24.58 sq ft
Distance to centroid:	16.60 ft
Pole shape coefficient:	0.9 (Table 2.6-3 in *ASCE Manual 74* [ASCE 1991])
Wind pressure on pole:	5.77 psf × pole shape coefficient
Pole bare weight:	486 lb
Specified ultimate moment capacity of pole at groundline:	57,830 ft-lb
Lower exclusion limit (% LEL) of specified strength:	1%
COV of pole strength:	0.10
Strength factor, ϕ:	1.07 (Table 2-2, for 1% LEL value and $COV_R = 0.10$).

The moment at groundline due to applied wire loads equals 45,162 ft-lb, and that due to wind on pole is calculated as:

$$1.0 \text{ (load factor } \gamma) \times 24.58 \text{ sq ft (wind area)} \times 5.77 \text{ psf (wind pressure)} \times 0.9\ (C_f) \times 16.60 \text{ ft (distance to centroid)} = 2{,}119 \text{ ft-lb},$$

yielding a total GLM of 47,281 ft-lb. Using a numerical modeling program to evaluate the nonlinear P-Δ effect, the resulting deflected unbalance is then determined to be 8,420 ft-lb. Thus, final required moment at groundline equals 55,701 ft-lb.

Because the specified moment capacity, or strength, represents the 1st percentile, a strength factor (ϕ) of 1.07 (Table 2-2) is applicable for design of the pole:

$$M_{5\%LEL} = 1.07 \times 57{,}830 \text{ ft-lb} = 61{,}878 \text{ ft-lb}.$$

Since the corresponding capacity of the pole exceeds the required capacity of 55,701 ft-lb, the pole design appears to be sufficient. Other points along the length are also checked to verify the local pole strength exceeds the corresponding local moment. *The selected round steel pole is considered acceptable.*

A.7 EXAMPLE 5: SPUN CONCRETE TRANSMISSION POLE

In this example, a spun concrete transmission pole will be selected to support the loads on the transmission pole described in Section A.2, for construction Grades B and C. Note that the modulus of elasticity for a cracked concrete section is different from the modulus of elasticity for an uncracked section. This characteristic makes the analysis of a prestressed concrete pole somewhat complex because the entire length of the pole does not necessarily become a cracked section when it is subjected to bending loads. Because the modulus of elasticity of a concrete pole is not uniform for its entire length (i.e., it varies depending on the strain at a particular cross section—cracked or uncracked), a computer program that accounts for this nonlinearity in modulus of elasticity was used to calculate the P-Δ effects for the poles in this example.

Grade B Construction or Grade C Construction. A spun concrete pole of the following dimensions and characteristics is considered:

Pole length:	75 ft (65.5 ft above ground)
Tip diameter:	7.75 in.
Tip wall thickness:	2.5 in.
Butt diameter:	21.25 in.
Butt wall thickness:	2.5 in.
Taper:	0.18 in./ft
Groundline diameter:	19.54 in.
Pole projected area:	74.48 sq ft
Distance to centroid:	28.03 ft
Pole shape coefficient:	0.9
Wind pressure on pole:	19.97 psf × shape coefficient
Primary reinforcement:	(12) ½-in., 270-ksi prestressing strands
Pole weight:	8,484 lb
Specified ultimate moment capacity of pole at groundline:	267,500 ft-lb
Lower exclusion limit (% LEL) of specified strength:	5%
COV of pole strength:	0.10
Strength factor, ϕ:	1.0 (Table 2-2 for 5% LEL value and $COV_R = 0.10$).

The moment at groundline due to applied wire loads equals 197,242 ft-lb, and that due to wind on pole is calculated as:

1.0 (load factor γ) × 74.48 sq ft (wind area) × 19.97 psf (wind pressure)
× 0.9 (C_f) × 28.03 ft (distance to centroid)
= 37,522 ft-lb,

yielding a total GLM of 234,764 ft-lb. Using a numerical modeling program to evaluate the nonlinear P-Δ effect, the resulting deflected unbalance is then determined to be 19,711 ft-lb. Thus, the final required moment at groundline equals 254,475 ft-lb.

Because the specified moment capacity, or strength, represents the 5th percentile, a strength factor (ϕ) of 1.00 (Table 2-2) is applicable for design of the pole:

$$M_{5\%LEL} = 1.00 \times 267{,}500 \text{ ft-lb} = 267{,}500 \text{ ft-lb}.$$

Since the corresponding capacity of the pole exceeds the required capacity of 254,475 ft-lb, the pole design appears to be sufficient. Other points along the length are also checked to verify the local pole strength exceeds the corresponding local moment. *The selected concrete pole is considered acceptable.*

A.8 EXAMPLE 6: FIBER-REINFORCED POLYMER DISTRIBUTION POLE

In this example, a fiber-reinforced polymer (FRP) distribution pole will be selected to support the loads on the distribution pole described in Section A.2, for construction Grades B and C.

Grade B Construction. An FRP pole of the following dimensions and characteristics is considered:

Pole length:	45 ft (38.5 ft above ground)
Tip diameter:	10.8 in.
Tip wall thickness:	0.38 in.
Butt diameter:	17.1 in.
Butt wall thickness:	0.34 in.
Taper:	0.14 in./ft
Groundline diameter:	16.2 in.
Pole projected area:	43.31 sq ft
Distance to centroid:	17.97 ft
Pole shape coefficient:	0.9
Wind pressure on pole:	5.77 psf × shape coefficient
Pole weight:	425 lb
Specified ultimate moment capacity of pole at groundline:	109,500 ft-lb
Lower exclusion limit (% LEL) of specified strength:	5%
COV of pole strength:	0.10
Strength factor, ϕ:	1.0 (Table 2-2 for 5% LEL value and $COV_R = 0.10$).

The moment at groundline due to applied wire loads equals 73,907 ft-lb, and that due to wind on pole is calculated as:

1.0 (load factor γ) $\times$ 43.31 sq ft (wind area) $\times$ 5.77 psf (wind pressure) $\times$ 0.9 (C_f) $\times$ 17.97 ft (distance to centroid) = 4,042 ft-lb, yielding a total GLM of 77,949 ft-lb. By modeling the structure to evaluate the nonlinear P-Δ effect, the resulting deflected unbalance is determined to be 24,127 ft-lb. Thus, the final required moment at groundline equals 102,076 ft-lb.

Since the specified moment capacity, or strength, represents the 5th percentile, a strength factor (ϕ) of 1.00 (Table 2-2) is applicable for design of the pole:

$$M_{5\%LEL} = 1.00 \times 109{,}500 \text{ ft-lb} = 109{,}500 \text{ ft-lb}.$$

Since the corresponding capacity of the pole exceeds the required capacity of 102,076 ft-lb, the pole design appears to be sufficient. Other points along the length are also checked to verify the local pole strength exceeds the corresponding local moment. *The selected FRP pole is considered acceptable.*

Grade C Construction. An FRP pole of the following dimensions and characteristics is considered:

Pole length:	45 ft (38.5 ft above ground)
Tip diameter:	10.7 in.
Tip wall thickness:	0.33 in.
Butt diameter:	17.0 in.
Butt wall thickness:	0.29 in.
Taper:	0.14 in./ft
Groundline diameter:	16.09 in.
Pole projected area:	42.98 sq ft
Distance to centroid:	17.96 ft
Pole shape coefficient:	0.9
Wind pressure on pole:	5.77 psf × shape coefficient
Pole weight:	395 lb
Specified ultimate moment capacity of pole at groundline:	87,600 ft-lb
Lower exclusion limit (% LEL) of specified strength:	5%
COV of pole strength:	0.10
Strength factor, ϕ:	1.0 (Table 2-2 for 5% LEL value and COV_R = 0.10).

The moment at groundline due to applied wire loads equals 45,162 ft-lb, and that due to wind on pole is calculated as:

$$1.0 \text{ (load factor } \gamma) \times 42.98 \text{ sq ft (wind area)} \times 5.77 \text{ psf (wind pressure)} \times 0.9\ (C_f) \times 17.96 \text{ ft (distance to centroid)} = 4{,}009 \text{ ft-lb},$$

yielding a total GLM of 49,171 ft-lb. By modeling the structure to evaluate the nonlinear P-Δ effect, the resulting deflected unbalance is determined to be 8,884 ft-lb. Thus, the final required moment at groundline equals 58,055 ft-lb.

Since the specified moment capacity, or strength, represents the 5th percentile, a strength factor (ϕ) of 1.00 (Table 2-2) is applicable for design of the pole:

$$M_{5\%LEL} = 1.00 \times 87{,}600 \text{ ft-lb} = 87{,}600 \text{ ft-lb}.$$

Since the corresponding capacity of the pole exceeds the required capacity of 58,055 ft-lb, the pole design appears to be sufficient. Other points along the length are also checked to verify the local pole strength exceeds the corresponding local moment. *The selected FRP pole is considered acceptable.*

A.9 CALCULATION OF *P-Δ* EFFECT USING THE GERE-CARTER METHOD

When using a first-order linear analysis to calculate a GLM for establishing the required size of pole, the deflected unbalance (P-Δ effect) should be taken into account. One method for doing this is to use the principles set forth in the paper "Critical buckling loads for tapered columns" (Gere and Carter 1962). This method, commonly referred to as the Gere-Carter method, has been widely used throughout this industry for determining the critical buckling (Euler column) loads for tapered columns and estimating the P-Δ effect for various types of tapered poles. The results using this method, however, tend to be more conservative than those obtained using more sophisticated FEA modeling methods. One of the reasons for the degree of conservatism of the Gere-Carter method is the assumption that all vertical loads are concentrated at the top of a fictitious pole, the height of which is somewhat arbitrary. Another limitation of the Gere-Carter method is that it is applicable only to poles of solid cross sections or hollow poles of constant thickness.

As illustrated in Fig. A-3 for a solid, round (i.e., wood) pole, the principles established therein can be used to develop an amplification factor that can be applied to a first-order linear moment to calculate a GLM that accounts for the deflected unbalance. This amplification factor is given by the following formula:

$$\text{Amplification Factor} = \frac{1}{1-\left(\dfrac{VL}{P_{CR}}\right)},$$

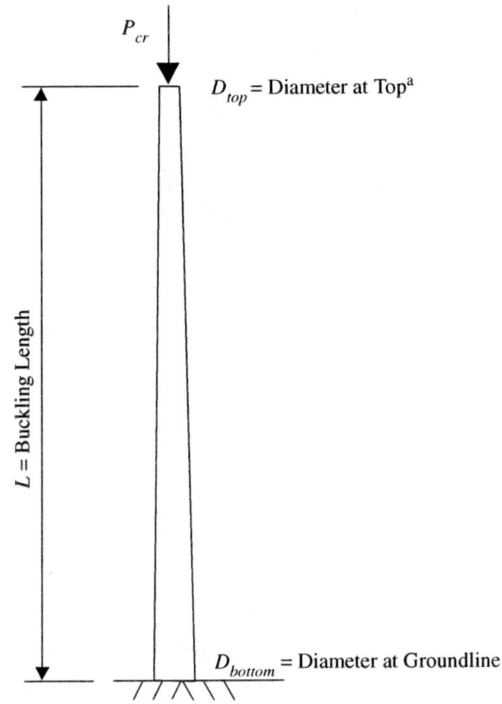

FIGURE A-3. Calculation of P-Δ *Effect.*

where

VL = total vertical load
P_{cr} = buckling load, as calculated below.

$P_{cr} = P_{cr}{'} * P^{*}$,

where,

$$P_{cr}{'} = \frac{\pi^2 * MOE * I_{top}}{[2 * L]^2 * 144}$$

and

$$P^{*} = \left(\frac{D_{bottom}}{D_{top}} \right)^{2.7}.$$

Note: The 2.7 exponent in the above equation is applicable only for solid, round cross sections.

As an example, in the Wood Transmission Pole (Grade B construction) example in Section A.3, the amplification factor of 1.112 was calculated as:

VL = factored weight of pole (above ground) + cross-arms + wires
= (3,551 + 0 + 3,111) × 1.10
= 7,328 lb

L = distance from groundline to centroid of horizontal loads
= 43.80 ft

D_{top} = 12.67 in.

D_{bottom} = 18.33 in.

MOE = mean modulus of elasticity for kiln-dried Douglas fir pole
= 2,376 ksi

I_{top} = moment of inertia for $D\phi_{\text{top}}$ of 12.64 in.
$= \pi \times D\phi_{\text{top}}{}^4 / 64$
$= 1{,}265 \text{ in.}^4$

Thus,

P_{cr}' = 26,845 lb

P^* = 2.71

P_{cr} = 72,750 lb, which yields an amplification factor = 1.112.

Note: This example uses the conservative approach of applying the full weight of the pole above ground (3,551 lb) in the calculation of the concentrated load that is applied to the top of the pole, VL. Since a large portion of this load is in the lower portion of the pole and thus has a lesser influence on P-Δ, a less-conservative approach would be to use only a portion of the pole weight to calculate the value, VL. In this case, using only one-third of the pole's weight would result in an amplification factor of 1.069 rather than 1.112, effecting a total GLM that is approximately the same as that obtained using nonlinear analysis techniques. Should such a reduction factor be used, it must be carefully selected to avoid producing nonconservative results.

Appendix B

EXAMPLES FOR CHAPTER 4: ASSESSING NOMINAL VALUE (R_n)

B.1 METHOD 1: EMPIRICAL ASSESSMENT OF R_n

B.1.1 Example 1: Wood Poles

In 1955, the American Society for Testing and Materials (ASTM) Committee on Wood sponsored a research program to characterize the probability density functions (PDFs) for distribution poles made of common wood species. Table B-1 and Fig. B-1 provide a summary of bending test results for the 106 untreated Southern pine poles tested in the ASTM program. These poles all met the American National Standards Institute (ANSI) *O5.1-2002* (ANSI 2002) minimum specifications. This sample comprised four subspecies in the proportions: 0.42 shortleaf, 0.40 longleaf, 0.10 shortleaf, and 0.08 loblolly pine. The poles ranged in length from 25 to 55 ft. The strength values were evaluated as the maximum groundline bending moment divided by a section modulus determined from the corresponding ANSI size class minimum circumference. Although this approach gives high estimates of groundline strength, when a design engineer multiplies these values by the specified pole class section modulus, the result is (on average) an accurate estimate of groundline moment (GLM) capacity. These values are therefore intended only for use with the ANSI class minimum dimensions. Table B-1 provides a summary of the test values. A student *t*-test was used to compare samples at the 95% confidence on the mean and shows no significant difference for the sample sizes and subspecies tested. The weighted average at the lower confidence bound and weighted variances give an overall mean of 10,480 psi and a coefficient of variance (COV) of 17%.

Table B-2 compares a point estimate of the 5% lower exclusion limit (LEL) with the 5% lower tolerance limit (LTL), 75% confidence limit derived using nonparametric, normal, log-normal, and three-parameter Weibull distribution functions. In this case, the normal distribution gave the most conservative estimate of a 75% tolerance on the 5th percentile and the nonparametric and log-normal distributions gave the highest. For the nonparametric estimate, the 5% LEL point estimate is based on the order statistic defined as $0.05 \times N$, or 5.3. This is 0.3 of the distance

TABLE B-1. Southern Pine Test Pole Evaluation from ASTM Wood Pole Tests

Subspecies	Sample	AMORGL[a]	COV	95% Confidence on Mean			
				t	SE	Low	High
Loblolly	9	10680	14%	1.86	498	9750	11610
Longleaf	41	11020	17%	1.68	293	10530	11510
Shortleaf	45	10920	18%	1.68	293	10430	11410
Short	11	11870	12%	1.81	429	11090	12646

[a] Adjusted modulus of rupture at groundline (AMORGL) is the groundline stress at failure, determined assuming minimum dimensions for the ANSI class and size.

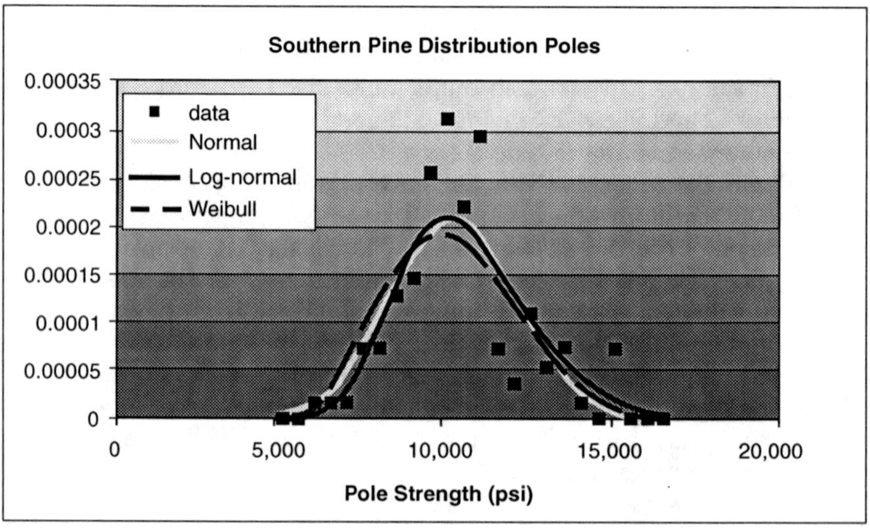

FIGURE B-1. Comparison of Normal, Log-Normal, and Three-Parameter Weibull PDF Fit to Test Data for Southern Pine Distribution Pole Bending Strength.

between the 5th-order (8,068 psi) and 6th-order (8,424 psi) statistics, or 8,170 psi. For the 75% confidence estimate, an interpolated order statistic would be 4.35. From Table 4-1 in Chapter 4, the 75% order statistic is 4 for a sample size of 102, and is 5 for a sample size of 125. This sample size was 106, which is 102 plus 4/23 of the difference between 102 and 125. The 75% confidence value is therefore 8,055. For the normal and log-normal distributions, the point estimate is 1.657 standard deviations below the mean, and the 5% LTL with 75% confidence value is 1.75 standard deviations below the mean in both cases. The Weibull value, corresponding to

TABLE B-2. Comparison of Nonparametric and Parametric Distribution Estimates of the 5% LEL Strength (psi) at 75% Confidence for a Sample of 110 Southern Pine Poles Tested Following the *ASTM Standard D 1036-99* (ASTM 1999b) Test Procedure

Nonparametric		Normal		Log-normal		Weibull	
4th-order statistic	8,052	Mean	10,480	λ	9.243	Shape	3.310
5th-order statistic	8,068	StDev	1,781	μ	–.1688	Scale	6,273
6th-order statistic	8,424	COV	17%			Location	5,403
7th-order statistic	8,553	K_{05}	1.645	K_{05}	1.645		
		K_{05}, K_{75}	1.752	K_{05}, K_{75}	1.752		
Point Estimate 5% LEL	8,170	7,980 (Eq. 4-4)		8,210 (Eq. 4-8)		7,960 (Eq. 4-11)	
75% Conf 5% LTL	8,055	7,786 (Eq. 4-4)		8,065 (Eq. 4-8)		7,805 (Eq. 4-11)	

5% of the area under the probability density function (PDF), varies with the shape scale and location parameters and is unique for each PDF. This can be calculated by downloading the computer program from Forest Products Laboratory (2005).

The 5% LEL and 5% LTL values shown in Table B-2 are derived as groundline strength from tests of green, untreated Southern pine poles. Established design standards for round timber poles and piles, *ASTM Standard D 2899-01* (ASTM 2001) and *AWPA Book of Standards* (American Wood-Preservers' Association [AWPA] 2004), recommend strength reductions for timbers subjected to pretreatment conditioning temperatures higher than 100 °F.

Strength change with height in wood poles is another factor that should be considered when evaluating pole strength. The importance of this adjustment varies with assumptions about pole shape and the method used to derive the groundline design value. Bohannan et al. (1974) concluded that linear interpolation between the *ANSI O5.1-2002* (ANSI 2002) 6-ft-from-the-butt and tip circumferences will underpredict section properties at any location in the pole. They concluded that the magnitude of this underprediction will generally offset any drop in the

TABLE B-3. Sample of 15 Steel Coupon Tests Used to Characterize the Strength in a Population of Steel Poles

Specimen	F_y	Specimen	F_y
1	64.98	9	65.27
2	65.14	10	67.81
3	66.87	11	67.6
4	65.71	12	68
5	67.35	13	65.71
6	68.2	14	68.16
7	65.81	15	67.84
8	67.6		

groundline strength if groundline values are derived to represent the average groundline stress at maximum load, regardless of failure location. The referenced groundline strength, however, can be derived as the actual stress at the failure location extrapolated to groundline using a strengthheight relationship such as that suggested in *ANSI O5.1-2002*, Section 9 (ANSI 2002). If this is done prior to adjusting for minimum class size, then this same change in strength with height must be used when checking for a critical stress location.

The wood utility pole industry standards for derivation of groundline strength have recently been modified and some published strength values are extrapolated values rather than stress-at-failure values. It is important for pole producers to document the method used to derive a groundline strength when supplying those values for use with reliability-based design (RBD).

B.1.2 Example 2: Evaluation of Yield Strength of Steel Using Material Test Data

A sample of 15 steel coupon tests is too small to provide an estimate of the 5th-percentile yield strength using either a nonparametric or a normal distribution assumption alone (Table B-3). Prior knowledge of steel strength PDFs, however, supports the normal assumption as being conservative. Table B-4 provides a comparison of the four PDF assumptions discussed in Chapter 4. In this case, all three parametric assumptions give similar estimates of the 5% lower exclusion value.

TABLE B-4. Comparison of Nonparametric and Parametric Distribution Estimates of the Yield Strength (psi) of 15 Steel Poles at a 5% LTL with 75% Confidence

Nonparametric		Normal		Log-normal		Weibull	
1st-order statistic	64.98	Mean	66.8	λ	4.202	Shape	1.0
2nd-order statistic	65.14	StDev	1.22	μ	0.018	Scale	1.823
3rd-order statistic	65.27	COV	1.82%			Location	64.98
		K	1.645	K	1.645		
		$K_{05},K_{.75}$	1.99	$K_{05},K_{.75}$	1.99		
Point Estimate	NA	64.76		64.78		65.0 7	
75% Conf	NA	64.39		64.41		64.87	

B.2 METHOD 2: MONTE CARLO SIMULATIONS WITH MECHANICS-BASED MODELS

B.2.1 Example 1: Custom-Designed Steel Poles (Range of Pole Sizes)

The following example for a tubular steel pole in bending serves to illustrate the concepts presented in Chapter 4. Note that the data presented in this example are hypothetical and are not based on actual tests or measurements. There are many applications of steel poles where the primary stresses are flexural. *ASCE Manual 72* (ASCE 1990) provides mechanical models for establishing the maximum allowable bending stresses for various cross-sectional geometries of these poles. In this example, a steel pole with a 12-sided cross section will be used to model the technique a supplier could use to develop the PDF for the bending strength of a particular-geometry pole made with a single type and grade of steel. The equations in *ASCE Manual 72* that define the allowable bending stress of a 12-sided pole are:

$$If\ \frac{w}{t} \le \frac{240}{\sqrt{F_y}} \qquad \text{(Eq. B-1)}$$

$$If\ \frac{w}{t} > \frac{240}{\sqrt{F_y}}\ and\ \frac{w}{t} \le \frac{365}{\sqrt{F_Y}},\ then\ F_a = 1.45F_y(1.0 - 0.00129\sqrt{F_y}\,\frac{w}{t}) \qquad \text{(Eq. B-2)}$$

where

w = width and t is the thickness of each side of the pole
F_y = yield stress
F_a = allowable (failure) bending stress. If plate width/thickness ratio (w/t) falls between $240/(F_y)^{0.5}$ and $365/(F_y)^{0.5}$, the pole capacity ($P/A + MC/I = F_a$) is limited by local buckling. Values greater than $365/(F_y)^{0.5}$ are uncommon and no allowable stress equations are provided for them.

Step 1: Obtain Statistical Properties of the Professional Factor, P. The first step is to collect data on the accuracy of the predictive model used in the design (i.e., Eqs. B-1 and B-2). This entails comparing the model predictions with full-scale pole test data, using the actual geometric and material properties extracted individually from each of the specimens tested. This model error is called the professional factor *P*. If we have a high correlation between predicted and measured, and a least-squares regression indicates a consistent, random pattern of residuals over the range of test values, it can be defined as the average ratio of test strength (f_{test}) to calculated strength (f_{calc}). Table B-4 provides a summary of verification test results. The professional factor $P = f_{test}/f_{calc}$ calculated for each of the five tested poles is given in the last column of Table B-5; f_{calc} is obtained from Eqs. B-1 and B-2 using the actual measured dimensions of each pole tested as well as the steel-yield stress data using coupons extracted from each pole.

These five tests indicate a mean yield strength of 69.6 ksi and a standard deviation of 2.69 ksi (COV = 3.87%). Because there are only five tests, the lower 5% tolerance factor for a confidence level of 75% is 2.46, corresponding to an estimated tolerance limit of 62.95 ksi. The standard error (SE), as calculated by Eq. 4-12 in Chapter 4, is then 2.64 ksi, which is 4% of the tolerance value. Thus, because of the low variability of this sample, five tests are sufficient to provide a tolerance estimate with an SE less than 10%.

TABLE B-5. Professional Factor *P* Extracted from Five Full-Scale Custom Pole Tests

Test No.	F_y (ksi)	w (in.)	t (in.)	w/t	S (in.^{3})	M_{ult} (in.-kips)	f_{calc} (ksi)	f_{test} (ksi)	$f_{\text{-ratio}}$
1	69.30	9.12	0.189	48.25	176.90	10,430	58.29	58.96	1.012
2	71.42	8.96	0.224	40.00	203.33	13,030	61.74	64.08	1.038
3	72.15	8.96	0.252	39.82	230.17	15,430	63.67	67.04	1.053
4	69.88	9.54	0.311	30.35	317.34	20,970	64.25	66.08	1.028
5	65.23	9.57	0.377	25.40	398.43	25,890	63.89	64.98	1.017

The professional factor P, estimated as the average ratio of measured to calculated failure stress, is 1.03. Individual values were close, with the measured value exceeding the calculated one in all cases. No relationship appears to exist between the *f-ratio* and the *D/t ratio*, with the highest *f-ratio* occurring at $D/t = 136$, where D is the pole diameter. This corresponds to a $w/t\ (F_y)^{0.5}$ of 295, which is on the border of the local buckling mode. Beyond this point, poles are limited by local buckling. A pattern such as this suggests that a professional factor derived as a function of w/t should be considered. These statistical properties of the professional factor P may be used with different pole sizes within the range of applicability of Eqs. B-1 and B-2.

Step 2: Collect Statistical Data on Geometric Tolerances and Material Strength Variability. The second step in this method requires (1) the statistical sampling of a large number of manufactured tubes to collect data on geometric tolerances, and (2) collection of data on steel-yield stress variability. This manual recommends separate sampling for each unique cross-sectional geometry and each grade/type of steel used for pole manufacture. For this example, which uses a 12-sided regular polygon, the sampling will be done on tubes manufactured from ASTM A572-Grade 65 steel.

The factors that influence the strength of a tube (Eqs. B-1 and B-2) include its yield strength (F_y), the tube diameter (D), the plate thickness (t), and the width of the flat farthest from the neutral axis (w). Measurements of these properties must be made on a large number of randomly selected tubes that represent the full spectrum of tube sizes manufactured. Furthermore, the quantities of tubes for the different size ranges used in this sampling process should be in direct proportion to the average number of tubes expected to be produced within the manufacturing process. For example, if 40% of tube production is expected to use a 3/16-in. plate, then approximately 40% of the tubes used for sampling should have 3/16-in. wall thickness. Table B-6

TABLE B-6. Nominal and Actual Data for Material and Geometric Properties of Samples

Iteration No.	F_y (ksi)	D_{nom} (in.)	W_{nom} (in.)	t_{nom} (in.)	$F_{y\text{-}act}$ (ksi)	D_{act} (in.)	w_{act} (in.)	t_{act} (in.)
1	65.0	22.73	5.59	0.18750	71.32	22.71	5.60	0.189
2	65.0	23.25	5.73	0.18750	69.36	23.25	5.71	0.181
3	65.0	21.62	5.29	0.18750	72.55	21.75	5.25	0.188
4	65.0	21.89	5.36	0.18750	74.11	21.84	5.41	0.190
...								
1,000	65.0	91.58	22.53	0.75000	65.89	91.65	22.51	0.756

indicates the information that would be collected in such a sampling process. In this example, 1,000 tubes were selected for sampling. The nominal properties were taken from the manufacturing drawings and the actual properties were the actual tube diameter, flat width, and wall thickness dimensions. The actual values for F_y were obtained from mill test reports.

In addition, ratios between the actual and nominal geometric property values were calculated and are shown in Table B-7. After all the data have been collected, appropriate PDFs must be defined for each of the individual variables based on the actual properties from Table B-6 ($F_{y\text{-}act}$, D_{act}, w_{act}, and t_{act}), as well as for each of the actual-to-nominal ratios of the different properties given in Table B-7. Although a log-normal distribution is usually preferred, it is often acceptable to assume a normal distribution. In the present example, a normal distribution is assumed for convenience. The final results are not significantly different from those obtained using a log-normal distribution.

To simulate a realistic set of geometric and material properties that is truly representative of manufacturing, correlations between the different variables and between these variables and the actual-to-nominal ratios must be defined. Since there are strong correlations between many of the variables, a proper Monte Carlo simulation cannot be run without them.

Step 3: Conduct Monte Carlo Simulations of Actual Bending Strengths for Each Pole Diameter and Thickness. The third step is to conduct a Monte Carlo simulation using these established PDFs, along with their corresponding correlation values, to create a large database of theoretical tube sizes and strengths. In this example, 5,000 iterations were made in order to generate enough pole strength values to accurately model the tail ends of the strength distribution curves and thus establish a reasonable 5% LEL. Table B-8 illustrates the data that were generated using this Monte Carlo technique.

For each iteration, both a theoretical tube strength and a nominal tube strength are calculated. The theoretical tube strength is calculated by using

TABLE B-7. Ratios of Actual-to-Nominal Geometric Property Values

No.	D_{act}/D_{nom}	w_{act}/w_{nom}	t_{act}/t_{nom}
1	0.9991	1.0019	1.008
2	1.0000	0.9968	0.965
3	1.0060	0.9921	1.003
4	0.9977	1.0086	1.013
. . .			
1,000	1.0008	0.9990	1.008

TABLE B-8. Values Generated from Monte Carlo Simulation

Iteration	F_y (ksi)	D (in.)	w (in.)	t (in.)	D_{act}/ D_{nom}	w_{act}/w_{nom}	t_{act}/t_{nom}
1	77.0690	13.5378	3.3734	0.1971	1.00004	1.004776	1.001151
2	73.8009	27.6386	6.8492	0.2744	1.000465	0.99783	0.998245
3	68.6695	45.6283	11.3003	0.3967	0.998851	0.999472	0.997051
4	66.3786	48.9046	12.1154	0.4223	0.997202	1.001781	1.005124
5	70.6206	36.9458	9.1406	0.3242	0.997005	0.997572	0.995524
6	72.7404	17.9316	4.4343	0.2195	1.002253	0.996617	1.002227
7	70.5583	30.5654	7.5615	0.2929	0.999316	1.001459	0.994127
8	69.1674	43.2234	10.6984	0.3669	0.999463	1.003988	1.008479
. . .							
5,000	66.3117	64.7032	16.0296	0.5136	1.00129	1.00037	1.004426

TABLE B-9. Resulting Nominal and Actual Tube Strengths

Iteration	S_{act} (in.^{3})	M_{act} (in.-kips)	S_{nom} (in.^{3})	M_{nom} (in.-kips)	M_{act}/M_{nom}
1	28.4	2,190	28.4	1,844	1.187
2	167.2	12,339	167.3	10,876	1.135
3	661.4	45,416	664.8	43,214	1.051
4	808.8	53,689	809.4	52,610	1.020
5	354.2	25,016	358.0	23,267	1.075
6	55.9	4,067	55.5	3,610	1.127
7	218.5	15,417	220.1	14,303	1.078
8	549.2	37,850	545.3	35,445	1.068
. . .					
5,000	1,725.7	111,528	1,713.8	108,882	1.024

the values from columns 2 through 5 in Table B-8 to establish a theoretical section modulus, S_{act}, and allowable bending stress value (using Eqs. B-1 and B-2), from which a theoretical bending moment M_{act} is calculated. The nominal tube strength is calculated by using the ratios of actual-to-nominal values to adjust the geometric properties from which a nominal section modulus, S_{nom}, is calculated. This nominal section modulus is then coupled with the nominal allowable bending stress value, which is based on a nominal yield stress value of 65 ksi, to calculate a nominal bending moment M_{nom} for the same tube. Ratios between the theoretical bending moment and nominal bending moment for each tube can then be calculated and an appropriate PDF defined. Table B-9 provides the resulting nominal and actual tube strengths.

In this example, the parameters that define the PDF of this strength ratio include a mean of 1.073 and COV of 3.68%. For an assumed normal distribution, the resulting 5% LTL is calculated as $1.073 \times (1 - 1.646 \times 0.0368) = 1.008$.

The COV of pole strength is also used to obtain the resistance factor ϕ from Table 2-2 in Chapter 2. The row corresponding to an LEL of 5% should be used.

B.2.2 Example 2: Commodity Steel Poles (Single-Size Round Pole)

The objective is to obtain bending strength properties (PDF, COV_R, and 5% LEL) for a round commodity pole section with single type and grade of steel, for a given nominal diameter and given nominal thickness. Since the procedures and principles are very similar to those illustrated in Example 1 (beginning on page 79), the steps and detailed results are outlined below. Again, the data presented in this example are hypothetical and are not based on actual tests or measurements.

Since this example applies to a specific steel pole design with a round section, the exercise must be repeated if either the nominal diameter of the pole, the nominal thickness, the type or grade of steel, or the cross-sectional shape changes.

The equations in *ASCE Manual 72* (ASCE 1990) that define the allowable bending stress of a round pole are shown below:

$$\frac{D}{t} \leq \frac{6{,}000}{F_y} \qquad \text{(Eq. B-3)}$$

$$\frac{6{,}000}{F_y} \leq \frac{D}{t} = \frac{12{,}000}{F_y} \quad F_b = 0.70F_y + \frac{1{,}800}{D/t} \qquad \text{(Eq. B-4)}$$

where

D = outer diameter and t is the thickness of the pole
F_y = yield stress and F_b is the allowable (failure) bending stress.

Step 1: Obtain Statistical Properties of the Professional Factor, P. The first step is to collect data on the accuracy of the predictive model used in the design (Eqs. B-1 and B-2, in this case). This entails comparing the model predictions with full-scale pole test data, using the actual geometric and material properties extracted individually from each of the specimens tested, within the range of applicability of Eqs. B-1 and B-2. Table B-10

TABLE B-10. Professional Factor *P* Extracted from Five Full-Scale Commodity Pole Tests

Test No.	F_y (ksi)	D (in.)	t (in.)	D/t	S (in.^{3})	M_{ult} (in.-kips)	f_{calc} (ksi)	f_{test} (ksi)	$\frac{f_{test}}{f_{calc}}$
1	69.3	34.7911	0.189	184.08	176.90	10,430	58.29	58.96	1.0115
2	71.4	34.3190	0.224	153.21	203.33	13,030	61.74	64.08	1.0379
3	72.1	34.4660	0.252	136.77	230.17	15,430	63.67	67.04	1.0530
4	69.8	36.4958	0.311	117.35	317.34	20,970	64.25	66.08	1.0284
5	65.2	37.2325	0.377	98.76	398.43	25,890	63.89	64.98	1.0171

is a summary of the five full-scale tests conducted to verify the accuracy of the model.

Based upon these tests, the following values of the mean and COV of the professional factor P are obtained as follows: $P_m = 1.030$ and $COV_P = 0.161$. As in Example 1, these results suggest that the theoretical model (Eqs. B-1 and B-2) is somewhat conservative and that a professional factor P of greater than 1.0 is justified.

Step 2: Collect Statistical Data on Geometric Tolerances and Material Strength Variability. For this example, which uses a round section, the sampling will be done on tubes manufactured from ASTM A572-Grade 65 steel, with one nominal wall thickness (0.170 in.) and one nominal diameter (17.50 in.). The factors that influence the strength of a tube in Eqs. B-1 and B-2 include its yield strength (F_y), the tube diameter (D), and the plate thickness (t).

Table B-11 shows the information that would be collected in such a sampling process. In Table B-11, 1,000 tubes were selected for sampling. The properties measured were the actual tube diameter and wall thickness dimensions on tubes manufactured for this nominal design.

In the case of the yield stress, F_y, the actual values are obtained from records on the corresponding mill test reports. An example of data for F_y is given in Table B-12.

Step 3: Conduct Monte Carlo Simulations of Actual Bending Strengths for Each Pole Diameter and Thickness. Table B-13 provides a sample of the

TABLE B-11. Geometric Tolerance Data Collected

Iteration	D_{act} (in.)	t_{act} (in.)
1	17.50	0.177
2	17.48	0.178
3	17.52	0.177
4	17.54	0.176
...		
1,000	17.49	0.176
Mean	17.50	0.176
COV	0.14%	0.64%

data that were generated using the Monte Carlo technique, based on the corresponding PDFs.

The simulated tube strength values are calculated by using the values from columns 2 through 4 in Table B-13 to establish a section modulus, S_{act}, and allowable bending stress value F_a (using Eqs. B-1 and B-2) and moment capacity M_{act} for each simulated pole. The nominal tube strength is calculated by using the nominal material and geometric properties. In this example, the nominal section modulus, S_{nom}, is 39.74 in.^{3}, the allow-

TABLE B-12. Material Tolerance Data (F_y) Collected

Wall Thickness	0.170 in.	
Yield Stress (ksi)	$F_{y\text{-}nom}$	$F_{y\text{-}act}$
Coupon 1	65	71.32
Coupon 2	65	69.36
Coupon 3	65	72.55
Coupon n	65	74.11
Mean		70.96
COV		2.68%

TABLE B-13. Simulations for a Round Pole with the Following Nominal Properties: F_{y-nom} = 65 ksi, D_{nom} = 17.50 in., and t_{nom} = 0.170 in.

Iteration	$F_{y\text{-act}}$ (ksi)	D_{act} (in.)	t_{act} (in.)	S_{act} (in.^{3})	$F_{y\text{-}act}$ (ksi)	M_{act} (in.-kips)
1	70.093	17.51	0.1714	40.10	66.69	2,674
2	71.166	17.43	0.1743	40.40	67.82	2,739
3	70.956	17.48	0.1723	40.19	67.41	2,708
4	72.121	17.46	0.1708	39.75	68.09	2,706
5	70.595	17.45	0.1744	40.52	67.40	2,731
6	69.376	17.50	0.1727	40.35	66.32	2,676
7	68.782	17.45	0.1818	42.18	66.90	2,822
8	75.249	17.47	0.1816	42.19	71.39	3,011
...						
5,000	67.496	17.454	0.1732	40.26	65.11	2,621

able bending stress, f_{nom}, is 62.99 ksi, and the nominal moment capacity, M_{nom}, is 2,503 in.-kips.

In this example, the parameters that define this PDF (assumed normal distribution) include a mean of 2,788 and COV_R of the pole strength is 0.0362 in.-kips. The resulting 5% LEL of the pole bending strength capacity is calculated as $2{,}788 \times (1 - 1.646 \times 0.0362) = 2{,}622$ in.-kips, which is higher than the nominal bending strength capacity of 2,503 in.-kips that would otherwise be used for design purposes.

Appendix C

REFERENCES

Aluminum Association, Inc. (AAI). (1982). *Aluminum electrical conductor handbook*, M. Walker, ed., Washington, D.C.

American Association of State Highway and Transportation Officials (AASHTO). (2002). *Standard specifications for highway bridges*, 17th Ed., Washington, D.C.

American Concrete Institute (ACI). (2002). "Building code requirements for reinforced concrete." *ACI 318-02*, ACI Committee 318, Detroit, Mich.

American Forest and Paper Association, American Wood Council (AF&PA). (1996). *Load and resistance factor design (LRFD) manual for engineered wood construction*, Washington, D.C.

American Institute of Steel Construction (AISC). (1999). *Load and resistance factor design specification for structural steel buildings*, Chicago.

American Iron and Steel Institute (AISI). (1996). "Appendix: tests for determining structural performance." *Specification for the design of cold-formed steel structural members*, Washington, D.C.

American National Standards Institute (ANSI). (1996). "American national standard for roadway lighting equipment-fiber-reinforced plastic (FRP) lighting poles." *ANSI C 136.20*, New York.

American National Standards Institute (ANSI). (2002). "Specifications and dimensions (for wood poles)." *ANSI O5.1-2002*, Alliance for Telecommunications Industry Solutions, Washington, D.C.

American Society for Testing and Materials (ASTM). (1992). "Standard specification for reinforced thermosetting plastic poles." *Standard D 4923-92*, West Conshohocken, Penn.

American Society for Testing and Materials (ASTM). (1993)."Standard specification for computing the reference resistance of wood-based materials and structural connections for load and resistance factor design." *Standard D 5457-93*, West Conshohocken, Penn.

American Society for Testing and Materials (ASTM). (1997). "Standard specification for spun cast prestressed concrete poles." *Standard C 1089-97*, West Conshohocken, Penn.

American Society for Testing and Materials (ASTM). (1998). "Standard test method for establishing clear wood strength values." *Standard D 2555-98*, West Conshohocken, Penn.

American Society for Testing and Materials (ASTM). (1999a). "Standard practice for evaluating allowable properties for grades of structural lumber." *Standard D 2915-99*, West Conshohocken, Penn.

American Society for Testing and Materials (ASTM). (1999b). "Standard test methods of static tests of wood poles." *Standard D 1036-99*, West Conshohocken, Penn.

American Society for Testing and Materials (ASTM). (2001). "Standard practice for establishing allowable stresses for round timber." *Standard D 2899-01*, West Conshohocken, Penn.

American Society of Civil Engineers (ASCE). (1990). "Design of steel transmission pole structures." *ASCE Manuals and Reports on Engineering Practice No. 72*, New York.

American Society of Civil Engineers (ASCE). (1991). "Guidelines for electrical transmission line structural loading." *ASCE Manuals and Reports on Engineering Practice No. 74*, Reston, Va.

American Society of Civil Engineers (ASCE). (1992) "Design of latticed steel transmission structures." *ASCE/ANSI 10-90*, New York.

American Society of Civil Engineers (ASCE). (2000). "Design of latticed steel transmission structures." *ASCE Standard 10-97*, Reston, Va.

American Society of Civil Engineers (ASCE). (draft). "Guidelines for electrical transmission line structural loading." *ASCE Manuals and Reports on Engineering Practice No. 74*, Reston, Va.

American Society of Civil Engineers (ASCE). (2002). "Minimum design loads for buildings and other structures." *ASCE Standard 7-02*, Reston, Va.

American Society of Civil Engineers (ASCE). (2003). "Recommended practice for fiber-reinforced polymer products for overhead utility line structures." *ASCE Manuals and Reports on Engineering Practice No. 104*, Reston, Va.

American Wood-Preservers' Association (AWPA). (2004) *AWPA Book of Standards*, Birmingham, Al.

Anderson, T. W., and Darling, D. A. (1952). "Asymptotic theory of certain goodness of fit criterion based on stochastic processes." *Ann. Math Stat.* 23, 193–212.

Balci, O., and Sargenti, R. G. (1984). "Validation of simulation models via simultaneous confidence intervals." *Am. J. Math Management Sci.*, 4:375–406.

Behncke, R. (1998). "500 Kv compact line design for the greater Bangkok area." CIGRE, Paris, 22\33\36.

Bodig, J., Goodman, J. R., and Brooks, T. (1986). "Wood pole properties, review and recommendations for design resistance data." *Western red cedar data and size effect*, Vol. 3, EPRI EP4109, Electric Power Research Institute, Palo Alto, Calif.

Bohannan, B., Habermann, H., and Lengel, J. E. (1974). "Taper of wood poles." *Gen. Tech. Rep. FPL-GTR-2*, U.S. Department of Agriculture, Forest Service, Forest Products Laboratory, Madison, Wisc.

Dagher, H. J., Kulendran, S., Peyrot, A. H., Maamouri, M., and Lu, Q. (1993). "System reliability concepts in the design of transmission lines." *J. Struct. Engrg.* ASCE, 119(1), 323–340.

Electrical Power Research Institute (EPRI). (1995). "Reliability-based design of foundations for transmission line structures." *Report TR-105000*, Palo Alto, Calif.

Ellingwood, B., et al. (June 1980). "Development of a probability-based load criterion for American National Standard A 58." *NBS Special Publication 577*, U.S. Department of Commerce, National Bureau of Standards, Washington, D.C.

Forest Products Laboratory (FPL). (June 2005). http://www1.fpl.fs.fed.us/nonpar.html

Gere, J., and Carter, W. (1962). "Critical buckling loads for tapered columns." *J. Struct. Engrg.*, ASCE, 88(1)1–11.

Guttman, I. (1970). *Statistical tolerance regions: classical and Bayesian,* Hafner Publishing Co., Darien, Conn., 88–93.

Hammersley, J. M., and Handscomb, D. C. (1964). *Monte Carlo methods,* Methuen, London.

Institute of Electrical and Electronics Engineers (IEEE). (1992). "Guide to the installation of overhead transmission line conductors." *IEEE Standard 524,* New York.

Institute of Electrical and Electronics Engineers (IEEE). (1996). "IEEE trial use guide for fall protection of the utility industry." *IEEE Standard 1307,* New York.

Institute of Electrical and Electronics Engineers (IEEE). (2002). *National electrical safety code, C2-2002,* New York.

International Electrotechnical Commission (IEC). (2002). "Design criteria of overhead transmission lines." *IEC 60826,* Technical Committee No. 11, Geneve, Switzerland.

Law, A. M., and Kelton W. D. (2000). *Simulation modeling and analysis,* 3rd Ed., McGraw-Hill, New York.

National Oceanic and Atmospheric Administration (NOAA). (1959–1995). *Storm data,* Washington, D.C.

Natrella, M. G. (1963). "Experimental statistics." *National Bureau of Standards Handbook, 91,* U.S. Government Printing Office, Washington, D.C.

Peyrot, A. H., and Dagher, H. J. (1984). "Reliability-based design of transmission line structures." *J. Struct. Engrg.,* ASCE, 110(11)2758–2777.

Precast/Prestressed Concrete Institute (PCI). (1997). "Guide for the design of prestressed concrete poles." Task Force/PCI Committee on Concrete Poles, *PCI Journal,* 42(6).

Precast/Prestressed Concrete Institute (PCI). (1999). "Specification guide for prestressed concrete poles." PCI Committee Report, *PCI Journal,* 44(2), 80–87.

Twisdale, L. A. (1982). "Wind loading underestimate in transmission line design." *Transmission and Distribution,* Dec., 40–46.

Wilkes, S. S. (1944). *Mathematical statistics.* Vol. XV, John Wiley & Sons, New York, 217.

Appendix D

NOTATION AND SI CONVERSION FACTORS

D.1 NOTATION

The following notation is used in this manual:

A	=	projected wind area (ft^2)
B	=	response term in calculating the wind force, F
C_f	=	shape factor (force, or drag, coefficient), *ASCE Manual 74* (ASCE 1991)
COV_X	=	coefficient of variation, measure of variability of the structural strength ($X = R$) or other paraeter of interest (for example, $X = P, O$); defined as the standard deviation divided by the mean value.
C&M	=	loads produced by construction and maintenance operations
D	=	outer diameter of steel pole
D_{bottom}	=	diameter at bottom (groundline) of pole
D_{top}	=	diameter at top of pole
DL	=	dead loads, such as weights of bare wires, hardware, insulators, and supporting structures
e	=	base of the natural logarithm = 2.718282
E	=	exposure factor in calculating the wind force, F
f_{act}	=	actual bending stress
f_{nom}	=	allowable bending stress
F	=	wind force, in pounds
F_a	=	allowable (failure) bending stress of multisided steel pole (Eqs. B-1 and B2)
F_b	=	allowable (failure) bending stress for round steel pole (Eqs. B–3 and B–4)
$F_{b5\%LEL}$	=	5% LEL bending stress
F_y	=	yield stress of steel

FA = variability in strength property introduced during fabrication

G = gust response factor as described in the forthcoming edition of *ASCE Manual 74* (ASCE 2002) or *National Electric Safety Code, C2-2002* (*NESC*) (IEEE 2002) for wires and structures

H = effective height in calculating K_z (ft)

I_{50}, I_{75} confidence level factors, nonparametric PDF (Table 4-1)

I_{top} = moment of inertia at top of pole

K = distance from the mean of a normal population to a target LEL in units of standard normal deviations (Eq. 4-4)

$K\xi$ K factor derived to provide confidence ξ that the sample LEL will be less than or equal to the target value of the parent population (Table 4-1)

k_N = log-normal factor relating mean strength to nominal 5% LTL (Eqs. 4-8, 4-9)

k_v = factor used in calculating the wind force F

K_z = velocity pressure exposure coefficient

L = distance from groundline to centroid of horizontal loads on pole

LEL = lower exclusion limit of distribution identifies the upper limit on a prescribed portion of the low tail of the distribution. For example, 5% of the area under a probability distribution function will lie below the 5% LEL and 95% will lie above it.

LL = legislated loads, which are any set of minimum loadings specified by law for the particular service area of the line; the most common of these are the *NESC* district loadings with the appropriate *NESC* overload and strength factors (IEEE 2002).

1n[ ] = natural logarithm (base e)

LTL = lower tolerance limit (of a strength property). The LTL strength is an estimate of the LEL strength for a finite sample size. The LTL is associated with a confidence value that it is less than or equal to the corresponding population LEL. For example, the 5% LTL of pole strength with 50% confidence calculated using a small sample has a 50% probability of being smaller than the population 5% LEL

$m_{[\,.\,]}$ = mean of variable in the subscript [.]

m_P = mean of professional factor, P

M = variability of strength property attributed to material variability

M_{nom} = nominal moment capacity
$M_{5\%\text{LEL}}$ = 5% LEL bending moment
MOE = modulus of elasticity
N = number of specimens in a sample
O = variability in strength property introduced by other effects such as deterioration, design error, and environmental risk. These effects must be combined to give an estimate of the pole strength variability.
P = professional factor representing the accuracy of a strength predictive model
P_{cr} = critical buckling load of pole
PDF = probability density function. Parametric or nonparametric function used to represent frequency of occurrence over a representative range of values of a random variable.
Q_{50} = loads produced by the wind velocity V_{50} or combinations of ice and wind that have a 50-year (approximate) return period (RP)
Q_{RP} = loads similar to Q_{50}, except that they have an RP-year return period
R = resistance or strength of a component or a structure
R_e = e% lower exclusion limit of strength, corresponding to the strength level for which there is an e% probability that a given pole or structural component will be less than or equal; thus, (100 – e)% of the poles or components exceed this strength level.
R_m = mean strength
R_{n} = nominal or characteristic strength of a component, calculated using an approved design guide, defined as the 5% lower exclusion limit (LEL), or as estimated by the 5% lower tolerance limit (LTL), at a specified confidence (e.g., 50% or 75%).
S = section modulus
S_{act} = actual section modulus
S_{nom} = nominal section modulus
s = sample standard deviation
SL = security loads such as anticascading loads that will result in a structure with sufficient strength to limit the consequences of failures from weather-related or accidental events
t = plate thickness of steel pole side or steel pole tube
T/D = transmission and distribution

V_{50} = 3-s gust wind velocity with a 50-year return period (RP)
VL = total vertical load on pole (Appendix A.9)
w = width each side (flat) of steel pole section
β = reliability index. (See Eq. 1-1.)
γ = load factor applied to the load effect Q_{50} under consideration
γ_{CM} = the load factor applied to all the loads in Eq. 2-3
θ = scale parameter for the Weibull distribution function (Eq. 4-10)
η = proportion (decimal value) of the range of values in a parent population included in a sample of N specimens
λ = mean of the natural logarithms of the data points
μ = standard deviation of the natural logarithm of the data points
ξ = assurance (%) that the range of values in a sample will encompass a prescribed portion of the parent population
σ = standard deviation for normal probability distribution function
σ_R = standard deviation of strength
ϕ = strength (or resistance or capacity) factor. The ϕ takes into account variability in material, dimensions, workmanship, and the uncertainty inherent in the equation used to calculate R_n as a predictor of the true strength of the component. It can also be used to coordinate the strengths of components and subsystems by adjusting their reliabilities. Procedures for selecting ϕ are provided in this document.
ϕ_{LL} = a strength factor to use with legislated loads (this factor is to be obtained from the appropriate code documents).
χ_0 = location parameter for the Weibull distribution function (Eq. 4-10)
Ω = coefficient of variation of the strength data (σ_R/R_m)
ω = slope or shape factor for the Weibull distribution function (Eq. 4-10)
$[\,.\,]m$ = mean of variable [.]

D.2 SI CONVERSION FACTORS

1 ft	=	0.305 meter (m)
1 ft-lb	=	1.36 m-N
1 in.	=	25.4 millimeters (mm)
1 pound (lb)	=	4.45 newtons (N)
1 lb/ft	=	14.6 N/m
1 lb/ft^2 (psf)	=	47.8 pascals (Pa) (N/m^2)
1 lb mass/ft^3 (pcf)	=	0.016 gram/cubic centimeter (g/cc)
1 mile per hour (mph)	=	0.45 meter/s (m/s)

To convert temperature, θ, from °F to °C, T (°C) = 5/9 [T(°F) − 32 °F].

INDEX

(e), equation; (f), figure; (t), table

Environmental Issues and Waste Management Technologies in the Materials and Nuclear Industries XII

Environmental Issues and Waste Management Technologies in the Materials and Nuclear Industries XII

Ceramic Transactions, Volume 207

A Collection of Papers Presented at the 2008 Materials Science and Technology Conference (MS&T08)
October 5–9, 2008
Pittsburgh, Pennsylvania

Edited by

Alex Cozzi
Tatsuki Ohji

The American Ceramic Society

WILEY

A John Wiley & Sons, Inc., Publication

Published by John Wiley & Sons, Inc., Hoboken, New Jersey.
Published simultaneously in Canada.

Library of Congress Cataloging-in-Publication Data is available.

ISBN 978-0-470-40848-3

Contents

Preface

The annual symposium on Environmental and Energy Issues took place in conjunction with MS&T'08, October 5–9, 2008 in Pittsburgh, Pennsylvania. Symposia for "Ceramics and Glass for Waste Minimization, Stabilization, and Disposition and Waste" and "Green Technologies for Materials Manufacturing and Processing" were sponsored by the Nuclear and Environmental Technology and Engineering Ceramic Divisions of the American Ceramic Society, the ASM International, and the International Commission on Glass Technical Committee on Nuclear and Hazardous Waste Vitrification. Several sessions with focused topics in nuclear, environmental, and green engineering were held, including a session of invited talks on Waste Glass Leach Testing and Modeling followed by a panel discussion. This volume documents a number of the papers presented at these symposia.

The success of the symposium and the issuance of the proceedings could not have been possible without the support of The American Ceramic Society and the other organizers of the program. The program organizers included Allen Apblett, Carol Jantzen and James Marra. Their assistance, along with that of the session chair's, was invaluable in ensuring the creation of quality proceedings.

Alex Cozzi
Kevin Fox
Liz Hoffman
Tatsuki Ohji
Mrityunjay Singh
Richard D. Sisson, Jr.

Ceramics and Glass for Waste Minimization, Stabilization and Disposition

GLASSES FOR IMMOBILIZING LANTHANIDE, ALKALI, AND ALKALI-EARTH FISSION PRODUCTS

J. V. Crum and J. D. Vienna

Pacific Northwest National Laboratory
K6-24 P.O. Box 999
Richland, WA 99354

ABSTRACT

A series of glasses were formulated for the immobilization of a potential waste stream from commercial nuclear fuel reprocessing, the combined lanthanide (LN), alkali, and alkaline earth (Cs/Sr) fission products. These glasses were formulated to meet repository disposal requirements while being processable in a cold-crucible melter. The glasses were fabricated and tested for product consistency test response, phase characterization, density, and glass transition temperature. The results suggest that the combined fission product waste forms are likely to meet repository requirements and generate less glass than if individual streams were vitrified.

INTRODUCTION

One major goal of the Global Nuclear Energy Program/Advanced Fuel Cycle Initiative (GNEP/AFCI) is to close the nuclear fuel cycle by recycling fissionable components of spent nuclear fuel. Non-fissionable components separated from the spent fuel will need to be immobilized in waste forms suitable for disposal. The program is in the process of addressing several questions related to wasteforms, treatment options, and disposal pathways. At the time of this study, the program was in the process of indentifying baseline and backup waste forms for each of the waste streams, individually. The reasoning behind treatment of individual streams is first to minimize the amount and toxicity of waste going to the high-level waste repository and to closely match the elements to be immobilized and the disposal environment through tailored waste forms. This study was undertaken to investigate combining the LN and Cs/Sr waste streams (shown in Table 1) into a single glass waste form. The waste stream contains 15 wt% alkali oxides, 17 wt% alkali-earth oxides, and 68 wt% lanthanide oxides.

Table 1 Composition of combined Cs/Sr and LN streams in terms of individual oxides

Component	Oxide, wt%	Component	Oxide, wt%
Cs_2O	0.13285	Gd_2O_3	0.01031
Rb_2O	0.01704	La_2O_3	0.07986
BaO	0.12687	Nd_2O_3	0.26366
SrO	0.04389	Pr_2O_3	0.07282
Ce_2O_3	0.15653	Sm_2O_3	0.05531
Eu_2O_3	0.00907	Y_2O_3	0.03178

The glasses in this study were designed to achieve a high concentration of glass modifiers (waste loading) in relation to glass formers using the lanthanide borosilicate glass (LaBS) system. LaBS glass was shown to successfully accommodate up to ~62 wt% lanthanides in glass with acceptable processing and performance properties [1]. There are also examples of glasses in literature that have high concentrations of alkaline earths or alkalis [2]. However there is a lack of data available in the literature for glasses with high concentrations of all three combined.

EXPERIMENTAL

Four glasses were formulated with loadings ranging from 40 wt% to 62 wt%. They were batched from oxides, carbonates and boric acid using reagent grade chemicals weighted out to the nearest 0.01 g (in 250g batches). Glass batches were then melted in a platinum crucible with a tight fitting lid at 1350-1550°C. Each glass was melted for 1 hr then cast onto a stainless steel plate. The glass was ground to a fine powder using a tungsten-carbide puck and chamber in a vibratory mill to homogenize the glass. The powdered glass was then remelted for additional hour and cast onto a stainless steel plate.

A sample of each glass was ground to a fine powder then analyzed by X-ray powder diffraction to characterize crystalline phases present in the quenched glass. The 7-day product constancy test (PCT) was performed on the quenched glass samples [3]. Densities were measured using helium pycometery. Lastly, the glass transition temperatures (T_g) were measured using a simultaneous differential temperature analyzer/thermogravinometric analyzer (DTA/TGA).

RESULTS

The glass composition, waste loading and measured properties XRD, PCT, Density, and T_g, are listed in Table 2. The XRD results confirm that quenched glasses 1C, 40 wt% loading, and 2C, 50 wt% loading, are crystal free. At waste loading of 55 wt% the quenched glass contains a minor amount of $CsAlSiO_4$. At waste loading of 62 wt% the quenched glass contains two major crystalline phases ($CsAlSiO_4$, $La_{4.67}Si_3O_{13}$) and one minor phase ($Ba_4Al_2O_7$) as shown in Figure 1.

Table 2. Targeted Glass Composition in mass fraction and Predicted and Measured Glass Properties

Glass ID	1C	2C	3C	4C
Al_2O_3	0.1700	0.1700	0.1700	0.1000
B_2O_3	0.1000	0.1000	0.0500	0.0500
BaO	0.0507	0.0634	0.0698	0.0786
Ce_2O_3	0.0626	0.0783	0.0861	0.0970
Cs_2O	0.0531	0.0664	0.0731	0.0824
Eu_2O_3	0.0036	0.0045	0.0050	0.0056
Gd_2O_3	0.0041	0.0052	0.0057	0.0064
La_2O_3	0.0319	0.0399	0.0439	0.0495
Nd_2O_3	0.1054	0.1318	0.1450	0.1635
Pr_2O_3	0.0291	0.0364	0.0400	0.0451
Rb_2O	0.0068	0.0085	0.0094	0.0106
SiO_2	0.3300	0.2300	0.2300	0.2300
Sm_2O_3	0.0221	0.0277	0.0304	0.0343
SrO	0.0176	0.0219	0.0241	0.0272
Y_2O_3	0.0127	0.0159	0.0175	0.0197
Total	1.0000	1.0000	1.0000	1.0000
Waste Loading (wt%)	40	50	55	62
Calculated Density (g/mL)	3.20	3.53	3.72	4.06
Measured Density (g/mL)	3.05	3.51	3.81	3.99
T_g (°C)	702	715	770	755
XRD Quenched	**1C**	**2C**	**3C**	**4C**
$CsAlSiO_4$			Minor	Major
$La_{4.67}Si_3O_{13}$				Major
$Ba_4Al_2O_7$				Minor
Normalized PCT release	**1C**	**2C**	**3C**	**4C**
r_B (g/m^2)	0.0214	0.0241	0.0185	0.0647
r_{Ba} (g/m^2)	0.0105	0.0167	0.0154	0.0373
r_{Si} (g/m^2)		0.0094	0.0073	0.0105
r_{Sr} (g/m^2)	0.0256	0.0356	0.0282	0.0491

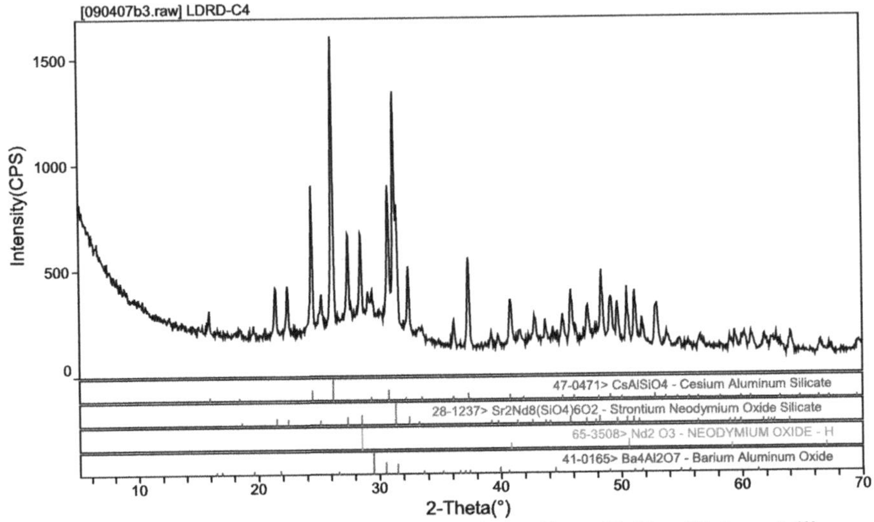

Figure 1. XRD scan of quenched glass at 62 wt% loading with identified crystalline phases.

Results of the 7 day PCT show, all of the analytes are at least two orders of magnitude below those of the environmental assessment (EA) glass [4]. The limited number of data indicates that as waste loading and/or crystallinity increases there is only a minimal negative impact on the PCT response.

Measured densities of the quenched glasses agree well with the model used for prediction [5]. The measured density increases (roughly linearly) with waste loading from 3.05 g/cm^3 at 40 wt% loading up to 4.06 g/cm^3 at 62 wt% loading, as shown in Figure 2.

The T_g is important for immobilization of highly radiolytic waste (from ^{134}Cs, ^{137}Cs, and ^{90}Sr in this case). Estimates of resulting temperatures are highly dependent on isotope concentrations, canister size, and cooling method. With these parameters yet unknown, we strive for high T_g to allow flexibility in the immobilization, storage, and transport system. The measured T_g of these glasses, shown in Figure 3, range from 700°C up to 770°C. T_g is plotted versus the ratio of glass formers (SiO_2 and $BO_{1.5}$) to glass modifiers (balance of the glass) on a mole fraction basis. The plot shows a sharp increase in T_g for glasses in the region rich in glass modifiers as compared to the smaller effect in the region rich in glass formers, with the change in slope occurring around a one-to-one ratio of former to modifiers (the inversion point). More data are needed to support and confirm the limited number of data.

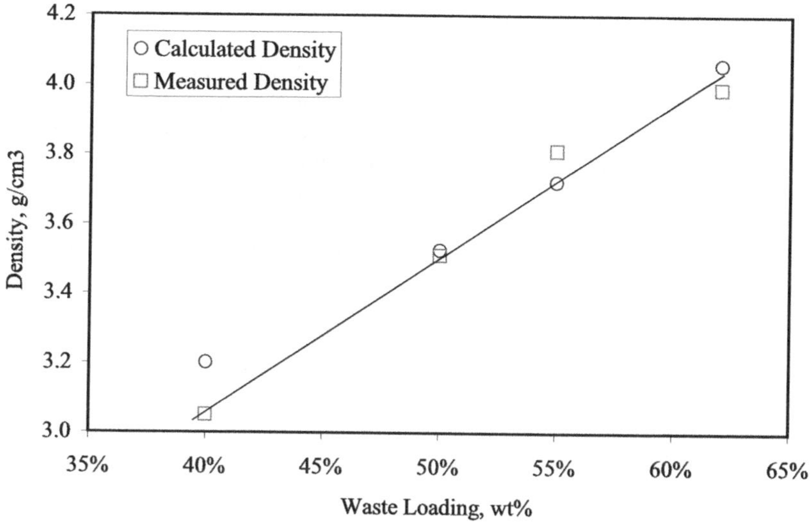

Figure 2. Linear relationship of density as a function of waste loading

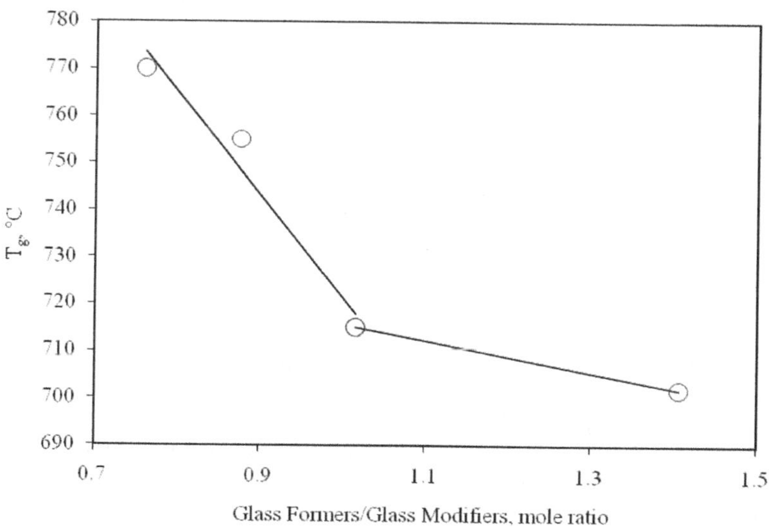

Figure 3. Measured T_g as a function of the ratio of glass formers to glass modifiers

DISCUSSION AND CONCLUSIONS

A series of glasses were prepared over a range of waste loadings of 40 wt% - 62 wt% with low PCT responses (two orders of magnitude below EA glass). At a waste loading of 55 wt%, crystallization begins to occur for the limited compositions tested. Density increases (from 3 to 4 g/cm$_3$) as function of waste loading, causing a compounding volume reduction of glass. Figure 4 shows glass volume as a function of loading. The effects of decreasing glass mass and increasing glass density are separated to demonstrate the compounding affect. The reduced glass volume affords the possibility to use smaller waste packages with increased heat transfer.

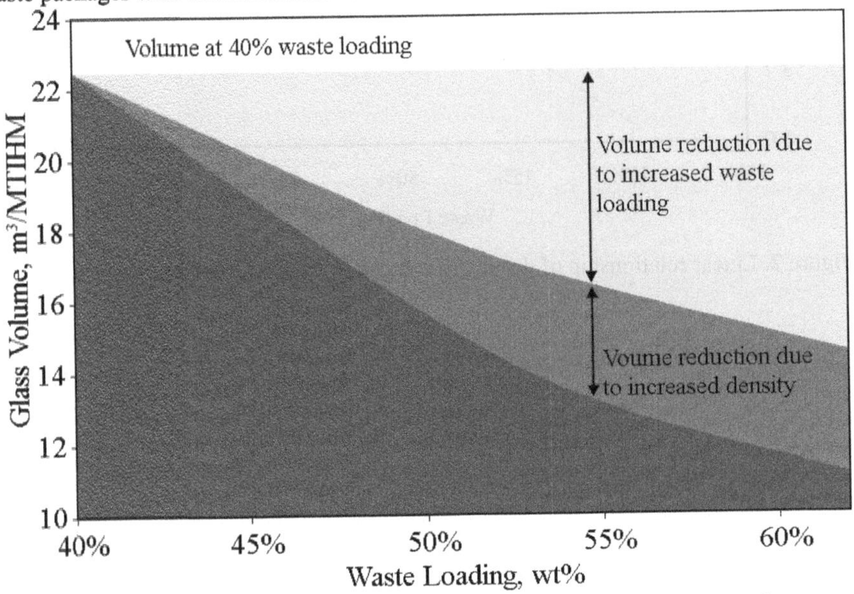

Figure 4. Glass Volume per MTIHM as Function of Waste Loading and Density.

REFERENCES

1. D. K. Peeler, T. B. Edwards, T. S. Rudisill, I. A. Reamer, J. D. Vienna, D. E. Smith, M. J. Schweiger, B. J. Riley, "Composition/Property Relationships for the Phase 2 Am-Cm Glass Variability Study (U)", WSRC-TR-99-00393, Rev. 0, 1999
2. SciGlass, version 6.6, ITC inc., 1998-2006
3. ASTM C 1285-97, Standard Test Methods for Determining Chemical Durability of Nuclear Waste Glasses: The Product Consistency Test (PCT), American Society for Testing and Materials, 1999
4. C. M. Jantzen, N. E. Bibler, D. C. Beam, C. L. Crawford and M. A. Pickett, "Charaterization of the DWPF Environmental Assesssment (EA) Glass Standard Reference Material", WSRC-TR-92-346, Rev. 1, 1993
5. J. D. Vienna, D. Kim, P. Hrma, "Database and Interim Glass Property Models for Hanford HLW and LAW Glasses", PNNL-14060, Rev. O, 2002

FULL-SCALE COLD CRUCIBLE TEST ON VITRIFICATION OF SAVANNAH RIVER SITE SB4 HLW SURROGATE

A.P. Kobelev, S.V. Stefanovsky, V.V. Lebedev, M.A. Polkanov, V.V. Gorbunov, A.G. Ptashkin, O.A. Knyazev
SIA Radon, Moscow, Russia

J.C. Marra
Savannah River National Laboratory, Aiken, SC, USA

K.D. Gerdes
US DOE Office of Environmental Management, Washington, DC, USA

ABSTRACT

A full-scale cold crucible vitrification test on sludge batch 4 (SB4) Savannah River Site HLW surrogate using a 418 mm inner diameter stainless steel crucible was carried-out for 66 hrs. Commercially available Frit 503-R4 (8 wt % Li_2O, 16 wt % B_2O_3, 76 wt % SiO_2) was used as a glass forming additive at a calcine to frit ratio of 1:1 (50 wt % calcine, 50 wt % frit). Three portions of slurry prepared from frit and mixture of chemicals simulating waste in amount of ~750 kg and from frit and waste surrogate prepared by the SRT-MST-2007-00070 procedure in amount of ~1,300 kg with water content of ~27 and ~50 wt %, respectively, was processed and ~875 kg of the vitrified product in total (~415 + 460 kg) was obtained. Average parameters were as follows: vibration power = 121.6 to 134.1 kW, feed rate (capacity) = 25.1 to 39.8 kg/hr; glass pour rate (productivity) = 14.0 kg/hr; specific energy expenses for feed processing = 4.8 to 3.4 kW×hr/kg; specific energy expenses for glass production (melting ratio) = 8.7 to 9.6 kW×hr/kg; specific glass productivity = 2453 kg/(m^2×d). The product was composed of major vitreous and minor spinel structure phases. No nepheline phase was found. The average degree of crystallinity was estimated to be ~12 vol %. Cesium was found to be the most volatile component (up to ~60 wt % of total added to the batch). Lithium, sodium and boron were less volatile. Other major feed constituents (Al, Si, Mg, Fe, Mn) were not volatile but their carry-over with gas vapor flow occurred.

INTRODUCTION

The successful vitrification demonstration of sludge batch 2 (SB2) waste surrogate using a cold crucible inductive melting (CCIM) process [1-8] allowed for the progression to the next step of SRNL – Daymos Ltd./Radon collaborative work on CCIM vitrification of SB4 waste surrogate. As followed from SRNL research work on selection of frit and glass composition and Radon lab-scale melting tests in alumina crucibles in a resistive furnace, a frit 503-R4 with chemical composition (wt %): B_2O_3 – 16, Li_2O – 8, SiO_2 – 76, was an appropriate glass forming additive to vitrify the SB4 waste at high waste loadings in the CCIM. Tests in small (50 ml) alumina crucibles showed that the waste/frit mixtures at waste loading of 40 to 60 wt % yielded almost homogeneous glass. Traces of spinel type crystalline phase were observed only at the highest waste loading of 60 wt % in glass.

Larger melt volumes in the cold crucible may cause enhanced crystallization of glass yielding higher degree of crystallinity in the vitrified waste product [1,5]. In initial testing, two tests with U-bearing SB4 waste surrogate in a lab-scale (56 mm inner diameter) cold crucible were performed. According to the work scope, two 216 mm diameter CCIM tests with U-free SB4 waste surrogate in a bench-scale (236 mm inner diameter) cold crucible and one test in a large-scale (418 mm inner diameter) cold crucible were conducted to optimize process parameters to achieve maximized waste loading (60 wt % target) and throughput for the SB4 composition. Specific emphasis was placed on minimizing cesium, halide and sulfate volatility from the melt.

EQUIPMENT AND FEED PREPARATION

The large-scale test was conducted at the Radon full-scale liquid waste vitrification plant (Figure 1) consisting of waste concentration system, feed preparation system, vitrification unit, off-gas system, energization unit, and auxiliary equipment. The cold crucible melter is energized from a VCI-1.60/1.76 high frequency generator with vibration power of 160 kW and operation frequency of 1.76 MHz.

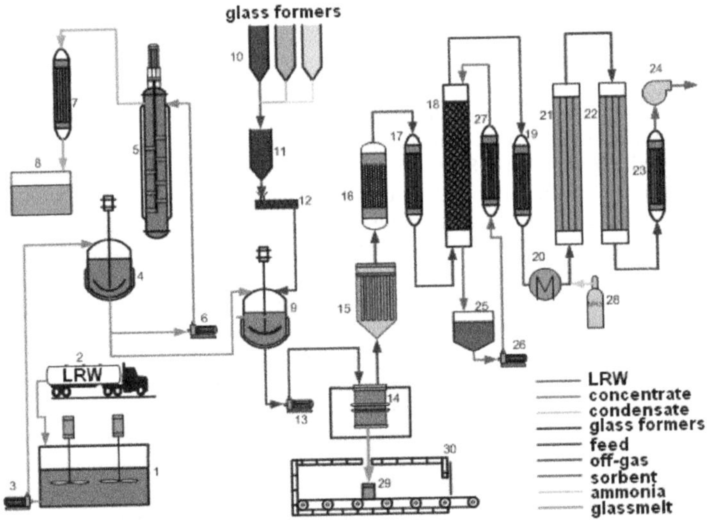

Figure 1. Schematic flowsheet of the Radon LILW vitrification plant (US Patent 6058741). 1 - storage tank, 2 – LRW transportation vehicle, 3, 6, 13, 26 - pumps, 4 – concentrate collector, 5 – rotary film evaporator, 6 - collecting tanks, 7, 17, 19, 23, 27 – heat-exchangers, 8 – condensate collector, 9 - mixer, 10 – glass formers bins, 11 – hopper, 12 – screw feeder, 13 - annealing tunnel furnace, 14 – cold crucible, 15 – coarse filter, 16 – glass fiber filter, 18 - absorbing columns, 20 - heater, 21 - catalytic reactor for NO_x reduction, 22 – catalytic reactor for NH_3 oxidation, 24 – ventilator, 25 – vessel with sorbent, 28 – cylinder with ammonia, 29 – canister, 30 – annealing furnace.

During routine facility operations, liquid radioactive waste (LRW) from an interim storage tank (1) is pumped into a concentrate tank (4) and concentrated in a rotary film evaporator (5) to a solids concentration of 1000 to 1100 kg/m^3. Vapor phase is cooled in a heat-exchanger (7) and collected in a tank (8). The waste concentrate is fed into a mixer (9) where it is intermixed with glass-formers fed from bins (10) and hopper (11) using a screw (12). The slurry with a water content of ~25 wt % is fed using a peristaltic pump (13) into the cold crucible melter (14). Molten glass is poured into canisters (29) and annealed in a tunnel furnace (30) to relax mechanical stress. Canisters with annealed glass are sent to a repository. Off-gas from the melter is purified from entrainment in a coarse filter (15) and from aerosols and fine solids in a glass fiber filter (16), cooled in a heat-exchanger (17), and passes an absorption column (18). Gas containing traces of NO_x passes a heater (20) and is purified from residual NO_x followed by NO_x reduction with NH_3 and oxidation of residual ammonia using catalytic reactors (21) and (22), respectively. The purified gas after cooling to room temperature in a heat-exchanger (23) is discharged to the atmosphere.

In our test we excluded the first step of the process (LRW concentration) preparing the feed immediately in the mixer (9). We also used a simplified off-gas purification scheme where the off-gas after purified on filters (15) and (16) was cooled in a heat-exchanger (17) and condensate was collected in a separate tank. The final off-gas was discharged to the atmosphere.

The cold crucible design was the same as in our previous tests and described in detail elsewhere [1,5,9]. The 418 mm inner diameter and 860 mm height cold crucible was manufactured from 12 mm outer diameter stainless steel pipes and equipped with a cover having feeding and off-gas ports, observation window, and a hole for the screw used to open the pouring unit (Figure 2).

Figure 2. View of the cold crucible (left), pouring unit installed in the cold crucible (middle) and cold crucible coated with a protective putty surrounded by inductor coils (right).

Three types of feed with a target waste loading in glass of 50 wt % were prepared. The first portion of the feed to produce a starting melt was prepared from chemicals: Al_2O_3, H_3BO_3, $CaCO_3$, $Ce(NO_3)_3 \cdot 6H_2O$, Cr_2O_3, Cs_2CO_3, Fe_2O_3, Li_2CO_3, MgO, MnO_2, Na_2CO_3, NiO, $PbSO_4$, Na_2SO_4, SiO_2. Water content in the feed was ~27 wt % i.e. similar to the conventional Radon LRW vitrification process.

The second portion of the feed was prepared from waste surrogate composed of chemicals (see above) and Frit 503-R4. In total 750 kg of feed with a water content of ~27 wt % was prepared.

The third portion of the feed in amount of about 1,300 kg was prepared from waste surrogate obtained in accordance with procedure SRT-MST-2007-00070 "Sludge and Glass Compositions for Cold Crucible Induction Melter (CCIM) Testing – Sludge Batch 4" [10] including 5 phases:

- Phase 1: Precipitation of MnO_2 [by the $KMnO_4$-$Mn(NO_2)$ reaction] followed by the precipitation of Fe and Ni as hydroxides.
- Phase 2: Analytical confirmation that the Mn, Fe and Ni meet the targeted specification.
- Phase 3: Washing the slurry prepared in Phase 1 with alkaline "inhibited wash water" until the soluble solids concentration in the aqueous fraction of the slurry is below 0.20 wt%. Note: inhibited wash water is water where NaOH and $NaNO_2$ (equimolar quantities) have been added to obtain a pH of 10.5±0.25.
- Phase 4: Addition of the remaining waste components and final adjustment of the total solids concentration.
- Phase 5: Analytical confirmation that the final slurry batch meets specifications described in Table I.

Table I. Expected Concentration of Anion/Cation Content of Sludge Feed Simulants (Final Product)

Cation	wt% total solids	wt % calcined solids	Cation	wt% total solids	wt % calcined solids	Anion	Concentration, g/L
Al^{3+}	10.381	14.830	Mg^{2+}	1.287	1.839	SO_4^{2-}	1.386
Ba^{2+}	0.048	0.069	Mn^{2+}	3.447	4.924	PO_4^{3-}	0.284
Ca^{2+}	1.521	2.172	Na^{+}	10.679	15.256	CO_3^{2-}	7.449
Ce^{3+}	0.141	0.201	Ni^{2+}	1.004	1.434	SO_4^{2-}	1.319
Cr^{3+}	0.104	0.149	Pb^{2+}	0.273	0.390	NO_3^{-}	5.595
Cu^{2+}	0.031	0.045	Si	0.975	1.393	Cl^{-}	3.550
Fe^{3+}	15.602	22.289	Zn^{2+}	0.031	0.044	F^{-}	0.045
K^{+}	0.043	0.062	Zr^{4+}	0.052	0.074	I^{-}	0.085
						NO_2^{-}	13.762
						OH^{-}	5.158
Specific Gravity, g/mL			1.14	Total Organic Carbon (TOC) [wet]			<0.05
Total Solids, wt %			20	Total Solids, g			273.6
Soluble Solids, wt %			4.72%	Estimated Calcine Factor			0.7

A feature of the sludge prepared according to the procedure [10] was the formation of magnetite-type spinel and nepheline phases at low temperatures (Figure 3). The waste surrogate was intermixed with Frit 503-R4 (produced by Ferro Corp., USA) having a chemical composition (wt %): 8 Li_2O, 16 B_2O_3, 76 SiO_2 and particle size distribution: ≤2.0 wt % retained on a U.S. Standard 80 mesh sieve, ≤10 wt % passing through a U.S. Standard 200 mesh sieve. Final water content in the feed was ~50 wt %. In total 1300 kg of feed was prepared. In order to evaluate Cs volatilization, $CsNO_3$ was added to the feed in amount to achieve 0.64 wt % Cs_2O in the final glass.

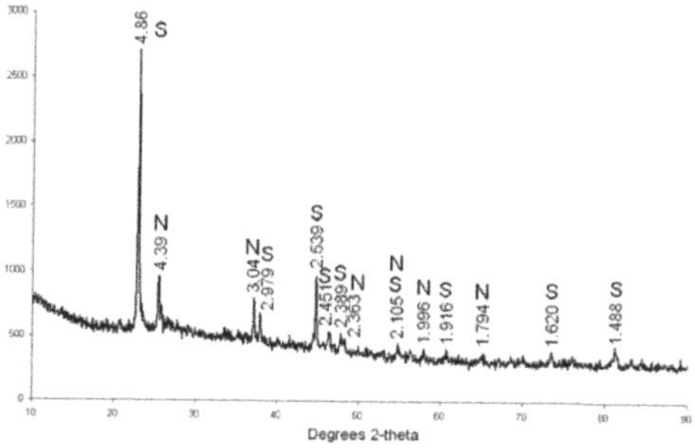

Figure 3. XRD Pattern of the sludge prepared by SRT-MST-2007-00070 Procedure and air-dried at 115 °C. N – nepheline, S – spinel.

CCIM PROCESS DESCRIPTION

Major steps of the run were as follows:

1) Starting melt production from chemicals;
2) Vitrification of waste surrogate with Frit 503-R4 at water content of ~27 wt %;
3) Vitrification of waste surrogate prepared using SRT-MST-2007-00070 procedure [10] with Frit 503-R4 at water content of ~50 wt %;
4) Process stopping and melter cooling.

In step 1, to start-up the CCIM process, the cold crucible was charged with the glass cullet produced in our previous test in the bench-scale unit (~50 wt % waste loading) and about 6.5 kg of dry batch produced from chemicals at the Radon vitrification plant. The starting melt was initiated through step-by-step increases of vibration power from 29.5 to 149 kW for 2 hrs-25 min. Molten glass was poured into canister #1.

In step 2, a mixture of chemicals with water content of ~27 wt % was fed for 29 hrs-40 min. Slurry was fed into the crucible in amount of 46 to 90 kg followed by holding the melt to complete melting of feed components and melt homogenization. Then the pour gate was opened and a portion of the melt was poured into a ~12 L canister. During this step the melt was poured into canisters #2 to #15 and ~415 kg of glass was produced (Table II).

Table II. Major ICCM Process Variables under Steady-State Conditions

Process Variables	Step 2	Step 3
Test Duration, hr:min	29:40	32:50
Average Vibration Power, kW	121.6	134.1
Weight of Processed Feed, kg	745.8	1307.9
Weight of Produced Glass, kg	414.3	458.6
Average Feed Rate (Capacity), kg/hr	25.1	39.8
Average Glass Pour Rate (Productivity), kg/hr	14.0	14.0
Specific Energy Expenses for Feed Processing, kW×hr/kg	4.8	3.4
Specific Energy Expenses for Glass Production (Melting Ratio), kW×hr/kg	8.7	9.6
Specific Glass Productivity, kg/(m^2×d)	2453	2453

There was variation in input power during feeding and holding the melt to homogenize the glass (Figure 4). The average value of input power at the step 2 was found to be ~122 kW (Table II).

Major process variables given in Table II demonstrate high efficiency of the cold crucible during the melting process: average feed rate (capacity), glass pour rate (productivity) and specific glass pour rate achieved ~25 kg/hr, ~14 kg/hr, and ~2450 kg/(m^2×d), respectively even with the variable process conditions.

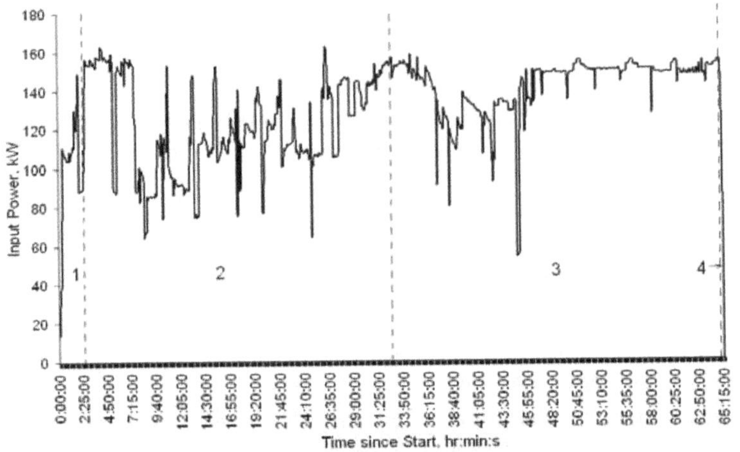

Figure 4. Variation of input power during the starting melt production (1), feeding of the mixture of chemicals and frit (2), waste surrogate prepared by SRT-MST-2007-00070 Procedure (3) and process stopping (4).

During step 3, the feed with Frit 503-R4 with increased water content (50-52 wt %) was processed. In total, slurry in amount of ~1300 kg was fed and ~460 kg of glass was produced. Under these conditions, both glass productivity and specific glass productivity remained similar to the values for the step 2 at lower water content (~27 wt %). This demonstrated advantages of the use of prepared waste surrogate and frit rather than chemicals as glass forming additives to increase melting rate. In spite of higher energy expenses on water evaporation, application of the feed prepared from waste surrogate and Frit 503-R4 compensates these expenses due to higher homogeneity of the feed and reduction in power expenses for silicate forming reactions. As a result, the feed rate during step 3 was even higher (~40 kg/hr) than that during step 2, specific glass productivity remained nearly the same [~2450 kg/(m^2×d)], and melting ratio was increased by only a factor of ~1.1 (~10%).

Energy parameters of the CCIM process may be summarized as a heat power distribution diagram (Figure 5). It is seen that about 60% of total power was spent on cooling of the generator's tube and cold crucible. Minor heat (up to ~4% of total) was removed with the off-gas. About 35% of total power was consumed on feed heating, water evaporation, calcine melting and melt homogenization.

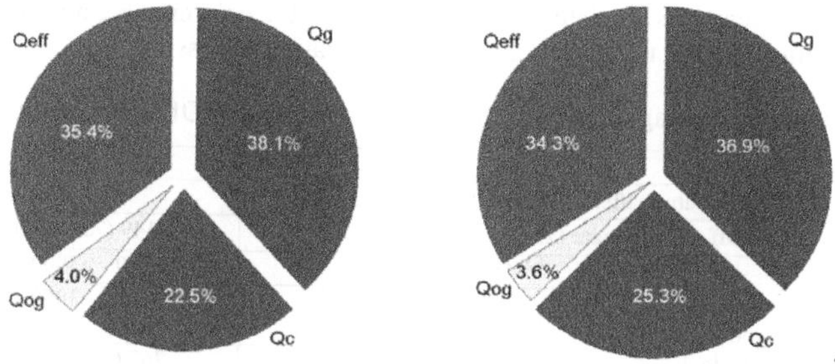

Figure 5. Heat power distribution for the CCIM process for treatment of slurries with water contents of ~27 wt % (left) and ~50 wt % (right). Qg is power spent on cooling of the HF generator, Qc is a power spent on cooling of the cold crucible, Qog is power removed with the off-gas, Qeff is power spent on batch heating, melting and melt homogenization.

OFF-GAS ANALYSIS

The off-gas sampling system schematic is shown in Figure 6. Major results are summarized in Table III.

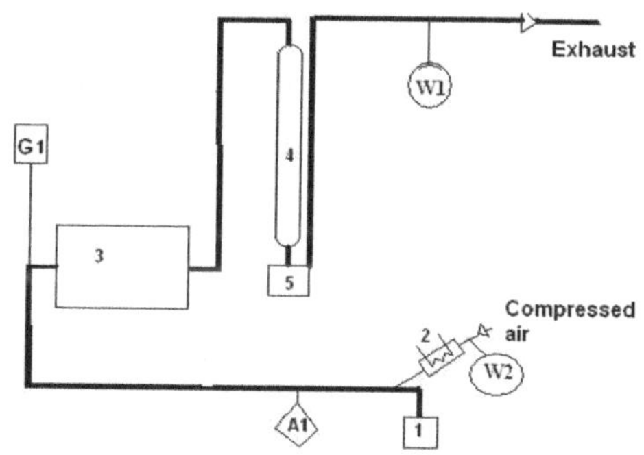

Figure 6. Schematic diagram of aerodynamic measurements and off-gas sampling points. 1 – melter, 2 – air heater, 3 – coarse filter, 4 – cooler, 5 – condensate collector. Measurement and sampling points: A1 – aerosol sampling, G1 – gas components sampling, W1 – off-gas flow rate measurement, W2 – compressed air flow rate measurement.

VOLATIZATION OF FEED CONSTITUENTS

Volatilization of feed constituents was evaluated using two methods: 1) by measuring of elemental concentrations in the off-gas using aerosol filters and 2) by difference between amounts of constituents (recalculated to oxides) fed into the cold crucible and found in the glass. Both give significant error, therefore, only rough evaluation is possible.

Table III. Average Concentrations and Variability of Off-gas Constituents

Parameters		Values
Off-gas temperature after dilution with compressed air, °C		150 - 207
Off-gas flow rate (25 °C, 1 atm), m^3/h		149 (72 - 234)
Concentration of components in off-gas	[1] NO. mg(N)/m^3	~0 - >3000
	[1] NO_2. mg(N)/m^3	~0 - >3000
	[1] CO. mg(C)/m^3	~0 - 50.0
	[2] HCl. mg/m^3	262 (26.2 - 846)
	[2] F^-. mg/m^3	1.32 (0.67 - 3.44)
	[2] SO_2. mg/m^3	132 (2.5 - 271)
	[3] Aerosols. mg/m^3	4410 (430 - 24100)

[1] "Kaskad-512" gas analyzer, [2] chemical analyses, [3] weighing procedure.

Major sources of error in determination of losses using the first technique are the variable off-gas flow rate and variation in elemental concentrations due to the semi-periodic mode of operation consisting of slurry feeding, melting/homogenization and melt pouring steps. Significant error in the second method of determination arises from differences in chemical composition of the glasses both in canisters and "dead volume" of the cold crucible and difficulties in precise determination of the amount of the slurry fed in the cold crucible and amount of glass poured into canisters (unknown amount of glass was poured into a catcher under transportation system in the tunnel furnace). Results are given in Table IV.

Table IV. Oxide Losses from Cold Crucible by Various Methods of Measurement

Oxides	Loss, wt %			
	XRF	EDS	ICP-AAS	OGA*
Li_2O	-	-	-	8.0
B_2O_3	12	-	7.3	4.5
Na_2O	3.5	7.9	2.6	6.7
Al_2O_3	0.1	-	0.3	0.3
SiO_2	-	-	-	0.1
CaO	0.5	-	4.3	2.4
MnO	3.9	5.1	7.1	5.3
Fe_2O_3	7.4	2.7	4.5	2.2
Cs_2O	56.3	48.4	71.9	61.6

* Determined by ICP-MS analyses of off-gas deposits on aerosol filters.

Considering data presented in Table IV note, that analyses of aerosol filters in the off-gas system indicated much lower boron volatilization as compared with the value obtained from analyses of glass. Boron loss occurs in two ways: mechanical carry-over mainly in aerosol form and vaporization from the melt surface at high temperatures. Appreciable glass depletion with boron is due to both boron vaporization from the open melt surface in the cold crucible and mechanical carry-over of fine-grained frit. These mechanisms were responsible for the predominant fraction of the carry-over deposits on the inner cold crucible walls and off-gas pipe (Table V). Only a minor fraction of boron as vapors and aerosols is volatilized.

Sodium losses determined from both off-gas and glass analyses are similar. XRF spectroscopy shows lower sodium loss than EDS but the value obtained from EDS is closer to that of the off-gas analysis.

Table V. Chemical Compositions (wt %) of Deposits on the Inner Surfaces of Cold Crucible Wall and Cover

Oxides	Wall Deposit	Cover Deposit	Oxides	Wall Deposit	Cover Deposit
Li_2O*	4.00	4.00	MnO	2.26	1.31
B_2O_3**	5.80	10.40	Fe_2O_3	22.37	13.52
Na_2O	8.55	22.94	NiO	3.85	2.40
MgO	1.29	1.37	CuO	0.00	0.00
Al_2O_3	20.52	4.31	ZnO	5.24	3.41
SiO_2	9.95	2.13	ZrO_2	0.07	0.02
P_2O_5	0.25	0.10	Cs_2O	1.45	4.59
SO_3	2.77	7.37	BaO	0.00	0.00
Cl	3.38	18.14	La_2O_3	0.00	0.00
K_2O	0.19	0.49	Ce_2O_3	0.00	0.00
CaO	1.58	2.25	PbO	0.75	0.51
TiO_2	0.00	0.00			
Cr_2O_3	0.34	0.31	Total	90.51***	95.57***

* Li and B are not determined by XRF spectroscopy, ** B was determined by potentiometric titration, *** analytical sum (without Li_2O), OH-groups may be present.

Cesium is the most volatile component. The values obtained from both off-gas (~60 wt %) and glass analyses (~50-55 wt %) are in rather good agreement. Some volatility was observed even for iron and manganese (Table IV and V). Probably, mechanical carry-over as aerosols was the major contribution to their volatilization. Nevertheless, as it has been shown in our previous work [11], appreciable losses of these and other iron group elements may be due to volatility of their chlorides that is consistent with analytical data for the wall and cover deposits (Table V).

As follows from Table V, if wall deposits are due predominantly to the mechanical carry-over, then vaporization of easily volatile constituents yields significant contribution to the cover deposit formation. The latter is strongly enriched with boron, sodium, cesium, sulfur, and chlorine. XRD data (Figure 7) show that the wall deposits consisted of major spinel type phase and minor sodium chroride (halite, NaCl) and sulfate (Na_2SO_4 – hexagonal). Phase composition of the cover deposit is much more complex. It contained major sodium sulfate and chloride, with spinel secondary in abundance and minor sodium zinc sulfate - β-$Na_2Zn(SO_4)_2$ and possibly cesium sulfate. Cs ion is too large and unable to substitute for Na ion in the sodium zinc sulfate crystal lattice.

Spinel might occur in the deposits due to mechanical carry-over of the slurry droplets. XRD pattern (Figure 3) demonstrates that the spinel (variety different from that occurred in the vitrified product) is present in the air-dried waste surrogate prepared by SRT-MST-2007-00070 procedure. Other volatiles (sodium chloride, sulfate, sodium-zinc sulfate, cesium sulfate) are the products of both aerosol carry-over and vaporization during calcination and melting and from open surface volatilization of the melt in the cold crucible.

Other feed constituents (magnesium, aluminum, silicon and calcium compounds) are not volatile although minor mechanical carry-over with gas-vapor flow occurs as well.

As follows from X-ray diffraction and scanning electron microscopy data, the product of SB4 waste vitrification is composed of major vitreous and minor (~10 vol %) spinel structure (trevorite and zincochromite) phases. Vitrified waste characterization details will be presented in a separate paper.

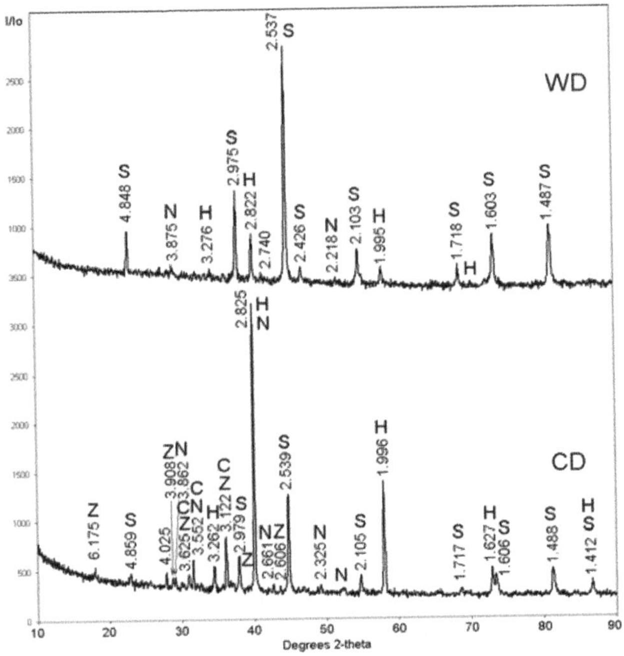

Figure 7. XRD patterns of the sublimates deposited on the inner surface of the cold crucible wall (WD) and cover (CD). C – CsSO4, H – NaCl, N – Na2SO4, S – spinel, Z – Na2Zn(SO4)2.

CONCLUSIONS

The full-scale cold crucible test using a 418 mm inner diameter stainless steel crucible was carried-out for 66 hrs. Commercially available Frit 503-R4 (8 wt % Li_2O, 16 wt % B_2O_3, 76 wt % SiO_2) was used as a glass forming additive at a calcine to frit ratio of 1:1 (50 wt % calcine, 50 wt % frit). Two portions of slurry prepared from frit and mixture of chemicals simulating waste in amount of ~750 kg and from frit and waste surrogate prepared by the SRT-MST-2007-00070 procedure in amount of ~1,300 kg with water content of ~27 and ~50 wt % was processed and ~415 kg and ~460 kg of the vitrified products was obtained. Major process variables achieved were as follows:

- Average Vibration Power, kW: 121.6 to 134.1;
- Average Feed Rate (Capacity), kg/hr: 25.1 to 39.8;
- Average Glass Pour Rate (Productivity), kg/hr: 14.0;
- Specific Energy Expenses for Feed Processing, kW×hr/kg: 4.8 to 3.4;

- Specific Energy Expenses for Glass Production (Melting Ratio), kW×hr/kg: 8.7 to 9.6;
- Specific Glass Productivity, kg/(m^2×d): 2453.

The product was composed of major vitreous and minor spinel structure phases (trevorite and zincochromite). No nepheline phase was found. The average degree of crystallinity was estimated to be ~12 vol %. Cesium was found to be the most volatile component (~60 wt % of total added to the bacth). Lithium, sodium and boron were less volatile. Other major feed constituents (Al, Si, Mg, Fe, Mn) were not volatile but mechanical carry-over with gas-vapor flow occurred.

ACKNOWLEDGEMENT
Authors thank N.P. Penionzkiewich for recording of XRD patterns of the samples.

REFERENCES

1. A.P. Kobelev, S.V. Stefanovsky, O.A. Knyazev, T.N. Lashchenova, A.G. Ptashkin, M.A. Polkanov, E.W. Holtzscheiter, J.C. Marra, "Results of a 50% Waste Loading Vitrification Test Using the Cold Crucible Melter for Savannah River Site," *Waste Management '06 Conf.* Tucson, AZ, February 26 – March 2, 2006. CD-ROM. ID 6127.

2. A.P. Kobelev, S.V. Stefanovsky, O.A. Knyazev, T.N. Lashchenova, E.W. Holtzscheiter, J.C. Marra, "Cold Crucible Vitrification of Defence Waste Surrogate and Vitrified Product Characterization," *Mat. Res. Soc. Symp. Proc.* Vol. 932, 2006, p. 353-360.

3. S.V. Stefanovsky, K.I. Maslakov, A.Y. Teterin, Y.A. Teterin, J.C. Marra, "Uranium Speciation in Borosilicate Waste Glass," *Proc. 36^{iemes} Journees des Actinides*. Oxford, UK, April 1-4, 2006. ID P-38.

4. S.V. Stefanovsky, A.G. Ptashkin, O.A. Knyazev, B.S. Nikonov, Y.A. Teterin, "Uranium Speciation in Borosilicate Glass for LRW Immobilization," *The Fifth Russian Conference on Radiochemistry "Radiochemisry-2006"* (Russ.), October 23-27, 2006. Dubna, Russia. Abstracts. 2006, p. 248.

5. S.V. Stefanovsky, and J.C. Marra, "The Effect of Waste Loading on the Structure and Leach Resistance of Borosilicate Glass for Savannah River Site SB2 Waste Immobilization," *Waste Management '07 Conf.* Tucson, AZ, 2007. CD-ROM. ID 7132.

6. K.D. Gerdes, J.R. Harbour, J.C. Marra, D.K. Peeler, T.B. Calloway, J.A. Roach, J.D. Vienna, A.S. Aloy, S.V. Stefanovsky and M.D. Bondarkov, "The U.S. Department of Energy – Office of Environmental Management's International Program," // *Ibid.* ID 7514.

7. S. Stefanovsky, A. Shiryaev, I. Zubavitchus and J. Marra, "XANES/EXAFS of U L_{III} Edge in Borosilicate High-Ferrous Nuclear Waste Glasses," *Proc. 37^{iemes} Journees des Actinides*. Sesimbra, Portugal, 24-27 March 2007. Abstracts. 2007, p. 61-62.

8. S.V. Stefanovsky, A.G. Ptashkin, J.C. Marra and A.A. Shiryaev, "Uranium Speciation and Structure of Vitrified Savannah River Site Sladge Batch 2 (SB2) Waste Surrogate," *Proc. XXI*[st] *International Congress on Glass*, July 1-6, 2007, Strasbourg, France, 2007. CD-ROM.

9. A.P. Kobelev, S.V. Stefanovsky, V.N. Zakharenko, M.A. Polkanov, T.N. Laschchenova, A.G. Ptashkin, V.A. Perepelkin, O.N. Tsveshko, E.A. Kachalova, A.S. Sorokaletova, O.A. Knyazev, M.A. Zen'kovskaya, N.P. Penionžkiewicz, B.S. Nikonov, Y.A. Oskolkov, L.A. Kalashnikova, A.V. Efimov, E.W. Holtzscheiter, J.C. Marra and B. Calloway, *Task 1: Maximizing Waste Loading for Application to Savannah River High Level Waste - 2006 & 2007 Scope*. Interim Report on Contract Work for FY 2006/2007. Subtask 1: CCIM Testing to Demonstrate Maximized Waste Loading with SB2, Moscow, 2007.

10. J.C. Marra. *Sludge and Glass Compositions for Cold Crucible Induction Melter (CCIM) Testing – Sludge Batch 4*. SRT-MST-2007-00070 (2007).

11. S.A. Dmitriev, I.A. Knyazev, S.V. Stefanovsky and F.A. Lifanov, "Volatilization of Radionuclides at Plasmachemical Treatment of Radioactive Waste," *Phys. Chem. Mater. Treat.* (Russ.) No. 4, 1993, p. 74-82.

PROCESSING FLY ASH FROM COAL BURNING POWER STATION IN A VARIABLE RADIOFREQUENCY FIELD

L. Barbieri, I. Lancellotti, F. Andreola, A. Corradi, C. Leonelli
Dipartimento di Ingegneria dei Materiali e dell'Ambiente, Università degli studi di Modena e Reggio Emilia, Via Vignolese 905, 41100 Modena, Italy

M. La Robina
Institute of Materials Engineering, Australian Nuclear Science and Technology Organisation ANSTO, New Illawarra Road – Lucas Heights, NSW 2234 - Australia

ABSTRACT

Experiments with fly ash obtained from a coal-burning power station in NSW, Australia, were conducted in the cold crucible facility at ANSTO using a 30kW variable frequency generator. The generator was configured as a standard Hartley oscillator with the tank capacitor able to vary from 110 pF to 4100 pF giving a frequency span of 5MHz to 18MHz. The applicator consisted of a coil with a water cooled base and top plate that formed an LC resonant cavity, which coupled to the RF generator by a single turn coil. A dark glass, confined inside the applicator, was formed by applying a high frequency electromagnetic field between two copper plates through a sample of fly ash. The temperatures reached were in excess of 1000 °C causing the powder to fuse into glass in a very brief period of a few minutes. The purpose of this experiment was to verify if dielectric and/or induction heating could melt the fly ash in a more rapid manner than conventional (resistive/joule) heating.

INTRODUCTION

The complexities of processing various waste streams both in the nuclear industry[1, 2] and other hazardous waste such as fly ash originating from electric power generation and municipal incinerators has lead to the development of equipment and methodologies to generate extreme environments for processing these materials in order to consolidate them for safe disposal.

Fly ash is a by-product of combustion from a variety of sources but predominantly from coal-based fuels burnt for the production electricity in high-efficiency power plants and from municipal waste incinerators.

In Australia the majority of fly ash produced is categorized as Class F – being mainly silica and alumina (80-85%) and >10% CaO[3]. The ash used in these experiments has less than 1% CaO, these ash constituents incorporate other elements in minor and trace quantities including various transition metal oxides [Table I].

The chemical composition of fly ash from municipal and industrial incinerators is dependent on its origin, e.g. what type of rubbish is incinerated. Fly ash from power stations is relatively similar with minor variation of oxide composition, dependent on the type of coal burnt and firing conditions of the power station and varies daily.

In Australia this waste originates mainly from power stations producing approx 8.6 million tonnes of fly ash annually[4], of which 7 million tonnes is disposed of as landfill. The remainder of the ash is processed as an additive for producing building materials by companies like Rocla Ltd., Australia that used to process fly ash using "solid flow method involving milling and sizing fly ash" for cements and geopolymers, while other companies use melting furnace technology which melts fly ash into molten slag to simultaneously achieve a triple purpose of:

a) Volume reduction

   Bottom ash and fly ash from incinerators is reduced in volume by converting to a molten slag, thus extending the life of the final disposal sites.

b) Detoxification

Ash melting decomposes more than 99% of dioxins, bringing it down below the detectable level.

c) Resource recycling

Slag can be used as roadbed and road surfacing materials. It also can be used as aggregate in concrete blocks, tile and so on.

In Australia, at least, the quantities produced far exceed the demand so an alternative use of fly ash needs to be found.

Experiments with fly ash obtained from a coal-burning power station in NSW, Australia, were conducted in the cold crucible facility at ANSTO using a 30kW variable frequency generator. During the R&D work conducted at ANSTO to melt ceramics and later glass and glass ceramics for immobilizing nuclear waste, it became apparent that this technology was advantageous for non conductive materials and avoided crucible contamination[5].

Nevertheless after a series of experiments the use of a susceptor to initiate induction melting for electrically nonconductive materials was proved to be an inefficient and potentially risky method, as was reported during trials conducted to establish operating parameters for the ANSTO cold crucible radio frequency (RF) generator[6]. For non-nuclear waste the problem was clearly only an inconvenience, however, with the need to have a cold crucible in a hot cell and operations being carried out with remote manipulators, the use of a conductor to initiate the melt was clearly a major problem. A better method needed to be found whereby these materials can be heated to their melting point without using a susceptor.

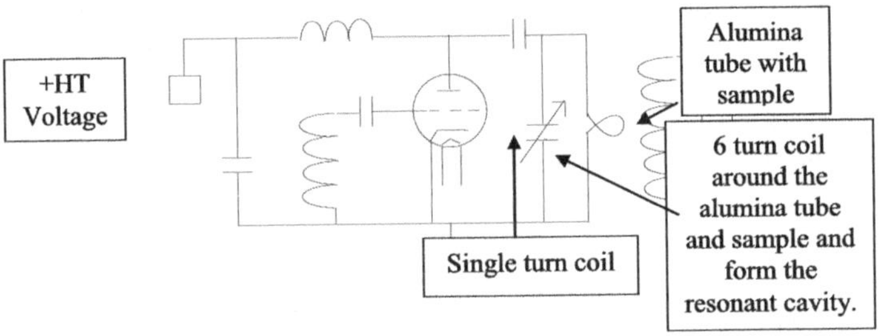

Figure 1. Hartley oscillator circuit used in the variable frequency generator

The solution to this problem came from a commercial variable frequency RF generator purchased together with an applicator that was a combination of an inductor and a capacitor that formed an LC resonant cavity that coupled to the RF generator by a single turn coil[7,8]. The circuit used was a tank loading Hartley oscillator similar to that shown in Fig.1, with the applicator consisting of a patented[9] water-cooled coil plate combination that formed a resonant LC circuit. This combination is capable of being used for generating the high energy electric field which subsequently generates plasma and then a discharge.

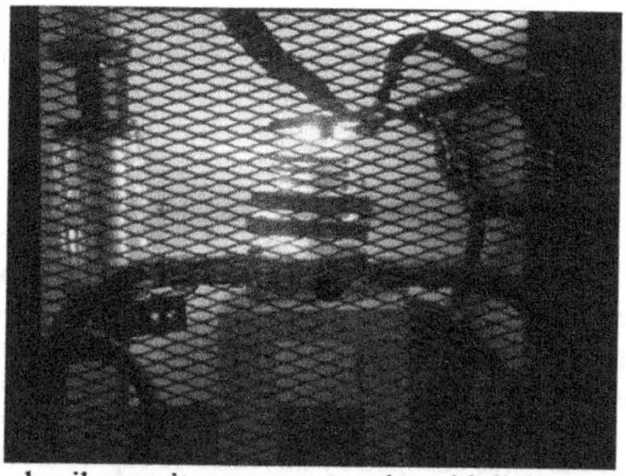

Figure 2. Combined coil capacitor resonant cavity with light emission during the melting of the fly ash. The intensity of the plasma formed generated light seen through the alumina tube that acted as the confinement for the powder and insulated the coil from the molten material.

EXPERIMENTAL

Sample Characterization

The fly ash powder used in this experiment was supplied to ANSTO by Australia's largest power station at Eraring, New South Wales. The Chemical analysis was also supplied by the power station and it is shown in Table I, second column. Eraring Power Station, located just north of Sydney, is one of the largest power stations in Australia, comprising four 660MW coal-fired units. The power station has operated reliably since 1981.

Table I. Chemical analysis (wt%) of the fly ash generated during production of electricity. (*ICP data from this work)

Oxide	As received fly ash	Vitrified fly ash*
SiO_2	66.80	66.13
Al_2O_3	23.60	23.25
Na_2O	0.57	0.49
K_2O	1.54	0.85
CaO	0.96	0.41
MgO	0.66	0.65
Fe_2O_3	3.83	2.66
TiO_2	0.88	1.02
P_2O_5	0.13	0.29
MnO	0.06	0.04
L.O.I	1.13	0.00

Since the fly ash from Eraring PowerStation is to be used to test the technology rather than the process, no additional analyses were conducted to more clearly understand which components are lost on ignition (L.O.I) as shown in Table I. This is out of the scope of these experiments.

The glassy sample obtained after the heating process was characterised by the chemical analysis, reported in Table I, third column, (inductively coupled plasma ICP-Varian Liberty 200); mineralogical analysis (diffractometer Philips PW 3710, Ni-filtered CuKα radiation) on the powdered samples (30 µm particle size) in the 5-70° 2θ range; thermal analysis (Netzsch TG/DTA STA 409) with 10°C/min heating rate, in the range 20-1400°C on powdered sample. Microstructure and qualitative chemical characterisation was performed by using a scanning electron microscopy (SEM Philips XL 30) coupled with energy dispersion spectroscopy (EDS) EDAX 9900) on a bulk sample.

Sample Melting

The high energy electric fields were generated by using combined copper coil and water-cooled plates that were coupled to the RF generator by a single turn coil to form a step up transformer capable of generating electric field intensity of 230kV/m between the two plates (Fig. 3), calculated based on the coil voltage of 4600V stepped up to 27,600V across the two plates and acting over a distance d of 12 cm (E=V/d). An amount of approximately 150 g of sample was put in an alumina tube placed inside the coil and in-between the two water-cooled plates. The copper coil generated the inductive component of the field; the frequency was changed by the variable capacitor, shown in the circuit (Fig. 1), within the range of 5 MHz to 18 MHz, as specified by the manufacturer. The principle of operation of this combined inductor-capacitor resonant cavity is that the capacitor acts as a dielectric heater at the high end and the coil acts as an induction heater at the low end of the frequency range.

At the low frequency, induction heating was dominant and as charge built up between the capacitor plates (Fig 3) a discharge occurred generating plasma and subsequently a breakdown, resulting in melting the material. Once it was known that the heating process is greatly enhanced by the discharge, the experiment was repeated using the same fly ash in the same experimental setup. The plasma was purposely initiated by using a tungsten rod of 1mm diameter to create a short circuit between the top and bottom plate. The discharge was propagated down the length of the alumina tube through the powder by slowly removing the tungsten rod but maintaining contact with the top plate so that the arc formed by the initial short circuit was not extinguished. Once the material began to melt, the conductivity increased so significantly that eddy currents induced by the coil were the dominant mechanism for the continued heating. At this point, the frequency was slowly reduced so that induction would dominate in the heating process.

The coil had the added advantage that the intense magnetic field focused the plasma and confined it to the interior of the alumina tube consequently increased the heating by the Joule effect (or arc melting). The resulting sample (shown in Fig 4.) obtained after a treatment time of approximately 10 minutes, was characterized and the results are shown below.

During this experiment, temperature was neither controlled nor measured. The geometry of the coil/capacitor and the extreme intensity of both the electric and magnetic fields were such that it is difficult to measure temperature with either a thermocouple or a fiber optic pyrometer. Previously experiments[10] were conducted to optimize the circuit of the oscillator and it was found that when capacitive coupling was dominant, i.e. dielectric heating was occurring, a plasma formed very rapidly and the sample melted in the arc. The measurement and control of temperature in the equipment is a significant problem and is the subject of ongoing research to solve it. Nevertheless, an idea of the thermal level can be estimated by calculating the theoretical liquidus temperature from the additive factors approach of Lakatos[11] by using the oxides content of the vitrified waste. Such calculated temperature results to be 1435°C.

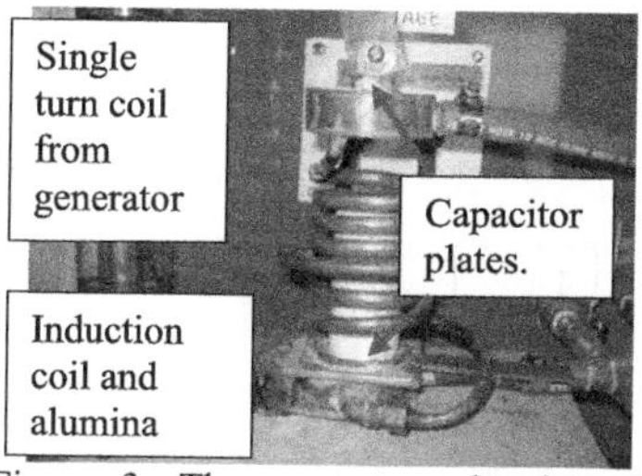

Figure 3. The apparatus showing the single turn coil connected to the generator and the LC capacitor-coil combination with the alumina tube in the centre. (See also Figure 1 showing the schematic circuit.)

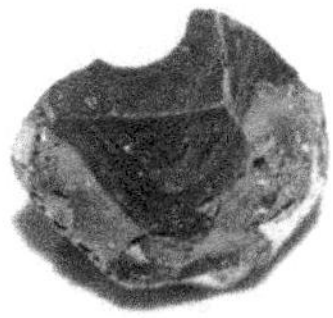

Figure 4. Sample glass produced in this experiment

RESULTS AND DISCUSSION

The characterization performed was aimed at confirming the efficiency of the dielectric heating in order to obtain a glassy material with extremely rapid heating cycle. The chemical analysis on both the as-received and vitrified fly ash (Table I) shows the elemental composition to be very similar. This suggests that this technique reduces volatilization and avoids crucible contamination.

SEM analysis (Fig. 5) shows a homogeneous glassy matrix with a uniform composition of Si and Al with minor amounts Ca, K, Fe and Ti. The Si/Al ratio is around 2.6 (detected by analyzing 20 areas throughout the sample) confirming the chemical composition obtained by ICP, where a ratio near 2.8 was detected. In addition few un-melted particles are dispersed throughout the glassy material.

Mineralogical study confirmed SEM data showing the broad band characteristic of short range order typical of a glass suggesting the treated fly ash has a glassy nature (Fig.6). The XRD pattern shows only a small peak which can be related to the presence of the un-melted particles dispersed in the glassy matrix.

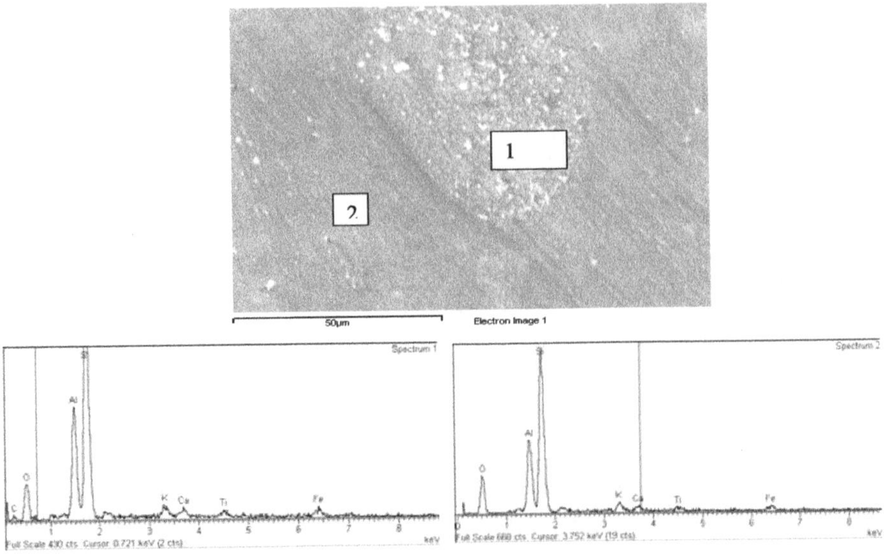

Figure 5. SEM micrograph and EDS spectra of vitrified fly ash.

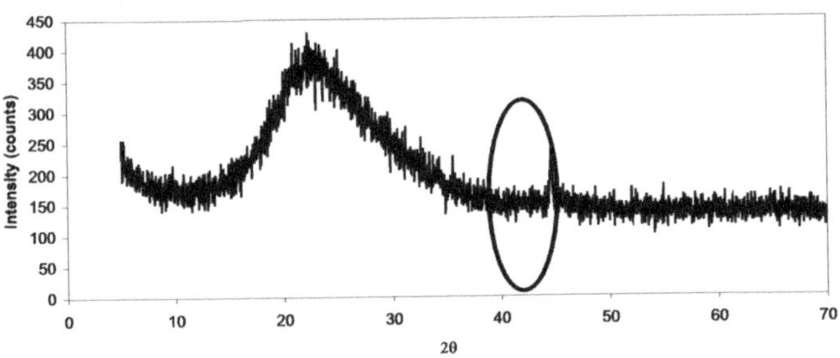

Figure 6. XRD pattern of vitrified fly ash

The study of thermal properties was conducted using thermo-gravimetric analysis (TG) and differential thermal analysis (DTA). These techniques permit to highlight physical and chemical changes in a sample subjected to a continuous and constant temperature variation. TG allows the determination of the material percentage weight loss as a function of temperature. DTA allows the evaluation of the enthalpy variation during the reaction.

DTA curve confirms the amorphous nature of the sample showing an inflexion point near 600 °C corresponding to glass transition temperature, Tg, further it shows an exothermic peak near 1050 °C associated to the formation of crystalline phases. All these temperatures are measured by a Pt-thermocouple located under the samples crucible.

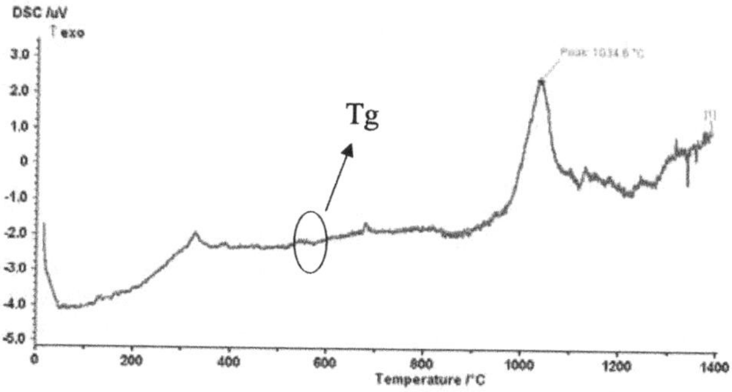

Figure 7. DTA curve of vitrified fly ash

The presence of the crystallization exothermic peak indicates that the vitrified fly ash can be transformed spontaneously into a glass-ceramic material by thermal treatment. In terms of inertization this transformation is advantageous because glass-ceramics, which are materials consisting of crystalline phases embedded in a glassy matrix, give show improved properties when compared to the starting glass. TG curves did not register any weight loss indicating that vitrification was completed.

CONCLUSIONS

In as far as the application of the variable frequency, the charge build-up that occurred in the applicator seems to suggest that circuit configuration behaves like a Tesla coil, resulting in an electrical breakdown. This break down resulted in the formation of a very intense electro/plasma beam which propagated through the body of the sample (the fly ash) resulting in the rapid fusion of the ash to form a glass. The aim had been to dielectrically heat the fly ash to reach a temperature at which it would undergo a transition to become electrically conductive. The plasma formation in the breakdown may be a result of the ash in fact becoming conductive and causing a short circuit within the coil and leading to the required fusion of the fly ash. However, a question mark still remains as to whether the break down was simply due to the enormous voltage build-up due to the step-up transformer effect of the oscillator coil (one turn) and the applicator coil (six turns) giving an electric field E between the capacitor plates of the applicator of the order of 230kV/m, which is more than sufficient to cause an electrical break down and formation of plasma. This is a technique common in arc melting. However one cannot deny that the rapid fusion of the fly ash was a success and gave a good result, as shown by the characterization of the sample, in a total of ten minutes treatment. The quality of the resulting glass would suggest that further improvements in the variable frequency generator and the applicator could give a device that will readily produce a high quality glass or glass ceramic form a waste material.

Although from point of view of the technology, the experiment requires to be optimised, it does show that once the technical difficulties are overcome in the equipment, the application of both dielectric heating and induction heating can be a viable method for producing high quality value added materials by melting fly ash in these extreme high energy environments. We can conclude that melting fly ash in such an extreme environment will allow temperatures to be reached that will produce high quality glass and glass-ceramics.

REFERENCES

[1]R. A. Day, M. La Robina, S. Moricca, T. Eddowes, N. Blagojevic, and M. L. Carter, ANSTO, "Influence of melting atmosphere on synroc-C microstructure and phase composition", INSTN – Saclay, France, 22-23 April 1999.
[2]M. La Robina, J. Ferenczy, The application of high energy electric and magnetic fields to process fly ash and ceramics. Microwave Applications for Engineering and Applied Science, MISA 2004, Ancona, Italy, 6-8 October 2004.
[3]Ash Development Association of Australia - http://www.adaa.asn.au/utilisations.htm
[4]J. Twining, P. McGlinn, E. Loi, K. Smith, R. Gieré, Risk Ranking of Bioaccessible Metals from Fly Ash Dissolved in Simulated Lung and Gut Fluids, *Environ. Sci. Technol.*, **39**(19), 7749 -7756 (2005).
[5]J. Ahearne, J.A. Gentilucci, D. Pye, E.T. Weber, F. E. Woolley, High-Level WasteMelter Review Report, July 2001, Prepared for the U.S. Department of Energy, under Contract DE-AC06-76RL01830.
[6]M. La Robina, Cold crucible experiments at various frequencies, Internal report, Materials Division, ANSTO (1998).
[7]Davies E J, Conduction and induction heating, IEE POWER ENGINEERING SERIES II Peter Peregrinus, London, UK, (1990).
[8]S. Zinn, S. L. Semiatin, Elements of Induction Heating, Design, Control, and Applications, Electrical Power Research Institute, Inc. (1998).
[9]M. Tarnawski, "Method and apparatus for heating refractory oxides", Australia, Patent WO2004/047495 A1, June. 3, 2004.
[10]M. La Robina, J. Ferenczy "Creating An Extreme Environment For Processing By Applying Variable Frequency High Energy Electric And Magnetic Fields", 10th International Conference on Microwave and High Frequency Heating Modena - Faculty of Engineering, 12-15 September 2005
[11]Lakatos, T., Johanson, L.G and Simmingsköld B., 1972. Viscosity temperature relations in the glass system SiO_2-Al_2O_3-Na_2O-K_2O-CaO-MgO in the composition range of technical glasses, Glass Technology, 13(3), 88-95.

NETEC COLD CRUCIBLE INDUCTION MELTER DEMONSTRATION FOR SRNL WITH SIMULATED SLUDGE BATCH 4 DWPF WASTE

Michael E. Smith, Allan B. Barnes, Alex S. Choi, and James C. Marra
Washington Savannah River Company
Aiken, SC, USA

ABSTRACT

Cold Crucible Induction Melter (CCIM) Technology is being considered as a possible next generation melter for the Defense Waste Processing Facility (DWPF). Initial and baseline demonstrations that vitrified a Sludge Batch 4 (SB4) simulant at a waste loading of 50 weight percent (versus about 38 weight percent in the current DWPF Melter) were performed by the Nuclear Engineering and Technology Institute (NETEC) in South Korea via a subcontract from the Washington Savannah River Company (WSRC). This higher waste loading was achieved by using a CCIM which can run at higher glass processing temperatures (1250°C and higher) than the current DWPF Melter (1150°C). Higher waste loadings would result in less canisters being filled and faster waste throughput at the DWPF. The main demonstration objectives were to determine the maximum melt rate/waste throughput for the NETEC CCIM with a Sludge Batch 4 simulant as well as determine the viability of this technology for use in the DWPF.

INTRODUCTION

The Cold Crucible Induction Melter (CCIM) offers several advantages over the ceramic-lined, Joule-heated melter currently used in the Defense Waste Processing Facility (DWPF). These advantages include; 1) increased waste loading (WL) of 50+ weight percent versus 34 to 38 weight percent, thus resulting in less glass canisters produced, therefore reducing the lifecycle costs for canister production and storage, 2) possible extended melter service life, 3) higher waste throughput with a smaller footprint and lower costs, and 4) higher tolerance for troublesome metals (aluminum and noble metals).

Due to the above potential benefits, Washington Savannah River Company (WSRC) awarded a contract (funded by DOE-EM 20) to the Nuclear Engineering and Technology Institute (NETEC) to perform two pilot-scale (550mm diameter) CCIM demonstrations of up to 72 hours each at the NETEC facility in South Korea. Unlike other CCIM's which use an agitator to help increase the melt rate, the NETEC CCIM uses 12 bubblers. The NETEC demonstrations were performed with simulated Savannah River Site (SRS) High Level Waste (HLW) Sludge Batch 4 (SB4) slurry feed at a waste loading of 50 weight percent and a melt pool temperature of 1250°C. The main objectives of the demonstrations were to enable WSRC to: 1) determine the viability of the CCIM technology for use with SRS HLW slurry feed in the DWPF, 2) aid in determining the required CCIM sizing for application at the DWPF to achieve the desired melt rate/waste throughput, and 3) establish additional information needed for future CCIM testing and project planning.

The initial and baseline demonstrations were performed in October and November, 2007, respectively. WSRC personnel witnessed the baseline demonstration. The purpose of the initial demonstration was to access, and optimize where necessary, the electrical parameters during operation of the CCIM, the nominal operating conditions, the general behavior of the system, and to identify any off-normal or operational anomalies (if any) encountered during this initial demonstration. Based on findings from the initial demonstration, the purpose of the baseline demonstration was to acquire process data that would support sizing of a CCIM that could be deployed in the DWPF (primarily melt rate flux and glass waste loading).

CCIM TECHNOLOGY

This section gives a general description of CCIM technology, as well as details of the NETEC CCIM equipment. Electromagnetic induction is applied into the CCIM by placing a conductor such as molten glass in a variable magnetic field. As shown in Figure 1, an AC potential difference is applied to the inductor. This creates a variable magnetic field both inside and outside of the inductor. If the electrically conductive glass melt is placed inside of the inductor, the variation in the magnetic field causes a change in the magnetic flux passing through the glass melt. Per Lenz's law, this induces an electromotive force (EMF) which causes induced currents. These induced currents in the glass are converted into heat due to the Joule effect in the glass melt.[1,2]

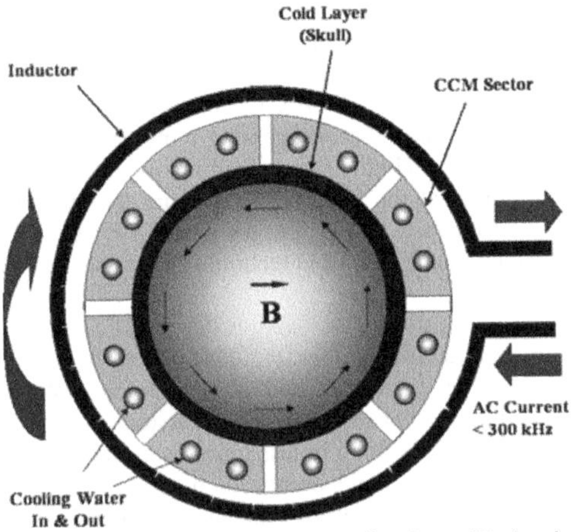

Figure 1. Sketch of CCIM Showing Application of Induction Heating

As shown in Figure 2, the CCIM is a simple and robust structure made of plain stainless steel and cooled by water. The electromagnetic field penetrates into the CCIM through the gaps between the stainless sectors (see Figure 1). For startup purposes, the CCIM is designed to inductively heat an electrically conductive susceptor that in turn oxidizes and heats the glass in the crucible. The molten glass is then directly heated while the crucible is water cooled. This cooling causes a thin layer of solidified glass (skull) to coat the inside surface of the crucible, thereby protecting it from the corrosive melt. Since the energy source is outside of the melter, electrode corrosion is avoided.

In addition to increasing the melter lifetime, this technology allows for much higher operating glass temperatures (> 1500°C) than conventional melters with either Inconel or molybdenum electrodes. The glass melt is batch poured through a cooled bottom drain valve as shown in Figure 2. Twelve bubblers are symmetrically located at the bottom of the CCIM to increase glass convection and thereby insure glass homogeneity and higher melt rates. Feed is delivered into the CCIM and the resultant off-gas is treated by an off-gas system that includes a

High Temperature Filter (HTF), HEPA Filter, Post Combustion Chamber, Off-Gas Cooler, Scrubber, Reheater A, Activated Carbon/HEPA Filter, Blower, Reheater B, and DeNOx System.

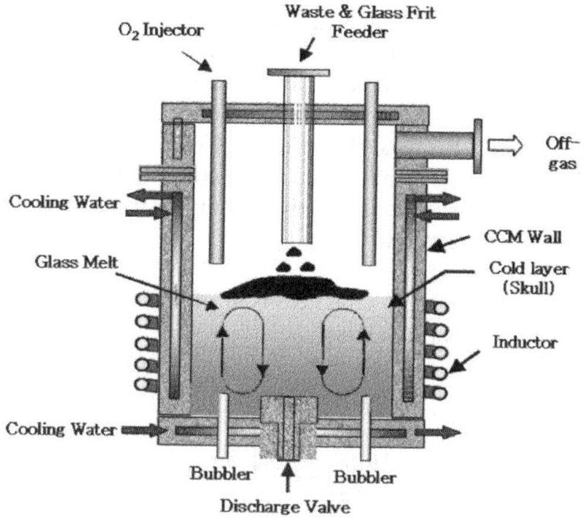

Figure 2. Cross-Section View of NETEC CCIM

FEED AND STARTUP GLASS DETAILS

The non-radioactive simulant used for the demonstrations was the SRS HLW SB4 which contains a high level of aluminum. The targeted and analyzed compositions of this simulant (calcined at 900°C) are given below in Table I.

Table I. Compositions of Calcined SB4 Simulant Sludge

Element	Calcined Sludge	
	Target Wt. %	Analyzed Wt. %
Al	14.904	14.40
Ca	2.183	1.18
Cr	0.150	0.11
Cu	0.045	0.03
Fe	22.40	17.7
K	0.06	0.33
Mg	1.848	0.22
Mn	4.948	3.52
Na	15.332	20.30
Ni	1.441	1.20
S	0.320	0.20
Si	1.40	1.12
Zn	0.044	0.04
Zr	0.074	0.06

The SB4 analyses show that the targeted weight percents of the major elements were met with the exception of sodium, which was higher than targeted. The impact of this high sodium content on the glass will be discussed later. The sludge slurry (about 32.64 weight percent sludge and the remainder water) was mixed with glass Frit 503-R3 to target a waste loading of 50% on a calcined solids basis. No reductant such as formic acid (used in the current DWPF feed preparation process) was added to the feed. This should result in highly oxidized feed and glass product. For reference, 10 liters of slurry should produce about 5.65 kg of glass. For the initial startup, startup glass SB4-503-R3 was placed in the CCIM. The composition of the SB4-503-R3 startup glass targets the composition of the SB4 sludge mixed with Frit 503-R3 at 50 weight percent waste loading. The targeted and analyzed compositions of the startup glass SB4-503-R3 and the glass Frit 503-R3 are given in Table II.

Table II. Compositions of Startup Glass SB4-503-R3 and Glass Frit 503-R3

Oxide	SB4-503-R3 Startup Glass		Glass Frit 503-R3	
	Target Wt %	Analyzed Wt %	Target Wt %	Analyzed Wt %
Al_2O_3	14.14	14.90	-	-
B_2O_3	8.00	10.60	16.00	18.00
CaO	1.53	1.53	-	-
Ce_2O_3	0.12	0.06	-	-
Cr_2O_3	0.11	0.11	-	-
Fe_2O_3	16.08	17.20	-	-
Li_2O	4.00	3.75	8.00	7.88
MgO	1.54	1.11	-	-
MnO_2	3.21	1.45	-	-
Na_2O	11.37	10.40	2.00	1.67
NiO	0.92	0.78	-	-
SO_3	0.48	0.40	-	-
SiO_2	38.50	38.70	74.00	73.90

The viscosity and electrical conductivity of the SB4-503-R3 startup glass were estimated using the computer code GlassForm v.1.1 to be about 50 poise and 0.45 S/cm, respectively, at 1,250°C. Both of these properties were deemed acceptable by NETEC.

INITIAL STARTUP

To start up the CCIM for the initial demonstration, 50 kg of the SB4-503-R3 startup glass was first added. A quarter-cut titanium ring (40 grams) was next loaded on top of the startup glass and then an additional 20 kg of startup glass was loaded on top of the ring. The High Frequency Generator (HFG) was started at 30 KW and incrementally increased by 10 KW per minute up to 80 KW. After that increases to the HFG of 5 KW/30 seconds were performed to achieve the targeted glass temperature of 1250°C. The nominal frequency for the HFG is 300 KHz and it can provide up 400 KW of power. Figure 3 shows the titanium ring with glass melting around it during the initial startup.

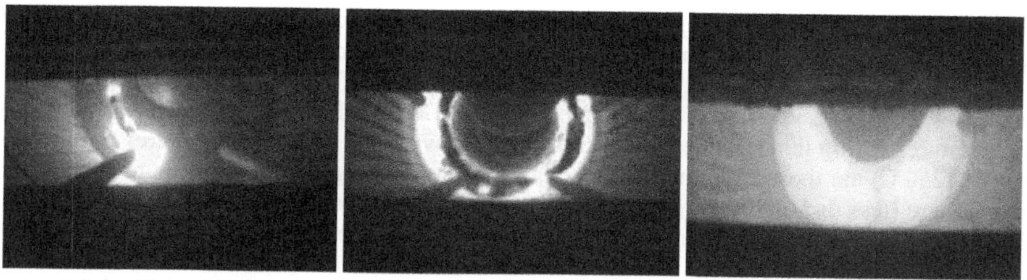

(a) Ignition of Ti Ring (b) Completion of Ignition (c) Preparation of Glass Melt

Figure 3. Initial Melting of Glass in CCIM Using a Titanium Susceptor

INITIAL AND BASELINE DEMONSTRATION DETAILS

The initial demonstration was started on 10/17/07 and finished on 10/19/07. The melter startup took only 1.5 hours to reach 1250°C. The HFG power was then increased to 240 KW to thoroughly mix the glass melt at 1350°C for a few minutes. The HFG power was then reduced to 200 KW to maintain a glass temperature of 1250°C.

The feed rate was started at 15 liters/hour and then incrementally increased to as high as 50 liters/hour. Pouring was done in a batch mode versus continuous pouring. From 15 to 30 liters/hour, the cold cap coverage was about 60 percent or less. Figure 4 gives a series of pictures of the cold cap at various feed rates. At 50 liters/hour, the cold cap coverage was about 80 percent. Unfortunately, the pressure drop in the off-gas HTF began to increase such that the demonstration had to be ended. The melter performance was deemed acceptable, including the various melter parameters being monitored.

Figure 4. Photographs of Cold Cap at Various Feed Rates

The baseline demonstration was started on 11/27/07 and completed on 11/30/07. The heatup of the CCIM was performed the same as for the initial demonstration. The CCIM feed was doped with nonradioactive cesium (fed as $CsNO_3$) to target a 0.5 weight percent Cs_2O concentration in the glass to determine a rough estimate of cesium volatility in the CCIM. The CCIM was fed at 40 liters/hour for about 70.5 hours of steady operation. The feed rate was then increased to 50 liters/hour for about 50 minutes without any issues. After this, the feed rate was increased to 60 liters/hour for about 30 minutes. Some irregularities were noticed in the cold cap, but vent holes were still present in the cold cap. At this time the test was ended after 72 hours of feeding. As with the initial demonstration, pouring was performed in a batch mode for the baseline demonstration.

As the demonstration progressed, it became a little more difficult to pour the glass via the bottom pour discharge valve. The reason for this was believed to be that glass in the drain valve had crystallized to some extent. Overall, the performance of the CCIM was deemed to be acceptable. Figure 5 shows the profiles of the CCIM electrical parameters and glass temperatures during the baseline demonstration. The two higher glass temperatures on the plot (about 1250°C) are those from the thermocouples located at heights of 8 and 12 cm. The lower temperature (less than 400°C) are those from a thermocouple located 4 cm above the bottom of the CCIM. This lower temperature is due to the close proximity of the water cooled melter floor.

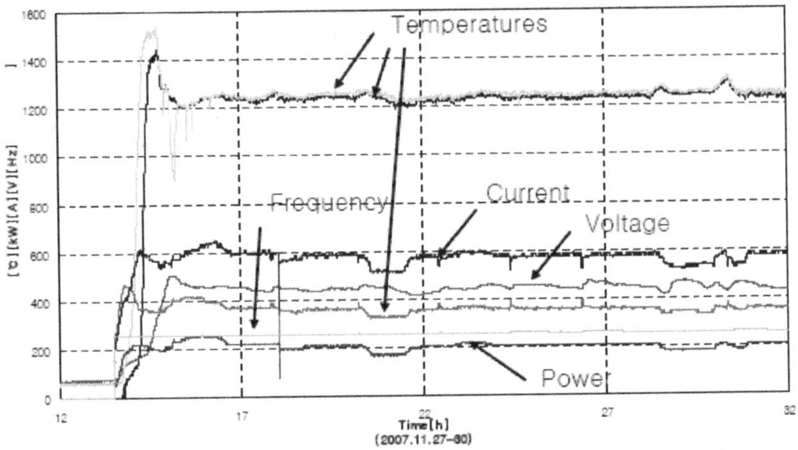

Figure 5. CCIM Melter Electrical Parameters and Temperatures during Baseline Demonstration

DISCUSSION OF BASELINE DEMONSTRATION RESULTS

Melt Rate and Waste Throughput

Table III summarizes the maximum feed rate, melt rate, waste throughput (WT), melt flux, and WT flux determined in the baseline demonstration at a waste loading of about 50 percent. The melt surface areas of the DWPF Melter and the 550mm CCIM are 28.3 ft^2 ($2.63m^2$) and 2.557 ft^2 (0.2376 m^2), respectively. It is noted that the calculated melt flux of the CCIM at 1,250 °C is 3 times higher than that of the DWPF melter design basis at 1,150 °C despite having an order of magnitude smaller melt surface area.

Table III. Summary of Maximum Rates Determined in NETEC CCIM Baseline Demonstration

Melt Temp (°C)	Max. Feed Rate gal/hr (l/hr)	Max. Melt Rate lb/hr (kg/hr)	Max. WT lb/hr (kg/hr)	Max. Melt Flux lb/hr/ft^2 (kg/hr/m^2)	Max. WT Flux lb/hr/ft^2 (kg/hr/m^2)
1250	13.2 (50)	62.4 (28.3)	31.3 (14.2)	24.4 (119.6)	12.2 (59.8)

Estimated Rates for 1.1m CCIM for DWPF

The required scale-up from a pilot 550mm CCIM to a larger CCIM that would potentially be installed in the DWPF is unknown, especially since this technology (bubblers, induction heating) is much different than the DWPF melter. However, if one assumes that the melt flux would be the same for a larger CCIM, an estimate of the melt rate and waste throughput for this particular feed can be made. If a CCIM with a diameter of 1.1m (3.609 ft) is used (per estimates by the NETEC staff), then for the same SB4 simulant feed processed in these demonstrations, the estimated melt rates and waste throughputs for this larger CCIM would be as given below in Table IV. The estimated waste throughput would therefore be about twice that of the existing DWPF melter. However, this is data from just one demonstration with one simulated feed. Still, the results are promising with regards to melt rate and waste throughput.

Table IV. Comparison of Estimated Maximum Melt Rate and WT for a 1.1m NETEC CCIM using Baseline Demonstration SB4 Simulant Feed.*

Melt Temp (°C)	Max. Melt Rate lb/hr (kg/hr)	Max. WT lb/hr (kg/hr)
1250	245.2 (111.2)	122.6 (55.6)

* For the SB4/Frit 510 feed, the highest melt rate and waste throughput of the DWPF melter at 35% WL are about 175 lb/hr (79.4 kg/hr) and 61 lb/hr (27.7 kg/hr), respectively.

Demonstration Glass Properties

The Savannah River National Laboratory (SRNL), part of WSRC, ran elemental analyses on the six glass pour samples from the baseline demonstration, and a review of the results was performed by the NETEC and SRNL personnel. It was determined that the measured glass composition was close to target with the exception of higher than expected Na_2O values (about 15 weight percent versus a target of 11.31). The higher than targeted Na_2O values were probably due to insufficient sludge washing. NETEC made the sludge per instructions from SRNL. Higher sodium in the feed can increase the amount of nepheline ($NaAlSiO_4$) crystals that are formed in the glass product. Nepheline is known to negatively impact glass durability.[3]

Per analysis of these same six samples, the glasses were found to be all fully oxidized with the exception of the sample taken from the pour made at the end of the demonstration (72 hours into the demonstration). Even this sample was almost fully oxidized with a $Fe^{2+}/\Sigma Fe^{total}$ ratio of 0.05 (where 0 is fully oxidized and the current DWPF target is approximately 0.20). These values agree with the fact that no reductant such as formic acid was used in the preparation of the feed for these demonstrations.

Samples from the same six glass pours during the baseline demonstration were submitted to SRNL for the Product Consistency Test (PCT) analysis. The PCT is used to determine the durability of waste glasses. A portion of each pour sample was also subjected to the modified Canister Centerline Cooling (CCC) thermal treatment before being subjected to the PCT. This modified CCC treatment is based on an estimate of the cool down rate of glass in the centerline of a DWPF canister filled at DWPF pour rates with a melt pool temperature of 1250°C (versus 1150°C for the current DWPF melter). It should be noted that the actual glass cool down rate of glass processed in a CCIM using batch pouring (versus the current DWPF mode of continuous pouring) had not been determined. A summary of the results is given in Table V. In the glass sample ID's, xxHR denotes the hour in which the pour sample was taken during the baseline demonstration. CCC indicates that the pour sample was submitted to the modified CCC thermal treatment prior to the PCT analysis. This thermal treatment increases the chance of crystal formation in the glass. EA glass is the HLW Environmental Assessment glass. The EA glass is the benchmark that DWPF glass must meet to be acceptable for repository disposal. The results showed that all of the quenched ("as is") glasses were more durable than the EA standard by about an order of magnitude. However, four of the six CCC thermal treated glasses failed the PCT test as they were less durable than the EA benchmark.

Table V. Summary of PCT Results (Average Normalized Release in Gram Glass/Liter Leachant)

Glass Sample ID	B	Li	Na	Si
EA Standard Sample	17.8772	9.5461	13.2464	3.9202
EA Cited Values[4]	16.6950	9.5650	13.3460	3.9220
2HR	0.9304	0.8805	0.9127	0.4964
12HR	1.0372	0.8862	1.0148	0.5249
28HR	1.2561	1.0853	1.2535	0.5872
44HR	1.1214	0.9630	1.0704	0.4863
60HR	1.2145	1.1159	1.3805	0.6002
72HR	1.0732	0.9732	1.2102	0.5816
2HRCCC	6.1058	5.2693	2.6169	0.9387
12HRCCC	44.1089	30.9586	16.6204	4.3939
28HRCCC	64.7884	26.3512	23.9690	3.5537
44HRCCC	10.9921	8.7307	5.0010	1.0347
60HRCCC	49.9757	28.8588	21.5560	3.9051
72HRCCC	48.229	28.0886	19.9305	3.8385

X-Ray Diffraction (XRD) analysis was also performed on the six pour samples with and without the CCC thermal treatment. The six quenched samples were essentially amorphous with small amounts of magnetite (crystalline form of iron oxide, $FeO \cdot Fe_2O_3$) and trevorite (nickel-iron oxide, $NiO \cdot Fe_2O_3$). The CCC thermal treated glasses all had magnetite and trevorite as well as nepheline. As stated previously, nepheline has a negative impact on glass durability.[3] The formation of nepheline was due to the higher than targeted sodium levels in the feed. This conclusion is substantiated by the fact that the glass made by SRNL in support of this work prior to the CCIM demonstrations contained the targeted sodium level and passed the PCT even after being subjected to the modified CCC thermal treatment.

Cesium Volatility

Cesium nitrate was added to the feed in the baseline demonstration to target a concentration of about 0.5 weight percent Cs_2O in the glass product. Analysis of the six glass pour samples taken indicates an estimated cesium volatility ranging from 34 to 70 percent. In fact, cesium is not just volatilized but entrained as well along with other feed materials. The DWPF design basis cesium carryover to the off-gas system is 10 percent. Therefore, testing to determine those process parameters that may be contributing to these higher than desired cesium carryover rates is needed. In particular, the effects of temperature, bubbling rate, cold cap coverage and waste loading on cesium carryover need to be measured. Testing would involve collection of off-gas particulate samples and subjecting them to the chemical and physical analyses for a full characterization of both volatile and nonvolatile carryovers.

Off-Gas Concentrations

Off-gas concentrations were measured in both demonstrations. The concentrations of CO (limit < 300 ppm) and NO_x (limit < 150 ppm) were well below the detection limits for each. The concentration of SO_x (limit < 100 ppm) was about 5 ppm.

CONCLUSIONS

Two 550mm CCIM demonstrations were performed by NETEC for WSRC via a subcontract. The results of the demonstrations indicated that the CCIM appears to be a viable technology for use in the DWPF to increase waste throughput relative to the current DWPF melter. The objectives of the CCIM per the subcontract were all met. The test results indicate that further optimization of glass properties such as liquidus and melt rate can be achieved and verified via additional testing. Below are the major achievements of the demonstrations:

- The ability to process simulated DWPF slurry feed was proven.
- Glass with waste loadings of about 50 percent was produced and subsequent durability tests (PCT) showed that the "as is" glasses were more durable than the EA glass standard by about an order of magnitude. However, four of the six CCC thermal treatment glass samples failed the PCT. The cause of this failure was the formation of nepheline in the CCC thermal treatment glasses as indicated by XRD. Nepheline is known to negatively impact glass durability. The formation of nepheline was attributed to the higher than targeted sodium levels in the feed.
- Based on the demonstration's maximum melt fluxes for 1250°C and applying these to a larger CCIM (>1m diameter), much higher waste throughputs than currently demonstrated in the DWPF appear possible.

Additional issues/research needs were identified, some of which are given below.

- Scale-up of melt rates and waste throughputs to full-scale CCIM (larger than 1m diameter)
- Processing of the DWPF Chemical Processing Cell flowsheet products in a CCIM
- Off-gas sampling and cesium volatility (impact of bubbling, cold cap coverage, etc.)
- Long term operation of a pilot/full-scale CCIM with DWPF feed
- Assessment of potential off-gas flammability due to lower vapor space temperature of the CCIM than the DWPF Melter

ACKNOWLEDGEMENTS

The authors acknowledge the excellent contribution of NETEC for the performance of the two CCIM demonstrations via WSRC subcontract AC59545T.

REFERENCES

[1]Myung-Jae Song, The Vitrified Solution, *Nuclear Engineering International,* 22-26 (February 2003).

[2]Myung-Jae Song and San-Woon Shin, Heart of Glass, *Nuclear Engineering International,* 25-28 (December 2005).

[3]K. M. Fox, et al., Nepheline Formation Study for Sludge Batch 4 (SB4): Phase 3 Experimental Results, WSRC-TR-2006-00093, Revision 0, WSRC, Aiken, South Carolina (2006).

[4]C. M. Jantzen, N. E. Bibler, D. C. Beam, C. L. Crawford, and M. A. Pickett, Characterization of the Defense Waste Processing Facility (DWPF) Environmental Assessment (EA) Glass Standard Reference Material, WSRC-TR-92-346, Revision 1, WSRC, Aiken, South Carolina (1993).

ADSORPTION AND SEPARATION OF URANIUM USING TUNGSTEN OXIDES

Hamed Albusaidi and Allen W. Apblett
Department of Chemistry
Oklahoma State University
Stillwater, OK, 74078.

ABSTRACT

Nuclear energy is undergoing a renaissance because it does not contribute to global warming. However, ignoring the issue of radioactive waste, the production of nuclear energy has its own environmental impacts. The mining and refining of uranium produces tailings that leach uranium and other toxic metals into aquifers. Military use of depleted uranium also releases materials that contaminate water supplies. However, the more common route for human ingestion of uranium is from natural waters in contact with uranium-rich granitoids. A strong need exists for processes to prevent contamination of aquifers and purification of drinking water supplies due to uranium's health risks. It was found that tungsten trioxide reacts with dissolved uranium (in the form of uranium acetate) to produce an amorphous hydrated mixed oxide of approximate formula of $U_5W_3O_{14}\cdot 8.3H_2O$. The uptake of uranium was found to be 1.67 moles of uranium per mole of WO_3 or 171% by weight. As an alternative, tungstic acid was found to react significantly more rapidly with aqueous uranium.

INTRODUCTION

Uranium is a common contaminant of ground water and can arise from natural and anthropogenic sources. Uranium occurs naturally in the earth's crust and in surface and ground water. When bedrock consisting mainly of uranium-rich granitoids and granites comes in contact with soft, slightly alkaline bicarbonate waters under oxidizing conditions uranium will solubilize over a wide pH range. These conditions occur widely throughout the world. For example, in Finland exceptionally high uranium concentrations up to 12,000 ppb are found in wells drilled in bedrock [1]. Concentrations of uranium up to 700 ppb have been found in private wells in Canada [2] while a survey in the United States of drinking water from 978 sites found a mean concentration of 2.55 ppb [3]. However, some sites in the United States have serious contamination with uranium. For example, in the Simpsonville-Greenville area of South Carolina, high amounts of uranium (30 to 9900 ppb) were found in 31 drinking water wells [4]. The contamination with uranium is believed to be the result of veins of pegmatite that occur in the area. Besides entering drinking water from naturally occurring deposits, uranium can also contaminate the water supply as the result of human activity, such as uranium mining, mill tailings, and even agriculture [5, 6]. Phosphate fertilizers often contain uranium at an average concentration of 150 ppm and therefore are an important contributor of uranium to groundwater [7]. The Fry Canyon site in Utah is a good example of the dangers of uranium mine tailings. The uranium concentrations measured in groundwater at this site were as high as 16,300 ppb with a median concentration of 840 ppb before remedial actions were taken [8]. Depleted uranium ammunition has also been demonstrated as a source of drinking water contamination [9].

Animal testing and studies of occupationally exposed people have shown that the major health effect of uranium is chemical kidney toxicity, rather than a radiation hazard [10]. Both functional and histological damage to the proximal tubulus of the kidney have been demonstrated [11]. Little is known about the effects of long-term environmental uranium exposure in humans but there is an association of uranium exposure with increased urinary glucose, alkaline phosphatase, and ß-microglobulin excretion [12] as well as increased urinary albumin levels [13]. As a result of such studies, the World Health Organization has proposed a guideline value of 2 ppb for uranium in drinking water while the US EPA has specified a limit of 30 ppb.

Current municipal treatment practices are not effective in removing uranium. However, experimentation indicates, that uranium removal can be accomplished by a variety of processes such as modification of pH and/or chemical treatment (e.g. alum) [14]. Several sorbants have been shown to be useful for removal of uranium from water. Activated carbon, iron powder, magnetite, anion exchange resin and cation exchange resin were shown to be capable of adsorbing more than 90% of the uranium from drinking water. However, two common household treatment devices were found not to be totally effective for uranium removal [4].

Besides treatment of well water, there is also a strong need for prevention of the spread of uranium contamination from concentrated source such as uranium mine tailings. Commonly used aboveground water treatment processes are not cost-effective and do not provide an adequate solution to this problem. However, permeable reactive barriers (Figure 1) have been demonstrated to be economically viable and elegant alternatives to active pump and treat remediation systems. Such barriers composed of metallic iron, ferric oxyhydroxide, and bone char phosphate have been designed and proven effective for uranium [8]. Iron metal performed the best and consistently lowered the input uranium concentration by more than 99.9 percent after the contaminated groundwater had traveled 1.5 ft into the permeable reactive barrier.

Molybdenum hydrogen bronze (also called molybdenum blue), HMo_2O_6, has been demonstrated to be useful for removal of uranium from aqueous solution and possible use in a cyclic process for uranium recovery [15]. In the latter respect, it was thought that the protons would provide ion exchange sites on the material in its reduced form while oxidation would remove these sites and release the uranium in concentrated form. However, it was found that oxidation of the blue reagent occurred during the adsorption process so that the reagent turned from blue to yellow. At the same time, it was found that the uptake of uranium was 122 percent by weight and far exceeded the expected capacity based on the number of protons present (the proton concentration is 3.46 mEq/g while the uranium adsorption was 5.14 mEq/g). Analysis of the product from uranium uptake revealed that the bronze had reacted with uranium to produce the mineral iriginite, $UMo_2O_9 \cdot 3H_2O$. This reaction was found to be very selective for uranium, other actinides, and heavy metals. For example, no reaction was observed with aqueous calcium and the uptake of uranium was only slightly hindered by the presence of calcium ions at five times the concentration of uranyl ions. The oxidation of the Mo(V) centers in the bronze was found to be due to reaction with molecular oxygen as the layered structure was disassembled by reaction with uranyl ions. This result suggested that prior reduction of MoO_3 to HMo_2O_6 was unnecessary for uranium adsorption thus prompting an investigation of the ability of molybdenum trioxide to absorb uranium from water [16]. It was found that MoO_3 could absorb up to 165 % by weight of uranium via a chemical reaction that produces an insoluble uranium molybdenum oxide mineral oxide called umohoite, $UMoO6 \cdot 2H_2O$. The rate of reaction between MoO_3 and a slight excess (43 mole %) of 0.100 M uranyl acetate was found to be zero order with a rate constant of 0.42 mmol/hr. A cyclic process was developed whereby MoO_3 adsorbed uranium from aqueous solution and then the uranium and molybdenum trioxide were separated by treatment with aqueous ammonia. Solid ammonium uranate was isolated by filtration and the aqueous ammonium molybdate was converted back to MoO_3 by heating. The recovery of uranium from the separation was 98.9%.

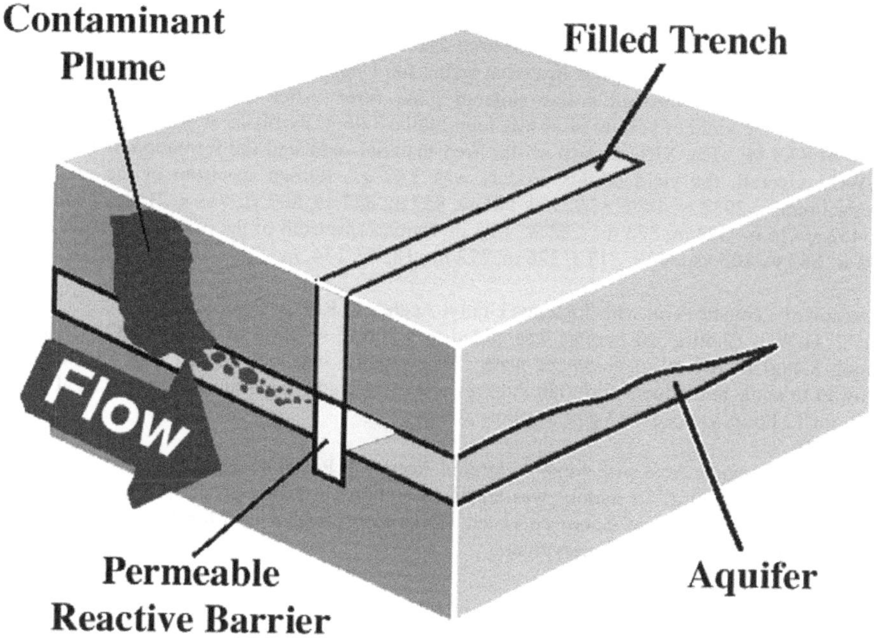

Figure 1. Operation of a permeable reactive barrier

Unfortunately, molybdenum trioxide has slight solubility in water and may have some toxicity issues if used to purify drinking water. Therefore, this investigation was undertaken to determine whether non-toxic tungsten trioxide and/or tungstic acid could be used for removal of tungsten from water.

EXPERIMENTAL

All reagents were commercial products (ACS Reagent grade or higher) and were used without further purification. Water was purified by reverse osmosis followed by deionization. Bulk pyrolyses at various temperatures were performed in ambient air in a digitally-controlled muffle furnace using approximately 1 g samples, a ramp of 10°C/min and a hold time of 4 hr. X-ray powder diffraction (XRD) patterns were recorded on a Bruker AXS D-8 Advance X-ray powder diffractometer using copper K_α radiation. Crystalline phases were identified using a search/match program and the PDF-2 database of the International Centre for Diffraction Data [17]. X-ray fluorescence spectroscopy (XRF) was performed on a Thermo electron QuanX XRD spectrometer. Uranium was quantified using standards produced by blending varying quantities of tungsten trioxide and uranium oxide (U_3O_8) Raman spectra were obtained using neat powders in a Nicolet NXF 9610 Raman Spectrometer.

Reaction of Tungsten Trioxide with Uranyl Acetate

WO_3 (1.16 g, 5.00 mmol) was added to a solution of uranyl acetate (8.48 g. 20 mmol) in 100 ml of water. The mixture was heated at reflux for 7 days. Upon cooling, a yellow solid was isolated by filtration through a fine sintered glass filter. After drying in vacuum at room temperature, the yield of yellow solid was found to be 3.96 g. Pyrolysis at 600°C gave a ceramic yield of 87.9 %. The XRD pattern of the fired material indicated the formation of UWO_6 and U_2WO_8. Overall, the yield of this product was 3.48 g. Raman spectrum of the uncalcined product (cm^{-1}): 2932 w, 1476 w, 973 m, 906 w, 857 m, 827 vs, 807 vs, 768 s, 715 m, 584 m, 490 w, 452 s, 415 m, 324 m, 273 m., 182 w, 137, m. Raman spectrum of the calcined product (cm^{-1}): 992 w, 883 w, 805 vs, 749 s, 717 s, 326 m, 272 m., 182 w, 134, m.

Reaction of Tungstic Acid with Aqueous Uranyl Acetate at Reflux Temperature

H_2WO_4 (2.50 g, 10 mmol,) was added to a stirred solution of $UO_2(C_2H_6O_2)\cdot 2H_2O$ (20 mmol, 8.48g) in 100 ml of deionized water. The mixture was refluxed for 6 hr and was then allowed to cool. Subsequent filtration through a fine sintered glass filter and drying in at 125°C oven for 12 hours yielded 7.13 g of a yellow solid.

Reaction of Tungstic Acid with Aqueous Uranyl Acetate at Room Temperature

H_2WO_4 (2.50 g, 10 mmol,) was added to a stirred solution of $UO_2(C_2H_6O_2)\cdot 2H_2O$ (20 mmol, 8.48g) in 100 ml of deionized water. The mixture was stirred for 7 days and was then filtered through a fine sintered glass filter,

RESULTS AND DISCUSSION

The reaction of tungsten trioxide with an excess of uranyl acetate in refluxing water is a slow reaction that eventually yields a yellow solid that was found to be amorphous by X-ray powder diffraction analysis, Raman spectroscopy showed that the solid consisted of both uranium (band at 807cm^{-1}) and tungsten oxides (bands at 768, 715, and 273 cm^{-1}) [18, 19] There were also weak bands at 2932 and 1476 cm^{-1} that can be attributed to acetate ions. The low intensity of these bands suggests that they are due to a small amount of chemisorbed acetate on the product surface and are not indicative that the product is a basic acetate. The infrared spectrum of the product demonstrated that it was a hydrate with broad absorptions centered at 3600 and 1600 cm^{-1} attributable to bound water or hydroxyl moieties. Upon heating, there was little change in the position of the tungsten-oxygen and uranium-oxygen vibrations, but the peaks attributable to functional groups other than the metal oxides disappeared. This result implies that the initially formed product is structurally quite similar to that of the uranium tungsten oxide phases in the calcined material. XRF spectroscopy (Figure 2) confirmed that the product contained both uranium and tungsten.

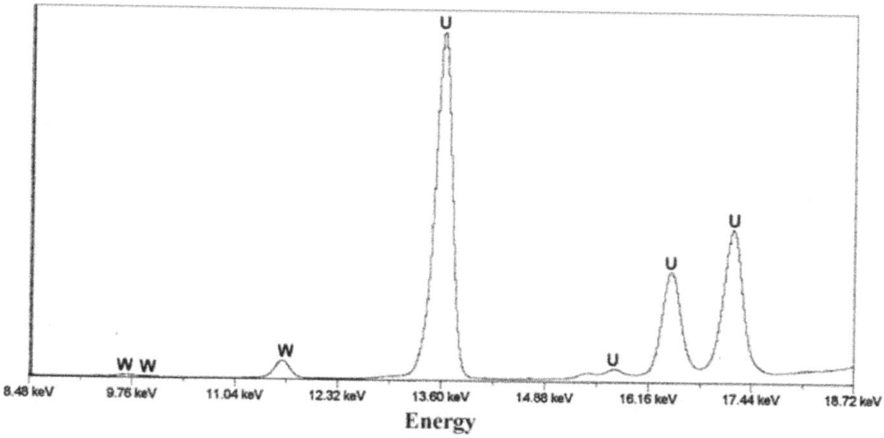

Figure 2. XRF Spectrum of Uranium Acetate/ Tungsten Trioxide Product

A thermal treatment was performed order to determine the amount of extraneous species such as water hydroxide, and acetate that was present and to demonstrate that the reaction of tungsten trioxide with uranyl acetate produced a mixed oxide compound. The weight loss upon heating to 600°C was 12.1 % by weight. The XRD pattern of the calcined product, shown in Figure 3, indicated that it was a mixture of UWO_6 and U_2WO_8. The amount of uranium oxide absorbed in the reaction is indicative of the formation of a 1:2 mole ratio mixture of these two species. This gives a uranium capacity of 1.67 moles of uranium per mole of WO_3 or 171% by weight. Also, the composition of the calcined product implies that the initially formed material has the composition $U_5W_3O_{14} \cdot 8.3H_2O$ assuming the acetate content is negligible. Therefore, on a mole: mole basis, WO_3 displays a greater capacity than MoO_3, which only absorbs one molar equivalent of uranium. This leads to almost the same capacity for uranium sorption by weight despite the significantly larger formula weight of tungsten trioxide as compared to molybdenum trioxide.

The rate of reaction of tungsten trioxide was determined in the presence of calcium ions as an example of competing ions found in ground water. The ratio of calcium ions to uranyl ions used was 3:1 on a mole basis. The reaction was found to be first order in uranium concentration (Figure 4) with a rate constant of 0.015 hr^{-1}. Thus, the reaction has a half-life of 2 days at 100°C. The reaction is slower than that of molybdenum trioxide with uranyl acetate and this may be due to the layered structure of molybdenum trioxide leading to faster reactions due to its ability to intercalate uranyl ions between its sheets. Tungsten trioxide's 3-dimensional structure does not permit this mechanism. It is also possible that the slight water-solubility of molybdenum trioxide contributes to the faster reaction kinetics

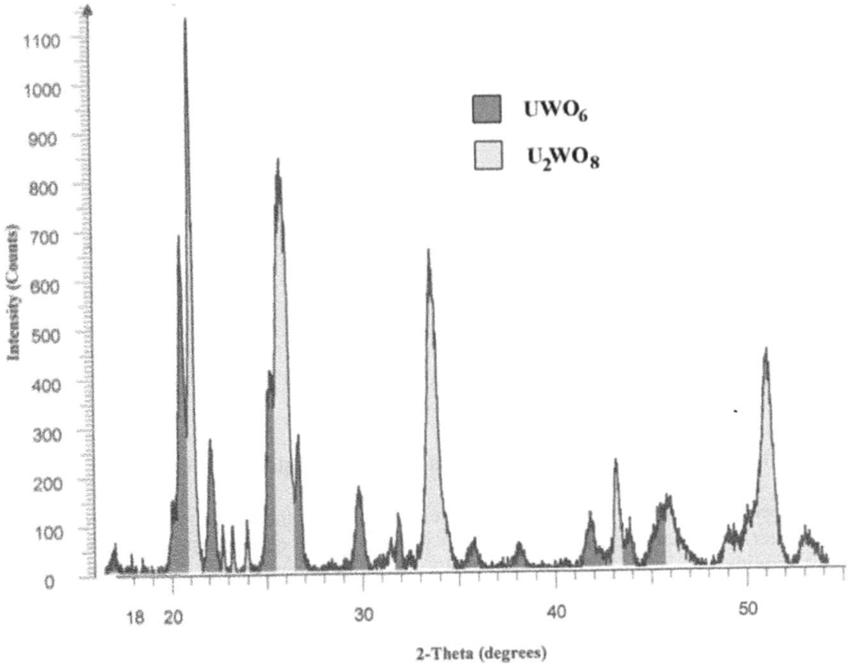

Figure 3. XRD pattern of Calcined Uranyl Acetate/Tungsten Trioxide Product

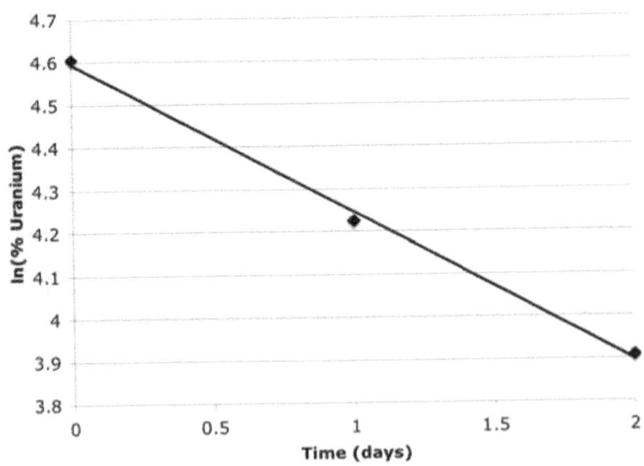

Figure 4. Reaction Kinetics for Uranyl Acetate/Tungsten Trioxide Reaction

The supposition that molybdenum trioxide reacts faster with uranyl ions because of its layered structure allowing intercalation suggests that tungstic acid, H_2WO_4, which also has a layered structure will react faster than tungsten trioxide. This was found to be true since a six-hour reaction of tungstic acid with refluxing aqueous uranyl acetate led to formation of a yellow solid that was 51% uranium by weight. As in the case of tungsten trioxide, the initially formed product was an amorphous hydrate but yielded a mixture of UWO_6 and U_2WO_8 when heated to 600°C (in a 91.8% ceramic yield). However, the mixture was richer in UWO_6 than the product from tungsten trioxide The promising kinetics at elevated temperature led to testing at room temperature which showed that H_2WO_4 absorbed 7% by weight of uranium in one week.

CONCLUSIONS

Tungsten trioxide and tungstic acid both react with dissolved uranium (in the form of uranium acetate) to produce amorphous hydrated mixed oxides. This reaction can form the basis of the means to remove uranium from drinking water or to treat contaminated ground water with a permeable reactive barrier.

ACKNOWLEDGMENTS

We are grateful for support provided for Hamed Albusaidi by the Ministry of Higher Education of Oman. The National Science Foundation is acknowledged for providing the funds for the X-ray powder diffractometer used in this investigation.

REFERENCES

[1] L. Salonen "^{238}U Series Radionuclides as a Source of Increased Radioactivity in Groundwater Originating from Finnish Bedrock. In: *Future Groundwater Resources at Risk* (Soveri J, Suokko T, eds):International Association of Hydrological Sciences, Wallingford, UK. pp. 71-84 (1994).

[2] M.A. Moss, R.F McCurdy, K.C. Dooley, M.L. Givner, L.C. Dymond, J.M. Slayter, and , M.M. Courneya "Uranium in drinking water - report on clinical studies in Nova Scotia". In: *Chemical Toxicology and Clinical Chemistry of Metals.* S.S. Brown and J. Savory (eds.): Academic Press, London. pp. 149-152 (1983).

[3] United States Environmental Protection Agency "Occurrence and Exposure Assessment for Uranium in Public Drinking Water Supplies" **EPA 68-03-3514** (1990).

[4] J.D. Navratil, "Advances in Treatment Methods for Uranium Contaminated Soil and Water" *Archive of Oncology*, **9**, 257-60 (2001).

[5] C.R. Cothern and W.L. Lappenbusch, "Occurrence of Uranium in Drinking Water in the U.S." *Health Phys.*, **45**, 89 (1983).

[6] D.R. Dreesen, J.M. Williams, M.L. Marple, E.S.Gladney, and D.R. Perrin, "Mobility and Bioavailability of Uranium Mill Tailings Constituents. *Environ. Sci. Technol.*, **16**, 702 (1982).

[7] R.F. Spalding and W.M Sackett "Uranium in Runoff from the Gulf of Mexico Distributive Province: Anomalous Concentrations" *Science*, **175**, 629 (1972)

[8] United States Environmental Protection Agency "Field Demonstration Of Permeable Reactive Barriers To Remove Dissolved Uranium From Groundwater, Fry Canyon, Utah" **EPA 402-C-00-001** (2000)

[9] United Nations Environment Programme "Depleted Uranium in Bosnia and Herzegovina Post-Conflict Environmental Assessment Revised Edition" Switzerland, 2003.
[10] D.M. Taylor and S.K. Taylor "Environmental Uranium and Human Health. *Rev Environ Health* **12**, 147-157 (1997).
[11] D.P. Haley. "Morphologic Changes in Uranyl Nitrate-Induced Acute Renal Failure in Saline- and Water-Drinking Rats. *Lab Investig* **46**,196-208 (1982).
[12] M.L. Zamora, B.L. Tracy, J.M. Zielinski, D.P. Meyerhof and M.A. Moss "Chronic Ingestion of Uranium in Drinking Water: a Study of Kidney Bioeffects in Humans. *Toxicological Science*, **43**, 68-77 (1998).
[13] Y. Mao, M. Desmeules, D. Schaubel, D. Bérubé, R. Dyck, D. Brûlé, and B.Thomas, Inorganic Components of Drinking Water and Microalbuminuria. *Environ. Res.*, **71**, 135-140 (1995).
[14] S.K. White, and E.A.Bondietti, Removing Uranium by Current Municipal Water Treatment Processes. *J. Am. Water Works Assoc.*, **75**, 374 (1983)
[15] B.P. Kiran, A.W. Apblett, and M. Chehbouni, Mohamed. "Selective Absorption of Heavy Metals and Radionuclides from Water in a Direct-to-Ceramic Process' Ceramic Transactions, **143**, 385-394 (2003)
[16] M, Chehbouni and A.W. Apblett Cyclic Process for Concentrating Uranium from Water, Ceramic Transactions. Vol. 168, 2005.
[17] "Powder Diffraction File (PDF-2)" (International Centre for Diffraction Data, Newtown Square, PA).
[18] M. Luisa and S.H. Taylor, Characterization of Uranium Oxides Using in Situ Micro-Raman Spectroscopy, *Applied Spectroscopy*, Vol 54 (No 9) 2000, p 1372.
[19] M.F. Daniel, B. Desbat, and J.C. Lassegues, Infrared and Raman Study of WO_3 Tungsten Trioxides and $W0_3{\cdot}xH_zO$ Tungsten Trioxide Hydrates, *Journal of Solid State Chemistry*; Vol 67, 1987, p 235-247.

ACCELERATED PROCESSING OF SB4 AND PREPARATION FOR SB5 PROCESSING AT DWPF

C.C. Herman
Savannah River National Laboratory
Aiken, South Carolina, USA

ABSTRACT

The Defense Waste Processing Facility (DWPF) initiated processing of Sludge Batch 4 (SB4) in May 2007. SB4 was the first DWPF sludge batch to contain significant quantities of HM or high Al sludge. Initial testing with SB4 simulants showed potential negative impacts to DWPF processing; therefore, Savannah River National Laboratory (SRNL) performed extensive testing in an attempt to optimize processing. SRNL's testing has resulted in the highest DWPF production rates since start-up. During SB4 processing, DWPF also began incorporating waste streams from the interim salt processing facilities to initiate coupled operations. While DWPF has been processing SB4, the Liquid Waste Organization (LWO) and the SRNL have been preparing Sludge Batch 5 (SB5). SB5 has undergone low-temperature aluminum dissolution to reduce the mass of sludge for vitrification and will contain a small fraction of Purex sludge. A high-level review of SB4 processing and the SB5 preparation studies will be provided.

INTRODUCTION AND BACKGROUND

The DWPF's mission is to immobilize the millions of gallons of High Level Waste (HLW) that is currently in storage at the Department of Energy's Savannah River Site. The HLW consists of two fractions that must be immobilized in DWPF. The sludge fraction consists of the insoluble solids in the HLW, which contain the long-lived radionuclides and inert components such as iron and aluminum hydroxide. The salt fraction destined for treatment in the DWPF consists of Cs, actinides, and sludge solids that have been treated through salt waste processing facilities. As of August 2008, the DWPF has processed 2.6 million gallons of the HLW sludge containing 17 million curies of radioactivity and a very small percentage of the salt stream. The HLW has been immobilized in glass that has been poured in over 2500 canisters that are currently stored in glass waste storage buildings awaiting disposal in the federal repository. Approximately 35 million gallons of HLW remain for treatment.

A schematic of the SRS HLW system and treatment processes is provided as Figure 1. The HLW sludge, along with the salt cake and concentrated supernate, is stored in 48 underground carbon steel tanks. Before the sludge and salt can be treated in the DWPF, several process steps must be performed to prepare the streams for treatment. Supernate is decanted off of the tanks for treatment in the site's three evaporators, where the waste is concentrated and stored until downstream processes are available for further processing. Currently, the salt is transferred to a storage tank for subsequent treatment in the salt waste processing facilities. The sludge must be transferred into the DWPF sludge preparation tank (Tank 51) to undergo Extended Sludge Processing (ESP).

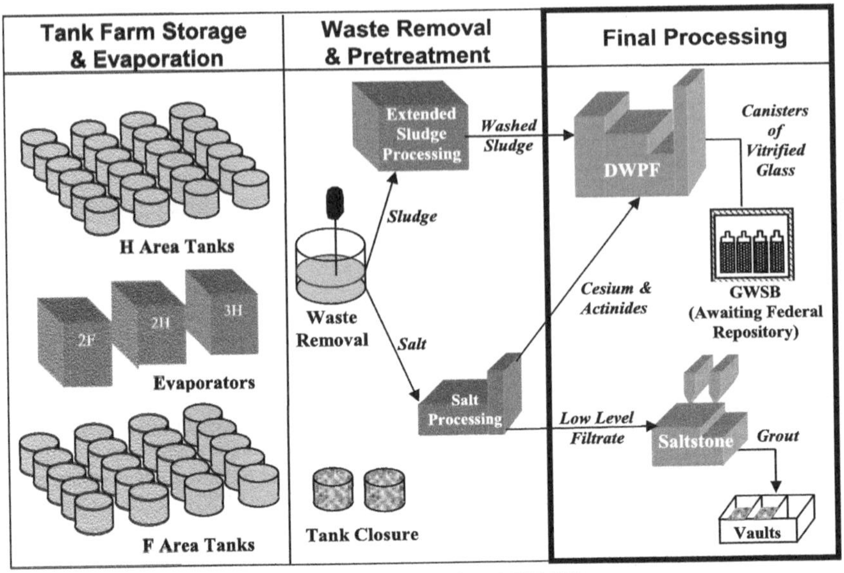

Figure 1. Flow diagram of SRS HLW Process

Sludge Processing and Qualification

Typically, sludge from more than one tank is transferred to the preparation tank to provide a blended sludge composition suitable for DWPF processing. The tanks are selected by the LWO based on necessary tank closure dates and expected sludge compositions. ESP is initiated with the introduction of inhibited water (water with a dilute concentration of sodium hydroxide and sodium nitrite) or decanting of supernate to reduce the high sodium salt content of the sludge. The iterative process is repeated until the final sodium supernate concentration target and the insoluble/total solids target are obtained. The final product constitutes a sludge batch for DWPF.

A key component of the sludge batch preparation process is the SRNL qualification testing that is performed to ensure that the sludge will meet DWPF processing and wasteform acceptance criteria. A sample of the sludge or sludges to be processed is taken early in the process and shipped to SRNL for characterization. Characterization includes the chemical constituents as well as the physical properties. This information is used to perform flowsheet testing, determine a frit for processing, and verify acceptability of the DWPF durability models over the expected glass composition region. Depending on when the sample is taken, multiple wash and decant cycles may need to be performed to complete the sludge batch preparation process. Replication of these cycles can identify potential problems with sludge settling and verify the accuracy of washing model predictions. All of this information is used by SRNL to recommend the sodium and solid target endpoints to LWO for each sludge batch.

In addition to characterization and washing steps, the radioactive sample is subjected to the chemical processing steps performed in DWPF to prepare the feed for melting. The chemical processing steps involve adding nitric and formic acids to the sludge at 93°C, heating the sludge to boiling, removing water to meet the target solids, and then boiling under reflux conditions to complete the necessary chemical reactions. These steps are performed in the Sludge Receipt and Adjustment Tank (SRAT) in DWPF and then the slurry is transferred to the Slurry Mix Evaporator (SME) tank. In the SME, frit is added at the target waste loading and the slurry is boiled further to concentrate to the melter feed solids target, which is typically in the range of 45 – 50 weight percent total solids. During the SRAT and SME processes, significant chemical reactions occur and hydrogen is generated, which requires monitoring to ensure that a flammable concentration is not reached. After verification of DWPF acceptability, a portion of the SME material is used to fabricate a glass at the nominal DWPF melter temperature of 1150°C. This glass is subjected to the ASTM Product Consistency Test (PCT)[1] to ensure that its durability performance is better than the HLW Environmental Assessment (EA) benchmark glass.

To support the radioactive testing, flowsheet testing with simulants is performed to refine the washing endpoint and to define the nominal acid addition quantities for the radioactive testing. These studies are typically completed before the actual waste testing and help define the width of the processing window. For the glass formulation development and melter processing, variability studies are performed considering the nominal sludge and frit compositions. The nominal sludge and frit compositions are combined over the waste loading interval of interest (typically 28 to 42 weight percent waste loading) and uncertainty in the compositions are considered to fully bracket the composition region. Glasses are then selected for fabrication representing points within the composition region. The PCT is performed on these glasses to verify the predictability of the DWPF durability models and to verify product acceptability. However, before these studies can be initiated, the candidate frit for that sludge batch must be determined. This involves both paper studies to determine the maximum processing window and laboratory studies to measure the melt rate of the sludge and frit compositions of interest. Initial laboratory testing focuses on dried feed testing and, therefore, is not able to discern differences in feed rheology and its subsequent impact on melter feed behavior. This Melt Rate Furnace (MRF) testing does allow a comparison between frits and waste loadings to identify relative melt rates for that system. This allows fine-tuning or down-selection of the frits for slurry fed melt rate furnace (SMRF) testing, which also provides relative melt rate data for DWPF processing.

Salt Processing Facilities and Associated DWPF Impact

To remove the salt, the salt cake is dissolved in the storage tank in preparation for treatment in one of the site's salt waste processing facilities. In the salt waste processing facilities, radionuclides (e.g., actinides, strontium, and cesium) are removed and prepared for transfer to DWPF. Removal of strontium and actinides is accomplished through the Actinide Removal Process (ARP) that entails sorption with monosodium titanate (MST). This produces a slurry that must be filtered to remove the MST and entrained sludge solids. The MST/sludge solids stream is subsequently transferred to DWPF for co-processing with the HLW sludge. Cesium removal is performed through a caustic-side solvent extraction (CSSX) process. In this process, the cesium becomes entrained in the organic solvent phase. After multiple extraction steps, the non-recovered solvent is transferred with the cesium as a dilute nitric acid stream to DWPF and the solvent is recycled. The decontaminated salt solution, on the other hand, is sent to the Saltstone Processing Facility for disposal as grout. Both of the salt treatment processes are currently being demonstrated on a pilot-scale to identify potential processing issues and to provide tank space until the larger capacity Salt Waste Processing Facility (SWPF) is brought on line.

In the DWPF, the MST/sludge solids stream is added to the SRAT at boiling to reduce the volume to be processed through the DWPF chemical process. After the stream is added, a sample is taken to determine the acid addition amounts and then typical processing is initiated. The MST/sludge solids stream itself has minimal impact on the SRAT processing chemistry, but does increase the SRAT processing time because of the additional boiling required. On the glass formulation side, more impacts are realized because of the high sodium and titanium content of the feed. The frit composition must be tolerant to significant sodium variation because in some cases none of the salt stream will be added, while in others a full tank addition may be necessary (which would be equivalent in volume to the nominal sludge processing volume). Nominally, 1500 to 2500 gallons of the stream would be processed with each SRAT batch. The titanium concentration must be considered because it is a known nucleating agent for crystals. Currently, the process is not creating a large enough volume to exceed the current 2 weight percent titania glass limit. However, once the SWPF begins processing, titania concentrations up to 6 weight percent may need to be accommodated. Due to the large volume of sample and associated radioactive dose that would be required to perform a demonstration with radioactive material, qualification is not performed on radioactive material as is done with the sludge. Qualification of the MST/sludge solids stream is performed by SRNL using simulants in flowsheet testing and surrogates in glass testing. The radioactive constituents are not believed to have a significant impact on either of these components of the qualification process and should have minimal impact on processing characteristics when the simulants are adequately formulated.

The strip effluent from the MCU process, and later the SWPF, is transferred to the DWPF in small batches. These batches are held in a hold tank until they can be transferred to the SRAT during the boiling process after all acid has been added. As with the MST/sludge solids stream, a SRAT batch can have no MCU stream or a full tank of the MCU stream (which would be equivalent in volume to the nominal sludge processing volume). The stream is a dilute nitric acid stream that contains small concentrations of the removed cesium and non-recovered solvent. In the SRAT, the stream has the potential to impact the feed redox because of the nitrate in the stream so this must be accommodated by the split of formic and nitric acids that are added at the beginning of the SRAT process. The cesium has no impact, while the residual solvent has resulted in flammability controls being implemented in DWPF to ensure that a flammable mixture was not created in the SRAT or SME cycle. Flammability is primarily controlled by the addition at boiling and insurance of purge gas during static and processing conditions. This has a net impact on the SRAT processing time, which will be further increased with the larger volume additions expected for SWPF. On the glass processing side, minimal impact is seen, as long as the redox is properly adjusted, because of the small concentration of all of the stream constituents. As with the MST/sludge solids stream, qualification is only performed using simulant streams, which are considered adequate to bound the potential processing impacts.

SLUDGE BATCH 4

Preparation of Sludge Batch 4 (SB4) was initiated in the Tank Farm in 2004 by transferring HM (high aluminum) sludge from Tank 11. The sludge exhibited settling problems soon after completion of the third transfer into the DWPF feed preparation tank (Tank 51). Good settling behavior is important during the sludge preparation process since ESP relies upon clean separation of the supernate and solids phase to remove the sodium salts during decanting and it needs to happen in a timely fashion to avoid a DWPF feed break. After completion of limited characterization of the Tank 51 samples, the settling problems were attributed to excessive shearing performed during sludge removal and the small boehmite particles that were the predominant compound in the insoluble solids.

The samples also indicated the presence of a much larger mass of sludge than originally projected. Due to the large mass of sludge solids, enhanced controls had to be enacted to ensure that hydrogen generated through reactions of the radioactive sludge components with the supernate did not result in the accumulation of a flammable concentration of hydrogen. The controls included limiting the mass to minimize the pathway for hydrogen to be released and stirring of the tank contents on a routine basis to limit accumulation. This caused problems with washing and sludge preparation, which were mitigated by performing less washing and extending the sludge preparation time. In addition, the sludge preparation plan was changed to limit the components of SB4; therefore, Purex sludge was not blended with Tank 51 resulting in a very high aluminum sludge content. The high aluminum coupled with higher sodium from limited washing created a durability issue for the glass wasteform. The combination of these two components can cause nepheline crystals to form upon slow cooling of the DWPF canister, and nepheline can have a significant impact on the product durability.

Glass Formulation and Melt Rate Observations

SRNL performed glass formulation testing to define a frit composition that would provide a reasonable operating window from a waste loading and processing perspective while also ensuring product durability. This testing concluded that increasing the B_2O_3 concentration in the frit could suppress the formation of nepheline in the SB4 system.[2] The testing also showed that the implementation of the nepheline constraint previously developed by Li et. al.[3] could adequately prevent the use of a glass formulation that was susceptible to nepheline formation. As an added benefit, increasing the B_2O_3 concentration was also shown to have a positive impact on melt rate and throughput.[4] This finding was equally important since initial SB4 melt rate testing had little success in finding a frit that would provide melt rate that was considered acceptable for DWPF processing.[5,6]

However, shortly after SRNL recommended the use of Frit 503, whoich appeared to have a throughput equivalent to that seen for the Sludge Batch 3 (SB3)-Frit 418 system[4], the LWO had to change the final composition of SB4. A confirmation sample of SB4 taken from Tank 51 near the end of sludge preparation indicated rheological properties that exceeded the transfer pump criteria. To solve this problem and to help increase the total solids content of the sludge in the DWPF feed tank (Tank 40, which would eventually be a blend of SB3 and the material qualified as SB4 in Tank 51 to be called SB4), supernate from Tank 40 was transferred to Tank 51. The net effect was a decrease in the total sodium content of the feed. This change could not be accommodated with Frit 503. Therefore, Frit 418 was recommended for initial processing due to the long lead times needed for frit procurement and the need for a SRNL variability study to verify the durability performance of the frit and sludge combination. Frit 418 satisfied these criteria since it was already being used for SB3 processing and had previously been tested in a variability study with SB4 using a wide compositional uncertainty region[7]. Therefore, DWPF initiated SB4 processing with Frit 418. Shortly after SB4 startup, SRNL performed additional melt rate testing with additional high boron containing frits with lower sodium content than Frit 503. The goal was to take advantage of the suppression of nepheline formation and the increased melt rate seen with the higher boron containing frits. Of these selected frits, Frit 510 provided an optimized melt rate in MRF testing and a reasonable operating window. Frit 510 was subsequently tested in the SMRF and shown to have an increased throughput compared to the

Frit 418 – SB4 system.[8] See Table I for the composition of the frits recommended for SB4 processing.

Table I. Frit Compositions for DWPF Processing

Frit #	B_2O_3 wt%	Li_2O wt%	Na_2O wt%	SiO_2 wt%
418	8	8	8	76
503	14	8	4	74
510	14	8	8	70

Initial DWPF throughput with the SB4-Frit 418 system was slower than that seen for the SB3-Frit 418 system. Waste loading was limited to ~34% in DWPF due to rheological properties of the sludge and slow throughput seen with the SB4 system. Canister pour times were near 30 hours per can with significant increases seen when the waste loading was increased. Feed rates were limited to ~0.5 gallons per minute, which was equivalent to the feed rate with SB3 at the end of processing. Once DWPF implemented Frit 510, DWPF increased their melter feed rates to upwards of 0.9 gallons per minute, which is near the upper limit of feeding capacity. While these high feed rates were utilized, pouring of DWPF canisters was accomplished in ~20 hours. This pour rate was the highest achieved in DWPF since startup and approached the design basis of the melter. However, waste loadings were still limited to ~34% since this appeared to provide the optimal throughput. Therefore, DWPF production verified the trends seen in the SRNL MRF and SMRF testing with regards to the frit composition.

Additional processing issues plagued SB4 due to leaking slurry pumps in the DWPF feed tank. This caused the feed to become diluted resulting in slower chemical processing since additional boiling to concentrate the feed was required. This caused a reduction in the melter feed and production rates because of the dilute feed being fed to the melter. To alleviate this problem, two decants of supernate were performed to concentrate the solids. However, before these decants could be performed, SRNL had to perform testing to ensure that these decants could be accommodated since sodium would be removed from the glass and the processing window might need adjustment due to the change in feed chemistry. SRNL flowsheet testing demonstrated minimal impact[9,10], but glass formulation assessments indicated a slight decrease in processing window with each decant[10,11]. To mitigate the decreases in the processing window and melt rate, sodium hydroxide was added to the feed tank to increase the Na_2O in the sludge by 3 weight percent on a calcined sludge oxide basis. Since the completion of the decants and sodium hydroxide addition, some of the processing rates have been recovered in the DWPF such that feed preparation remains the processing constraint in the facility.

Salt Processing Impacts

Transfers of the MST/sludge solids stream from the ARP facility were initiated to DWPF in the spring of 2008. The need to process this stream through the SRAT resulted in longer processing times for the feed preparation system, which in turn impacted the melter processing. With the high throughputs seen with Frit 510, the feed preparation process became the rate limiting step in the facility for the first time since startup. No other significant impacts have been seen with processing of the MST/sludge solids stream in DWPF.

Only small quantities of the MCU stream have been transferred to the DWPF facility. Therefore, its true impact cannot yet be assessed. However, simulants from the MCU facility have been processed through the DWPF to ensure that the transfer lines, hold tanks, and flammability controls are adequately operating. The true impact of this stream will not likely be known until the start of Sludge Batch 5 processing when higher volumes are processed in the facility.

SLUDGE BACTH 5

To try to broaden the operating window, reduce the sludge mass going to DWPF, and potentially improve processing for Sludge Batch 5, LWO decided to perform low temperature aluminum dissolution on the sludge remaining in Tank 51 after transfers were made for SB4. The decision was made before the high throughput data was obtained from DWPF processing with Frit 510. However, even with these high processing rates, it was believed that the potential for mass reduction (reduced canister production) and increased waste loading were sufficient reasons to pursue this technology. After completion of the aluminum dissolution process, the remaining sludge was blended with the Purex sludge previously selected for blending with SB4 (as stated above, this Purex sludge could not be blended with SB4 because the mass was too high). In addition, excess plutonium that had been dissolved and neutralized was transferred to Tank 51 for disposition as part of SB5. Upon completion of the qualification of SB5, the plan was to blend the SB5 material with the remaining SB4 heel in Tank 40 such that the blend would be processed as SB5. Final preparation of SB5 is underway in the Tank Farm and a sample will be used to confirm that the target composition has been met and to determine the target insoluble solids target to ensure acceptable rheology is obtained before the sludge is transferred.

Al Dissolution of Tank 51 Sludge

The goal of the low temperature aluminum dissolution process is to use caustic to dissolve the aluminum from the insoluble solids fraction of the sludge, whereby it becomes soluble in the supernate phase that can be decanted from the insoluble solids. In the Tank Farm, tankers of 50% sodium hydroxide were added to the remaining SB4 sludge in Tank 51 over a 4 week period. The slurry containing the sodium hydroxide was then mixed at a temperature between 55 - 65°C. The slurry pumps in Tank 51 provided mixing as well as the necessary heat to maintain the tank temperature over 46 days. After this period, the insoluble solids were allowed to settle for 29 days before the supernate was decanted to another HLW tank for treatment through the SWPF at a later date. Based on analyses of the sludge before and after dissolution, ~ 60% of the aluminum was removed through the dissolution process.[12] The oxide composition of the primary sludge components before (equivalent to the SB4 qualification material) and after the aluminum dissolution process are shown in Table II. The SB4 composition after blending with the SB3 heel (SB4 processing composition) and the washed SB5 qualification sample composition (post aluminum dissolution and Purex blending) are also given in Table II for comparison. The Tank 51 composition after aluminum dissolution is very high in sodium concentration because of the significant amount of sodium hydroxide that is added to perform aluminum dissolution. The impact of washing on the SB5 qualification sample is evident by the dramatic reduction in Na_2O concentration, while the impact of the Purex sludge addition on the Tank 51 sample both before and after aluminum dissolution is evident by the increase in Fe_2O_3 and U_3O_8 concentration.

Table II. Oxide Compositions of SB4 and SB5 Sludges of Interest

Oxide	Tank 51 before Aluminum Dissolution[13] (SB4 as Qualified)	SB4 as Processed (Blended with SB3 Heel)[14]	Tank 51 after Aluminum Dissolution[15]	Washed SB5 Qualification Sample[16]
Al_2O_3	42.9	24.1	20.6	20.6
CaO	1.65	2.64	1.30	2.24
Fe_2O_3	15.2	27.4	11.31	28.6
MgO	0.71	2.60	0.532	1.23
MnO_2	3.98	6.73	2.88	7.08
Na_2O	28.5	19.9	55.27	25.1
NiO	1.19	1.53	0.781	3.64
SiO_2	1.35	2.57	0.924	2.42
U_3O_8	2.80	8.38	2.03	7.72

Sludge Preparation Challenges

A 3-L sample of the post aluminum dissolution sludge blended with the Purex sludge material was taken to perform sludge batch qualification in the SRNL Shielded Cells. Approximately half of the plutonium was already in the sample at the time it was pulled. Since the qualification sample was taken very early in the sludge preparation process, the planned washing and concentration steps needed to be mimicked on the qualification sample. This included 6 washes with one being supernate material from Tank 40, 7 decants, and 1 plutonium stream/chemical addition. Performance of these process steps would result in a sludge with a final supernate concentration of roughly 1 M sodium and an insoluble solids concentration of approximately 10.5 weight percent. As part of the SRNL testing, chemical and physical behavior of the sludge was monitored throughout the processing steps to ensure that the final sludge would meet acceptable processing behaviors and to provide guidance to the Tank Farm on the effectiveness of their models for predicting sludge chemistry.

From the very beginning of processing of the sludge in the Tank Farm, problems were experienced with support equipment and with physical behavior of the sludge. Problems included limitations on the evaporator capacity due to the high hydroxide content of the feed, increased radiolytic hydrogen generation, and poor settling behavior. This resulted in changes to the sludge preparation plan on several occasions. In addition, SRNL testing showed that the insoluble solids might be slightly higher than projected due to the precipitation of some of the soluble species and sodium could be more effectively removed than the washing model indicated. The higher solids are advantageous from a DWPF processing perspective since it requires less evaporation in the facility during either the feed preparation or melter processing steps, while the lower sodium provides more flexibility to find a larger glass composition operating window. In the end, SRNL completed 5 washes of the qualification sample with one being a sample of Tank 40 supernate, addition of a plutonium stream and cold chemicals to properly balance the corrosion chemistry, and 6 decants to obtain an insoluble solids concentration of 11.3 weight percent. Therefore, SRNL recommended that a wash/decant cycle be eliminated and a higher solids endpoint targeted, which had the potential to reduce the preparation time by up to 1 month and would save valuable tank space. The primary oxide components of the SB5 qualification sample are shown in Table II.

Chemical Processing Cell Demonstrations

As discussed above, a routine step in the qualification process is demonstration of the chemical process cell SRAT and SME cycles. For SB5, a new challenge was introduced since the sludge simulant did not adequately represent the dissolved aluminum behavior. More specifically, the soluble aluminum hydroxide phase that was known to be present in the radioactive sludge could not be adequately represented by the aluminum hydroxide chemical that was added to the sludge simulant upon final trimming. The difference manifested itself in the acid demand for base equivalents when a titration or hydroxide measurement was performed. This made estimating the required acid challenging for the qualification run in the Shielded Cells. Testing with the simulant at several acid addition levels was performed. From this series of tests, a recommended acid addition amount was provided and was based on acceptable processing to meet DWPF defined objectives. Although no blended runs with radioactive sludge were planned, simulant testing was still necessary to ensure that a processing recommendation could be made to DWPF. Similar acid stoichiometric factors were used in the blend series testing; however, more problems were experienced with mercury reduction, which is one of the primary goals of the SRAT. As part of the flowsheet testing, a run where a simulant of the MST/sludge solids stream and the MCU strip effluent stream was added was also performed at the nominal acid stoichiometry. Minimal impact to processing was seen from these additions.[17]

When the testing was performed in the Shielded Cells, three processing problems were identified during demonstration of the SRAT cycle. Although below the DWPF process limits, significant hydrogen was generated, which was above the levels seen in the simulant runs. The sludge exhibited foaming throughout the process demonstration requiring the addition of multiple doses of antifoam above that typically required in DWPF processing. Finally, the rheology of the SRAT product was above DWPF design limits and yield stress was higher than before the start of processing, which is counterintuitive to the SRAT cycle goals. Although multiple problems were experienced in the SRAT demonstration, the SRAT chemical reaction requirements were met and all of the SME cycle parameters were met after diluting the SRAT product at the start of testing and adding antifoam with each frit or water addition.

An additional flowsheet demonstration will be performed with a 3-L sample taken from Tank 51. The original intent of this sample was to confirm the SB5 washing endpoint. However, a neptunium stream will now be added to the confirmation sample to qualify the stream for addition in Tank 40 since the decision to disposition this stream through DWPF was not made early enough to include it in the SB5 qualification process. As part of this second demonstration, the solids concentration target will be reduced slightly to try to alleviate the foaming and poor rheological properties seen in the first demonstration. It is anticipated that the lower solids targets should alleviate some of the rheological problems and may decrease the hydrogen generation rate. Once this test is completed, an updated recommendation will be made on the processing parameters to be used in DWPF.

Glass Formulation Challenges

Frit selection for SB5 was made more difficult due to uncertainty over the final composition of the SB5 sludge. This was partially driven by the performance of low temperature aluminum dissolution for the first time and the need to decant SB4-Tank 40 to concentrate the solids (which provided uncertainty in the composition to be blended with Tank 51). However, the problem was compounded by the typical uncertainty associated with the composition of a new sludge being blended (Purex-Tank 7 in this case) and definition of the final washing endpoint. To attempt to accommodate all of this uncertainty, several projections of the potential sludge composition were provided both for the material as prepared in Tank 51 (SB5 qualification) and after blending with Tank 40. The blended compositions addressed uncertainty in the heel mass that would remain in Tank 40 for blending with SB5.

The projections were holistically reviewed and then composition groupings were developed with uncertainty around the sodium concentration. Sodium concentration represented the greatest uncertainty because it was the component most strongly impacted by the washing endpoint and decisions on decanting the Tank 40 supernate. To provide the most flexibility and accommodate this large uncertainty, Frit 418 that had been used in initial processing of SB4 and to increase throughput for SB3 was recommended for initial SB5 processing.[18] This frit was used in a variability study to ensure that DWPF's durability model would adequately predict the performance of the DWPF glass over the anticipated compositional range when subjected to the ASTM PCT[1]. Glass performance data from this testing demonstrated that the models were acceptable. Included in the glass variability assessment were options to include up to half a tank volume of the MST/sludge solids stream.

Frit 418 was sufficiently robust that it also could be processed with the SB5 material before blending with SB4. Therefore, it was used as the frit in the SME testing of the qualification sample. A glass was fabricated from the SME product and was subjected to the ASTM PCT[1]. The data indicated that the glass was more durable than the EA glass and its durability performance was predictable by the DWPF product and process control models.

While Frit 418 provides tremendous compositional flexibility for SB5 processing, it may not be the best frit for optimal throughput in DWPF. Due to the delays in receiving the qualification sample and the numerous changes in the sludge preparation plan, melt rate testing to support the development of a frit that would optimize throughput for SB5 could not be developed in time to procure frit for DWPF and/or complete the variability study. Therefore, additional testing is currently underway to determine if another frit can be identified that exhibits enhanced throughput over Frit 418 with SB5. Once the melt rate testing is completed, a recommendation will be made to DWPF on the path forward for proceeding.

CONCLUSIONS

Each sludge batch that is qualified for DWPF continues to present new challenges that must be overcome. SRNL has worked successfully with DWPF to overcome these challenges as they have arisen during the sludge preparation or qualification processes. For Sludge Batch 4, problems with sludge preparation, glass acceptability, and melter processing were identified early on in the preparation process. These problems were attributed to the high aluminum content and particular compound in the HM sludge that was SB4. After performing testing to understand the potential processing issues, SRNL was able to recommend a frit that not only suppressed nepheline formation (therefore increasing glass durability) but also provided the highest throughputs seen in DWPF since startup. The higher throughputs were accomplished even with a dilute feed that required additional chemical processing time to accommodate the actinides and sludge solids from the interim salt processing facilities.

For Sludge Batch 5, low temperature aluminum dissolution was successfully performed to remove a significant mass of the insoluble solids destined for DWPF immobilization. After demonstration of this process, similar challenges to SB4 were realized and identified. These challenges were enhanced by the need to understand the behavior of the aluminum hydroxide component that had become soluble during the dissolution process, which included uncertainty in the final composition to be processed. Currently, SRNL is completing most of the testing and/or documenting the results of testing needed to support SB5 qualification. DWPF should initiate processing of SB5 in November 2008. When processing is initiated, Frit 418 will be used and will continue to be used unless a more optimized frit is identified in SRNL melt rate testing that will be performed in late 2008.

ACKNOWLEDGEMENTS

Testing reported was performed under the auspices of contract DE-AC09-96SR18500 between the Department of Energy and the Washington Savannah River Company. Qualification studies for both sludge batches were performed by J. Pareizs, C. Bannochie, D. Click, K. Zeigler, D. Lambert, B. Pickenheim, S. Reboul, N. Bibler, and Shielded Cells' facility personnel. Chemical processing studies for both sludge batches were performed by M. Stone, D. Lambert, D. Koopman, B. Pickenheim, M. Barnes, M. Williams, and technical analysts assigned to the Process Science & Engineering Section of the Environmental & Chemical Processing Technology directorate of the SRNL. Glass formulation and melt rate assessment studies for the sludge batches was supported by D. Peeler, T. Edwards, M. Smith, K. Fox, D. Miller, T. Jones, F. Raszewski, and technical analysts assigned to the Process Science & Engineering Section. Analytical characterization was supported by D. Best, W. Riley, technical analysts assigned to the Process Science & Engineering Section, and the Analytical Development organization of SRNL.

REFERENCES

[1]American Society of Testing Materials, Standard Test Methods for Determining Chemical Durability of Nuclear Waste Glasses: The Product Consistency Test (PCT), ASTM C-1285-2002 (2002).

[2]K. Fox, T. Edwards, D. Peeler, D. Best, I. Reamer, and R. Workman, Durability and Nepheline Crystallization Study for High Level Waste (HLW) Sludge Batch 4 (SB4) Glasses Formulated with Frit 503, WSRC-STI-2006-00009, Washington Savannah River Company, Aiken, South Carolina (2006).

[3]H. Li, P. Hrma, J. Vienna, M. Qian, Y. Su, and D. Smith, Effects of Al_2O_3, B_2O_3, Na_2O, and SiO_2 on Nepheline Formation in Borosilicate Glasses: Chemical and Physical Correlations, *Journal of Non-Crystalline Solids*, **331**, 202-216 (2003).

[4]M. Smith, M. Stone, T. Jones, D. Miller, P. Burket, and D. Immel, Sludge Batch 4 Melt Rate Furnace and Slurry-Fed Melt Rate Furnace Tests with Frits 418, 425, and 503, WSRC-STI-2006-00015, Washington Savannah River Company, Aiken, South Carolina (2006).

[5]M. Smith and D. Miller, Initial SB4 Melt Rate Furnace Testing, WSRC-TR-2005-00339, Washington Savannah River Company, Aiken, South Carolina (2005).

[6]M. Smith, T. Jones, D. Miller, and M. Stone, SB4 SMRF Runs with Frits 418 and 320, WSRC-TR-2005-00569, Washington Savannah River Company, Aiken, South Carolina (2005).

[7]K. Fox, T. Edwards, D. Peeler, D. Best, I. Reamer, and R. Workman, High Level Waste (HLW) Sludge Batch 4 (SB4) Variability Study, WSRC-STI-2006-00204, Washington Savannah River Company, Aiken, South Carolina (2006).

[8]M. Smith, T. Jones, and D. Miller, Sludge Batch 4 Baseline Melt Rate Furnace and Slurry-Fed Melt Rate Furnace Tests with Frits 418 and 510, WSRC-STI-2007-00450, Washington Savannah River Company, Aiken, South Carolina (2007).

[9]D. Koopman, D. Best, and M. Williams, Impact of SB4 Tank 40 Decant and ARP/MCU Additions with/without Added Caustic on DWPF CPC Performance, WSRC-STI-2008-00147, Washington Savannah River Company, Aiken, South Carolina (2008).
[10]F. Raszewski, M. Stone, T. Edwards, and D. Peeler, The Impact of a Second Tank 40 Decant on the Frit 510 – Sludge Batch 4 System: Projected Operating Windows and Chemical Process Cell Processing, WSRC-TR-2008-00141, Washington Savannah River Company, Aiken, South Carolina (2008).
[11]F. Raszewski, T. Edwards, and D. Peeler, The Impact of a Tank 40H Decant on the Projected Operating Windows for SB4 and Glass Selection Strategy in Support of the Variability Study, WSRC-STI-2008-00017, Washington Savannah River Company, Aiken, South Carolina (2008).
[12]J. Pike, Evaluation of Low Temperature Aluminum Dissolution in Tank 51, SRNS-STI-2008-00021, Savannah River Nuclear Solutions, Aiken, South Carolina (2008).
[13]J. Pareizs, C. Bannochie, D. Click, E. Hansen, D. Lambert, and M. Stone, Washing and Demonstration of the DWPF Flowsheet in the SRNL Shielded Cells Using Post Aluminum Dissolution Tank 51 Sludge Slurry, WSRC-STI-2008-00086, Washington Savannah River Company, Aiken, South Carolina (2008).
[14]C. Bannochie, Tank 40 Final SB4 Chemical Characterization Results, WSRC-STI-2007-00674, Washington Savannah River Company, Aiken, South Carolina (2008).
[15]M. Hay and D. McCabe, Characterization of Tank 11H and Tank 51H Post Aluminum Dissolution Process Samples, WSRC-STI-2008-00227, Washington Savannah River Company, Aiken, South Carolina (2008).
[16]C. Bannochie, J. Pareizs, and H. Ajo, Sludge Batch 5 Acceptance Evaluation: Chemical Characterization of the Tank 51 SB5 Qualification Sample Prepared at SRNL, SRNL-PSE-2008-00125, Savannah River National Laboratory, Aiken, South Carolina (2008).
[17]D. Lambert, M. Stone, B. Pickenheim, D. Best, and D. Koopman, Sludge Batch 5 Simulant Flowsheet Studies, SRNS-STI-2008-00024, Savannah River Nuclear Solutions, Aiken, South Carolina (2008).
[18]K. Fox, T. Edwards, and D. Peeler, Recommended Frit Composition for Initial Sludge Batch 5 Processing at the Defense Waste Processing Facility, WSRC-STI-2008-00338, Washington Savannah River Company, Aiken, South Carolina (2008).

EFFECT OF COMPOSITIONAL CHANGES ON THE STRUCTURE AND CRYSTALLIZATION TENDENCY OF A BOROSILICATE GLASS CONTAINING MoO_3

M. Magnin, S. Schuller
CEA, DEN, DTCD, Laboratoire d'études et Développement de Matrice de Conditionnement, Marcoule BP 17171, Bagnols sur Cèze, France

D. Caurant, O. Majérus
Ecole Nationale Supérieure de Chimie de Paris (Paris Tech), Laboratoire de Chimie de la Matière Condensée de Paris (UMR CNRS 7574), Paris, France

D. de Ligny
Université Claude Bernard Lyon1, Laboratoire de Physico-Chimie des Matériaux Luminescents (UMR CNRS 5620), Villeurbanne, France

C. Mercier
Université de Valenciennes et du Hainaut-Cambrésis, Laboratoire de Matériaux Avancés et Céramiques (UPRES EA 2443), Valenciennes, France

ABSTRACT

In borosilicate glasses, molybdenum combined with other elements such as alkali and alkaline-earth elements may form crystalline molybdates during melt cooling, among which sodium molybdate, Na_2MoO_4, is water-soluble. The solubility limit of MoO_3 was found to be 2.5 mol% in a simplified SiO_2–B_2O_3–Na_2O–CaO nuclear glass at about 1300°C. Higher MoO_3 concentrations induced liquid phase separation followed by crystallization of Na_2MoO_4 and $CaMoO_4$. These phenomena even occur for a cooling rate of 10^3 °C/min. The aim of this study was the impact of Na^+ and Ca^{2+} cations relative concentrations on the tendency of the melts to crystallize (nature of phases, intensity of crystallization). To determine the environment of molybdenum in the glass structure, ^{95}Mo MAS NMR and Raman spectroscopies were implemented on a series of quenched glasses with variable Rz concentrations ratio from 0 to 0.5, Rz=[CaO]/([CaO]+[Na_2O]) and with a fixed MoO_3 amount (2.5 mol%).

INTRODUCTION

In order to decrease the frequency of both refueling operation in nuclear reactors and spent fuel reprocessing, the use of high burn-up nuclear fuels (60 GW.day/ton) are planned in France in pressurized water reactors[1,2]. At the end of reprocessing (i.e. after Pu and U extraction by the Purex process), the remaining highly radioactive liquid wastes (HLW) will be enriched in fission products (such as molybdenum oxide) and minor actinides in comparison with current HLW solutions originating from 45 GW.day/ton spent fuels reprocessing. In order to decrease the number of glass canisters, a new glass matrix with higher level waste is under development at the CEA in France[2].

Molybdenum oxide (MoO_3) has a low solubility in borosilicate glasses. Beyond 2 or 3 wt% and following the melt cooling and the melting temperature, it can participate with alkali and alkaline-earth oxides to the formation of a yellow phase in the glass. This complex yellow phase is formed during the melting stage of the process by separation of molten salts from the borosilicate melt and may be responsible for an enhanced corrosion of the metallic glass melter[3]. On account of the high solubility of alkali molybdates in water, this phase separation may induce a decrease of the chemical durability of the vitrified nuclear waste material in aqueous environment[4,5].

Concerning the structural role of molybdenum in silicate and borosilicate glasses, extended X-ray absorption spectroscopy (EXAFS) showed that hexavalent molybdenum cations are located in tetrahedral (orthomolybdate) MoO_4^{2-} entities within alkali and alkaline-earth rich depolymerised regions outside the silicate or borosilicate network (i.e., the oxygen atoms of MoO_4^{2-} units are not bounded to Si or B atoms)[6-9]. The same kind of isolated molybdate tetrahedral entities are present in the Na_2MoO_4 and $CaMoO_4$ phases that can crystallize in nuclear borosilicate glasses. In spite of this structural similarity, the solubility of Na_2MoO_4 in water is more than three orders of magnitude higher than that of $CaMoO_4$. The similarity between the local environment of Mo^{6+} ions in glasses and in molybdate crystalline phases may explain the low solubility of molybdenum in glasses and their tendency to crystallize during slow cooling of melts or heat treatment of glasses. In the study of Caurant et al[10], it was shown that glass composition changes and more particularly the increase of B_2O_3 concentration has a strong impact on the nature of molybdates that may crystallize during the cooling of Mo-rich SiO_2-B_2O_3-Na_2O-CaO melts.

In this paper, MoO_3 (2.5-5 mol%) was added to a simplified SiO_2-B_2O_3-Na_2O-CaO composition representative of HLW glass in order to determine the solubility limit of MoO_3 (Mx glass series). Moreover, as sodium and calcium molybdates may crystallize in the melt during cooling, the impact of the relative proportions of sodium and calcium oxides on the solubility of MoO_3 and on the melt tendency to crystallize during slow cooling was studied by varying the R concentrations ratio from 0 to 0.5, Rz=[CaO]/([CaO]+[Na_2O]) (Rz glass series). For all glasses of this series, the evolution of molybdenum environment within the glass structure was followed by Raman and ^{95}Mo MAS NMR spectroscopies.

EXPERIMENTAL PROCEDURE

Glass Preparation

For this study, a non-radioactive simplified glass composition was chosen in the SiO_2–B_2O_3–Na_2O–CaO system derived from that of a UOX3 glass under development, keeping the same oxides ratio as in this complex composition. For instance, in this simplified composition, Na^+ and Ca^{2+} cations respectively simulate all the alkali (Li^+, Na^+, Rb^+, Cs^+) and alkaline-earth (Ca^{2+}, Ba^{2+}, Sr^{2+}) cations that will be present either in HLW UOX3 solutions or in the glass frit that will be added to the waste to prepare glass. In order to facilitate ^{95}Mo MAS NMR spectra recording, a small and fixed quantity of paramagnetic Nd_2O_3 (0.15 mol%) was added to all glass compositions prepared in this study. Two series of glasses were prepared for this study (mol%): first a Mx series (Table I) with increasing MoO_3 contents (1-x)[66.40 SiO_2-14.05 Na_2O-8.65 CaO-10.75 B_2O_3-0.15 Nd_2O_3] – x MoO_3 for x = 2.50, 3.00, 4.00, 5.00 mol% and secondly, a Rz series (Table II) derived from the M2.5 glass composition by changing Rz ratio and keeping constant MoO_3 concentration (2.50 mol%). In this last series Rz takes values of 0.00, 0.20, 0.38, and 0.50. No glasses were prepared with Rz >0.5 because too high CaO concentrations are known to induce glass-in-glass phase separation in the simple SiO_2-Na_2O-CaO system[11]. Glass R0.38 of the Rz series is the same M2.5 glass of the Mx series. The choice of the MoO_3 concentration for the Rz series (2.50 mol%) was made according to the crystallization results obtained for the Mx series shown in the result part of this paper.

Table I Nominal Glass Compositions of Mx Series (mol.%).

	M2.5	M3	M4	M5
% SiO_2	64.76	64.47	63.81	63.18
% B_2O_3	10.48	10.44	10.30	10.18
% Na_2O	13.68	13.63	13.47	13.32
% CaO	8.43	8.40	8.28	8.18
% Nd_2O_3	0.15	0.15	0.14	0.14
% MoO_3	2.5	3.00	4.00	5.00

TABLE II Nominal Glass Compositions of Rz Series (mol.%).

	Rz=0	Rz=0.2	Rz=0.38	Rz=0.5
% SiO_2	64.76	64.76	64.76	64.76
% B_2O_3	10.48	10.48	10.48	10.48
% Na_2O	22.11	17.69	13.68	11.06
% CaO	-	4.42	8.43	11.06
% Nd_2O_3	0.15	0.15	0.15	0.15
% MoO_3	2.5	2.5	2.5	2.5

Moreover a reference glass sample (SiCaMo0.82) containing only SiO_2, CaO and MoO_3 was also prepared for comparison with Mx and Rz glasses (^{95}Mo MAS NMR and Raman spectra). The composition of this glass was 58.88SiO_2 - 40.15CaO - 0.15Nd_2O_3 - 0.82MoO_3 (mol%). All glasses (~ 50 g batch) were prepared using reagent grade SiO_2, $CaCO_3$, Na_2CO_3, H_3BO_3, MoO_3 and Nd_2O_3 powders. The Mx and Rz glasses series were melted under air in platinum/gold crucibles at 1300°C for 3 h. The SiCaMo0.82 glass was also melted under air but a platinum/rhodium crucible has been used for its elaboration at 1400°C. Each melt was cooled by quenching between two copper plates with a speed cooling of about 10^4°C/min. The samples obtained were disc-shaped whose thickness was about 1mm.

Samples Characterization

The amorphous or crystallized character of all Mx samples was checked using X-ray diffraction (XRD) and the patterns were recorded with a Philips X'Pert Pro theta-theta X-ray diffractometer using monochromatized Cu Kα radiation (λ = 0.15406 nm). The crystalline phases were identified by comparison of XRD powder patterns with the crystallographic data base of the International Center of Diffraction data. All samples of the Mx and Rz series were also studied by electron paramagnetic resonance (EPR) to check the presence of paramagnetic Mo^{5+} ($4d^1$) or Mo^{3+} ($4d^3$) ions (Mo^{6+} ($4d^0$) ions are diamagnetic). EPR spectra (not shown) indicated the presence of only a very small quantity of Mo^{5+} ions ($[Mo^{5+}]/[Mo(Total)] < 1\%$) in all the samples. Consequently, almost all the molybdenum occurred as Mo^{6+} ions in our glasses which is in accordance with the oxidizing conditions (air) used during melting.

In literature, the environment of Mo^{6+} ions in glasses is frequently studied by EXAFS spectroscopy[12-14]. However, this technique remains relatively complicated and the average Mo-O distances of the MoO_4^{2-} units in glasses deduced from EXAFS spectra simulation does not seem to be very sensitive to the nature and to the composition of glasses (d(Mo-O)~1.76Å)[12-14]. Nevertheless, several papers indicated that the frequency of the Mo-O symmetric stretching vibration in MoO_4^{2-} tetrahedral units in crystalline phases are rather sensitive to their composition[15-21]. Moreover, other papers about structural characterisation of crystalline compounds containing Mo^{6+} ions showed that ^{95}Mo MAS NMR could also be a powerful technique to get information about the coordination and the evolution of molybdenum environment with composition changes[22,23]. Indeed, the position of the central (m_I=+1/2 ↔ m_I=-1/2) ^{95}Mo NMR transition is also rather sensitive to the nature of the crystals containing Mo^{6+} ions in 4- or 6-fold coordination. To the best of our knowledge, studies of ^{95}Mo MAS-NMR in glasses were only performed on the Mo-rich AgI-Ag_2O-MoO_3 system[22] but not for Mo in silicate or borosilicate glasses. Thus in this work, we decided to study the evolution of molybdenum environment in the glasses of the Rz series by both Raman and ^{95}Mo MAS NMR spectroscopies.

The molybdenum environment in glasses of the Rz series was studied by Raman and ^{95}Mo MAS NMR spectroscopies. For Raman experiments, an Aramis Jobin–Yvon confocal microRaman spectrometer equipped with a CCD detector was used. The 325 nm line of a coherent He-Cd laser was the excitation source and spectra were recorded at room temperature. ^{95}Mo is difficult detect by NMR because of both its low gyromagnetic ratio (-1.751 $10^7 radT^{-1}s^{-1}$), its low natural abundance (15.92 %) and its relatively high quadrupolar moment (22 mB). In the present case, these problems are exacerbated by the amorphous nature of the materials that broadens the lines (NMR parameters distribution) and by the low levels of Mo in the glasses of interest. Despite of these difficulties, ^{95}Mo MAS NMR enables to probe directly Mo^{6+} ions in our samples and may prove useful in understanding glass structure. ^{95}Mo MAS NMR experiments were performed at 18.8 T with an Avance II Brucker spectrometer using a 3.2 mm rotor at 20 kHz spinning frequency. All chemical shifts were referenced to a 2M Na_2MoO_4 reference solution at pH=12. In order to increase further the sensitivity of ^{95}Mo nuclei, two pulse sequences were used: QCPMG (Quadripolar Carr-Purcell Meiboon-Gill) and HS (hyperbolic secant).

RESULTS

Determination of MoO_3 Solubility Limit (Mx series)

The quenched glasses with increasing MoO_3 concentration (Mx series) were studied by XRD. M2.5 glass was transparent and totally amorphous according to XRD (Fig. 1). For x = 3 and 4, XRD patterns revealed the presence of powellite crystals (nominally $CaMoO_4$). Powellite has the same tetragonal structure as scheelite ($CaWO_4$) with nearly regular isolated MoO_4^{2-} tetrahedra and one site for Ca^{2+} ions in which these cations are surrounded by eight close oxygen atoms[24]. Figure 1 shows that the amount of powellite increases with x. For x = 5, new XRD lines that can be attributed to α-Na_2MoO_4 were also detected on the patterns. According to the literature[25,26], the cubic spinel-type α-Na_2MoO_4 is stable at room temperature up to 444°C[26] or 460°C[25]. α-Na_2MoO_4 crystals can also be detected in the yellow phase that may crystallize in nuclear glasses containing significant molybdenum concentration[27]. It thus appeared that for x >2.5 the solubility limit of molybdenum in Mx glasses was reached. The composition of the glasses of the Rz series was defined accounting for the solubility limit content of MoO_3 (2.5 mol%).

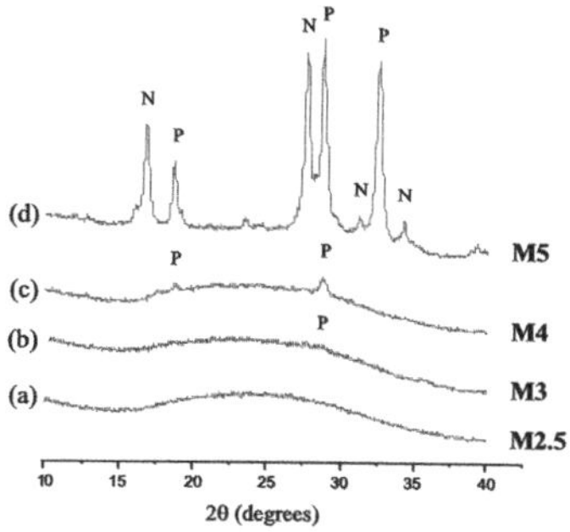

Figure 1. XRD patterns of quenched Mx samples: (a) M2.5, (b) M3, (c) M4, (d) M5. P: $CaMoO_4$ (powellite), N:, α-Na_2MoO_4.

Study by Raman spectroscopy (Rz series)

A study by Raman spectroscopy was recently performed by Magnin et al.[28] at room temperature and at higher temperatures (for melt and undercooled melt) on a glass composition belonging to the Mx series but with $[MoO_3]$ = 2 mol%. For this Mo-rich borosilicate glass, it was observed that the MoO_4^{2-} tetrahedra located in glass structure gave an intense and relatively wide vibration band on the Raman spectrum at about 900 cm^{-1}. A similar band was observed by Raman spectroscopy for another Mo-rich simplified borosilicate glass composition developed to immobilize other Mo-rich wastes solutions[29]. By comparison with Raman spectra of crystalline molybdates, this band was attributed to the symmetric stretching vibration of Mo-O bonds of the MoO_4^{2-} units. When these tetrahedral units are located in a crystallized environment such as in $CaMoO_4$ or Na_2MoO_4, this vibration band displaced to 880 or 892 cm^{-1} respectively[28].

For Rz=0 glass, a vibration band near 895 cm^{-1} is observed (Fig. 2 (e)). The position of this band is thus intermediate between that of the MoO_4^{2-} vibration band in a vitreous network containing both sodium and calcium (900 cm^{-1}) and that of the crystalline phase Na_2MoO_4 (892 cm^{-1})[28]. It would therefore appear that the glass is partially crystallized. Concerning the silicate glass without Na_2O (SiCaMo0.82) for which MoO_4^{2-} tetrahedra are only charge compensated by calcium ions, it appears that the vibration band has displaced to 919 cm^{-1} (Fig. 2 (a)). For Rz=0.2 and Rz=0.38 glasses, the MoO_4^{2-} vibration band seems to move to higher wave numbers corresponding to an environment with more Ca^{2+} ions (Fig. 2 (c) and (d)). The Rz=0.5 glass presents the particularity to be partially crystallized in $CaMoO_4$ (band at 880 cm^{-1}) inducing a decrease of Ca^{2+} ions concentration in the glassy matrix surrounding the crystals (Fig. 2 (b)). Whereas for this sample, the MoO_4^{2-} units remaining in the calcium depleted glassy phase give a vibration band at 896 cm^{-1} (i.e. in almost the same region of the spectrum as the Na-rich glass R0). Thus, the MoO_4^{2-} tetrahedra remaining in the glassy phase seem to be mainly charge compensated by sodium ions.

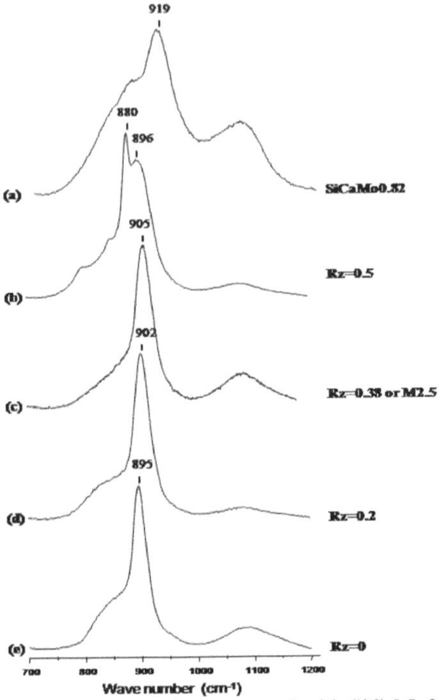

Figure 2. Raman spectra of Rz glass series compared with SiCaMo0.82 glass spectrum (a) SiCaMo0.82, (b) Rz=0.5, (c) Rz=0.38 or M2.5, (d) Rz=0.2, (e) Rz=0.

Study by ^{95}Mo MAS NMR

^{95}Mo MAS NMR spectra of Rz samples progressively moved towards lower chemical shift values when Rz increased (Fig. 3). The values pointed on the spectra correspond to the peak position δ in ppm. Rz=0 glass spectrum gave the chemical shift value of molybdenum when MoO_4^{2-} units are only surrounded by Na^+ cations (δ=- 4ppm). In contrast, SiCaMo0.82 glass spectrum shows that when MoO_4^{2-} units are only charge compensated by Ca^{2+} cations, the peak position decreased (δ=-80 ppm). By comparing the ^{95}Mo peak positions of these two particular glasses with those of Rz glasses (when Rz ≠ 0), it appears that the substitution of sodium by calcium in glass composition significantly influence molybdenum environment. As soon as we add calcium to the composition, molybdenum will prefer to surround himself by Ca^{2+} cations at the expense of Na^+ cations.

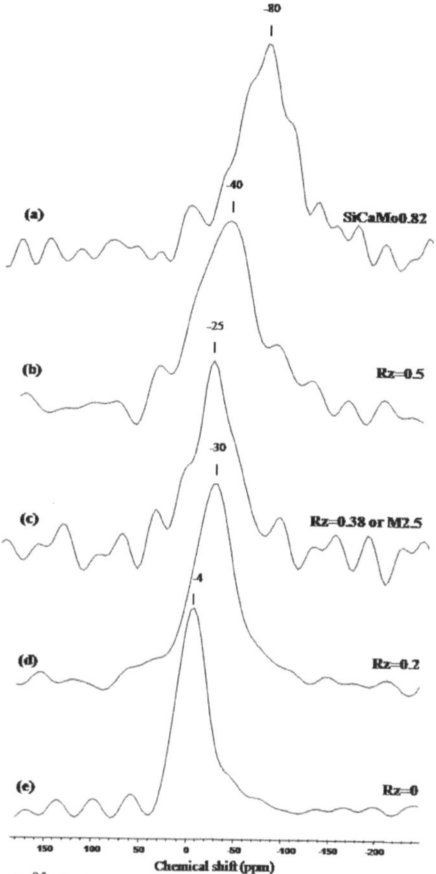

Figure 3. Comparison of ^{95}Mo MAS NMR spectra of Rz glasses with that of SiCaMo0.82 glass: (a) SiCaMo0.82, (b) Rz=0.5, (c) Rz=0.38 or M2.5, (d) Rz=0.2, (e) Rz=0.

DISCUSSION

The analysis of the Mx glass series by X-ray diffraction, allows us to determine the MoO_3 solubility limit (2.5 mol% MoO_3) in a simplified version of the UOX3 glass. Indeed, beyond 2.5 mol% with a melting processing at 1300°C and a cooling rate of 10^4 °C/min, $CaMoO_4$ and Na_2MoO_4 crystals can be detected. The knowledge of this solubility limit was then used to prepare glasses of the Rz series all containing 2.5 mol% MoO_3 by varying the R ratio.

For the Rz series, Raman spectroscopy results show that when the CaO content is increased at the expense of the Na_2O content, a shift of the band associated with the Mo-O symmetric stretching vibration (near 900 cm^{-1}) of the MoO_4^{2-} units located in glass structure is observed.

This vibration bands shift to higher wave numbers with R can be explained by the fact that the tetrahedral MoO_4^{2-} entities located in a glassy environment would tend to surround themselves preferentially by calcium ions.
From the ^{95}Mo MAS NMR study of glasses of the Rz series, a shift of the NMR peak position δ was observed towards that of the calcium silicate SiCaMo0.82 glass - as with Raman spectroscopy - when R increased (i.e. δ decreased when the CaO content increased). Thus, the ^{95}Mo MAS NMR results do not inform us about the crystallization tendency but showed that when R increased molybdenum environment changed: MoO_4^{2-} units progressively surround themselves by calcium cations.
In addition, we can see the influence of calcium on the tendency of $CaMoO_4$ and Na_2MoO_4 crystallization. On the one hand, when the composition on CaO is the same that Na_2O (Rz = 0.5 glass), we can observe the crystallization of $CaMoO_4$. Note that this phase has a strong chemical durability. On the other hand if we remove the calcium composition (Rz=0 glass), there is the crystallization of Na_2MoO_4. And account of Na_2MoO_4 strong solubility in water, this crystalline phase is unacceptable in nuclear glass canisters. The optimal Rz ratio ranges between 0.2 and 0.38 for which no crystallization phenomena are observed.

CONCLUSION

By increasing MoO_3 concentration (Mx glass series), we showed that the molybdenum solubility limit in a representative simplified version of UOX3 glass was about 2.5 mol%. For this MoO_3 content (Rz glass series), the role of the sodium and calcium elements on the solubility and the environment of molybdenum in glass structure has been studied by Raman and ^{95}Mo MAS NMR spectroscopies. Thus we were able to observe the changes of environment following the calcium and sodium. To pursue this study other glasses will be prepared and studied to refine the trend of molybdenum to surround itself Ca^{2+} cations. Further analysis by ^{29}Si, ^{23}Na and ^{11}B MAS NMR are in progress in order to determine the role of each elements towards the molybdenum crystallization tendency in borosilicate glasses.

REFERENCES

[1]R. Guillaumont, *C. R. Chimie*, **7**, 1129-1134 (2004)
[2]I. Bardez, D. Caurant, J.L. Dussossoy, P. Loiseau, C. Gervais, F. Ribot, D.R. Neuville, N. Baffier, and C. Fillet, *Nucl. Sci. Eng.*, **153**, 272-284 (2006)
[3]M. Le Grand, ‘‘Les platinoïdes et le molybdène dans des verres d'intérêt nucléaire: étude structurale’’; Ph.D. thesis, University of Paris VII, France (1999).
[4]R. J. Short, R. J. Hand, and N. C. Hyatt, *Mat. Res. Soc. Symp. Proc.*, **807**, 169–174 (2004).
[5]B. Camara, W. Lutze, and J. Lux, *Scientific Basis for Nuclear Waste Management II*, Edited by C. J. M. Northrup. Plenum Press, New York 93–102 (1979).
[6]G. E. Brown, F. Farges, and G. Calas, *Rev. Mineral.*, **2**, 317–410 (1995).
[7]G. Calas, M. Le Grand, L. Galoisy, and D. Ghaleb, *J. Nucl. Mater.*, **322**, 15–20 (2003).
[8]R. J. Short, R. J. Hand, N. C. Hyatt, and G. Mobus, *J. Nucl. Mater.*, **340**, 179–186 (2005).
[9]N. Sawaguchi, T. Yokokawa, and K. KawamuraA, *Phys. Chem. Glasses*, **37**, 13–18 (1996).
[10]D. Caurant, O. Majérus, E. Fadel, M. Lenoir, C. Gervais and O. Pinet, *J. Am. Ceram. Soc.*, **90**, 774-783 (2007).
[11]D. G. Burnett and R. W. Douglas, *Phys. Chem. Glasses*, **11**, 125-135 (1970).
[12]N. Sawaguchi, T. Yokokawa, K. Kawamura, *Phys. Chem. Glasses,* **37** (No. 1), 13-18 (1996).
[13]R.J. Short, R.J. Hand, N.C. Hyatt, and G. Möbus, *J. Nucl. Mater.* **340**, 179-186 (2005).

[14]G. Calas, M. Le Grand, L. Galoisy, and D. Ghaleb, *J. Nucl. Mater.*, **322**, 15-20 (2003).
[15]V. P. Mahadevan Pillai, T. Prodeep, and M.J. Bushiri, *Spectrochimica Acta Part A*, **53**, 867-876 (1996).
[16]S. J. A. Pope, and Y.D. West, *Spectrochimica Acta Part A*, **51**, 2011-2017 (1995).
[17]R. H. Busey, O. L. Keller, *J. Chem. Phys.*, **41** (No. 1), 215-225 (1964).
[18]F. D. Hardcastle, I. E. Wachs, *Journal of. Raman Spectroscopy*, **21**, 683-91 (1990).
[19]E. Sarantopoulou, C. Raptis, S. Ves, D. Christofilos, and G.A. Kourouklis, *J. Phys. Condens. Matter.*, **14**, 8925-8938 (2002).
[20]S.P.S. Porto, and J.F Scott, *Phys. Rev.*, **157** (No.3), 717-719 (1967).
[21]D. Christofilos, G.A. Kourouklis, and S. Ves, *J. Phys. Chem. Solids*, **56** (No.8), 1125-1129 (1995).
[22]N. Machida, and H. Eckert, *Solid State Ionics*, **107**, 255–268 (1998).
[23]J.-B. d'Espinose de Lacaillerie, F Barberon, et al., *J. Phys. Chem. B*, **109**, 14033-14042 (2005).
[24]O. Muller and R. Roy Eds. The Major Ternary Structural Families. Springer-Verlag, Berlin, (1974).
[25]Y. G. Petrosyan, E. V. Tkachenko, and V. M. Zhukovsii, *Inorg. Mater.*, **11**, 1381–1384 (1975).
[26]C. W. T. Pistorius, *J. Chem. Phys.*, **44**, 4532–4537 (1966).
[27]M. Le Grand, ''Les platinoïdes et le molybdène dans des verres d'intérêt nucléaire: étude structurale''; Ph.D. thesis, University of Paris VII, France (1999).
[28]M. Magnin, S. Schuller, D. Caurant, O. Majérus, D. de Ligny and T. Advocat, Proceedings of 3rd Atalante International Conference: Nuclear fuel cycles for sustainable future, May 13-23 (2008), Montpellier, France.
[29]D. Caurant, O. Majérus, E. Fadel, M. Lenoir, A. Quintas, C. Gervais, T. Charpentier and D. Neuville, *XXIst International Congress on Glass (ICG 2007)*, July 1-6 (2007), Strasbourg, France (extended abstract).

THE IMMOBILISATION OF A CHLORIDE CONTAINING ACTINIDE WASTE SURROGATE IN CALCIUM ALUMINOSILICATE GLASSES

J. M. Schofield, P. A. Bingham, R. J. Hand.
Immobilisation Science Laboratory, Department of Engineering Materials, University of Sheffield, Mappin Street, Sheffield, S1 3JD, U.K.

ABSTRACT

The vitrification of an actinide chloride containing waste stream, arising from the processing of impure Pu metal, in calcium aluminosilicate glasses is discussed.

Three studies have been undertaken all based on a glass with composition 51.40 mol % CaO, 7.17 mol % Al_2O_3 and 41.43 mol % SiO_2; firstly pure $CaCl_2$, as a simplified model waste stream was combined on a mole for mole basis for CaO in the glass batch (E#Cl). Secondly, a simulated waste stream which consisted of 93.34 mol % CaO, 5.91 mol % HfO_2 and 0.74 mol % Sm_2O_3 (as simulants for PuO_2 and Am_2O_3 respectively) were combined on a total waste mole for mole basis for CaO in the glass batch (E#o). Thirdly, a simulated waste stream which consisted of 93.34 mol % $CaCl_2$, 5.91 mol % HfO_2 and 0.74 mol % Sm_2O_3 was combined on a total waste mole for mole basis for CaO in the glass batch (E#).

These studies indicate that a surrogate waste loading (93.34 mol % $CaCl_2$, 5.91 mol % HfO_2 and 0.74 mol % Sm_2O_3) of up to 17.5 mol % can be achieved. Above 17.5 mol % waste, the potential for chlorine incorporation into these calcium aluminosilicate glasses, using this processing method, has been exceeded.

INTRODUCTION

Waste vitrification is a well established technology for dealing with radioactive wastes. The existing technology, based on mixed alkali borosilicate glasses, has primarily been developed to vitrify high level liquid wastes arising from fuel reprocessing [1]. As these wastes are not rich in actinides, (reprocessing is intended to remove both U and Pu from the waste) the low actinide solubilities in such glasses are not normally of concern. However radioactive waste streams arising in other contexts can contain significant amounts of actinides which need to be safely immobilised in the final wasteform. One such waste stream is an actinide containing waste arising from a process developed by the Atomic Weapons Establishment (AWE) that also contains significant quantities of inactive chloride ions [2, 3, 4]. Chlorine is also a difficult element to incorporate into conventional oxide glasses as it has low solubility and tends to form low–melting salt layers on the surface of the glass melt [5]; it has also been suggested that chlorine in glass is detrimental to durability [3].

One possibility for dealing with a waste that contains both actinides and chlorides is to utilise a process whereby the chlorine and actinide element are incorporated into a glass wasteform. In order for such a process to be viable it is essential that the actinide element is fully incorporated into the glass melt and chloride volatilisation is kept to a minimum. We are currently researching the possibility of utilising calcium aluminosilicate glasses to immobilise both the actinide and chloride components. The waste stream under consideration results from pyrochemical and aqueous processing of impure Pu metal and Pu oxide and consists primarily of calcium, plutonium and americium chlorides (92.66 mol % $CaCl_2$, 5.87 mol % $PuCl_3$ and 1.47 mol % $AmCl_3$).

A calcium aluminosilicate glass composition has been chosen because of the high calcium content of the waste hence the potential for high waste loading, whilst being a glass type with favourable properties for a wasteform [6]. To assess the viability of incorporating the active waste components into a 51.40 mol % CaO, 7.17 mol % Al_2O_3, 41.43 mol % SiO_2 glass, three surrogate waste streams were chosen; 100 mol % $CaCl_2$, a 93.34 mol % CaO, 5.91 mol % HfO_2 (as a simulant for PuO_2) and 0.74 mol % Sm_2O_3 (as a simulant for Am_2O_3), and a 93.34 mol % $CaCl_2$, 5.91 mol % HfO_2 and 0.74 mol % Sm_2O_3 waste stream. Each simulant waste has been added on a total waste mole for mole basis, for CaO (according to (x waste stream)+(CaO-x + Al_2O_3 + SiO_2) mol%).

EXPERIMENTAL PROCEDURES

Glass samples were prepared using calcium carbonate ($CaCO_3$) (98 % purity; Alfa Aesar, UK), aluminium hydroxide ($Al(OH)_3$) (extra pure; Acros Organics, UK), Loch Aline sand (SiO_2), calcium chloride ($CaCl_2$) (97 % purity; Alfa Aesar), hafnium oxide (HfO_2) (>97 % purity; Alfa Aesar) and samarium oxide (Sm_2O_3) (99.99 % purity; Alfa Aesar) according to Table I.

Table I. Nominal glass compositions

Name	Glass components (mol %)			Waste components (mol %)			
	SiO_2	Al_2O_3	CaO	$CaCl_2$			Total waste loading
E5Cl	41.43	7.17	46.40	5			5
E10Cl	41.43	7.17	41.40	10			10
E15Cl	41.43	7.17	36.40	15			15
	SiO_2	Al_2O_3	CaO	CaO	HfO_2	Sm_2O_3	
E0	41.43	7.17	51.40	0	0	0	0
E5o	41.43	7.17	46.40	4.7	0.296	0.037	5
E10o	41.43	7.17	41.40	9.3	0.591	0.074	10
E15o	41.43	7.17	36.40	14.0	0.887	0.111	15
E20o	41.43	7.17	31.40	18.7	1.183	0.148	20
	SiO_2	Al_2O_3	CaO	$CaCl_2$	HfO_2	Sm_2O_3	
E5	41.43	7.17	46.40	4.7	0.296	0.037	5
E10	41.43	7.17	41.40	9.3	0.591	0.074	10
E15	41.43	7.17	36.40	14.0	0.887	0.111	15
E17.5	41.43	7.17	33.90	16.3	1.035	0.130	17.5

Batches were mixed for ten minutes in a polythene bag, emptied into a recrystallised uncovered alumina crucible and then heated at 1.4 °C min^{-1} to 1450 °C in a muffle furnace in a fume cupboard and held at temperature for 3 hours. After the melting period the crucible was removed from the furnace and the molten glass cast into a pre-heated stainless steel casting mould on top of a stainless steel plate. The glass was immediately transferred to an annealing furnace at 780 °C for 1 hour and then cooled at a rate of 1 °C min^{-1}. Batches were considered to have been at the melt temperature for the required duration ± 6 min, due to the time taken to pour each melt and slight delays in ramp temperature due to the burden put on the furnace elements by the fume cupboard extractor fans. The melting furnace was calibrated to ± 2 °C. Once cooled to ambient temperatures the samples were transferred, using latex gloves to prevent any surface contamination, into a polythene sample bag and placed in a desiccator.

A Philips 500 scanning electron microscope with an Energy Dispersive Spectroscopy (EDS) system was used for analysis of the as–prepared glass. Glass samples for EDS were cut to ~ 15 × 15 × 4 mm, mounted in epoxy resin and polished to ~ 1 μm using silicon-carbide papers, water based diamond pastes and colloidal silica polishes. Three quantitative EDS measurements were made on regions near the top, in the middle and towards the bottom of the cross–sectioned, as-poured, glass samples. Estimated errors in EDS measurements are dependent upon elemental abundances: ± 0.5 wt % for major constituents (Ca, Al, Si, Cl) and ± 0.15 wt % for minor constituents (Hf, Sm), taking into consideration a variation of ± 1° in sample tilt angle and variations in the carbon coating thickness of the prepared sample.

Glass samples were cut on a diamond trim saw using water as a lubricant. Density measurements were repeated 10 times on ~ 40 g glass samples using the Archimedes method and distilled water as the measuring medium. Measurements were taken using a four-figure Oertling model R20 balance. Density measurements were assessed as being accurate to ~± 1×10^{-4} g cm^{-3}.

A Netzsch DIL 402C dilatometer was used to measure the thermal expansion coefficient over the range 50 to 300 °C ($\alpha_{50\text{-}300}$) and glass transition onset temperature, T_g. Dilatometer sample dimensions are ~ 15 × 4 × 4 mm. The thermal expansion coefficient is accurate to ± 2×10^{-7} taking into account the ± 1° variation in the parallel end surfaces of the sample, furnace temperature accuracy of ~ 0.5 °C, and sample thermocouple accuracy of ~ 0.5 °C. Glass transition temperature, T_g, is accurate to ± 2 °C. Samples were cut and polished to 45 μm on all surfaces using water as a lubricant, and heated at 5 °C min^{-1} to 20 °C above the dilatometric softening point, T_d. After an initial assessment of T_g onset of the glass samples using dilatometry, the glasses were then re–annealed at their recorded T_g onset temperatures ± 2 °C. For re–annealing, the samples were heated from room temperature to T_g onset at 1 °C min^{-1}, held for 1 hour at T_g onset, then cooled at 1 °C min^{-1} to room temperature. After the re–annealing step the samples have been tested using the dilatometer again; the values for $\alpha_{50\text{-}300}$ and T_g onset reported here have been obtained from this second dilatometric measurement.

A PerkinElmer Spectrum 2000 Fourier Transform Infra–Red (FTIR) spectrophotometer was used to obtain absorbance spectra between 400 and 1500 cm^{-1} on ~ 20 × 20 × 10 mm samples polished initially with silicon carbide papers to 45 μm then with cerium oxide to a 0.06 μm finish. Reflectance FTIR was used and the Kramers–Kronig transform utilized to calculate an absorbance index spectrum from the reflectance spectrum.

An accelerated static durability test has been used to assess the long term chemical durability of the glasses, similar to the American Society for Testing and Materials (ASTM) C1285–02 Product Consistency Test – B (PCT–B) [7]. The crushed glass samples were sieved to 75-150 μm. Adhering fines are removed, as specified in reference 7, using distilled water and methylated spirit (97 % pure). The powdered glass samples were dried overnight in an oven at 90 ± 2 °C. The dried glass powders were taken out of the oven, weighed to 3 ± 0.0005 g, transferred into Nalgene high density polyethylene (HDPE) 60 ml bottles then 30 ± 1.5 ml of distilled water added to give a volume of solution to mass of glass of 10 ± 0.5 ml g^{-1}. The average glass surface area to leachate volume ratio for this experiment was 1.908 m^2 L^{-1} with a range of 0.082 m^2 L^{-1}. The bottled samples were then sealed with their lids, agitated, weighed, and placed into an oven at 90 ± 2 °C for 7, 14 and 28 days ± 1 %. Blank samples containing only distilled water were also subject to the same experimental procedure. Samples and blanks were prepared in duplicate. After the test duration the bottled samples were removed from the oven, allowed to cool for 1 hr, weighed, ~18 ml of leachant removed via pipette, filtered, using Whatman ashless 542 (2.7μm particle retention) grade filter paper, and decanted into a HDPE bottle to be analysed. Leachates were analysed by inductively coupled plasma mass spectroscopy (ICP–OES), for measurement of chemical composition. Samples in the form of aqueous solutions were introduced as aerosols into the flame of a high temperature (10,000 K) Ar plasma flame. The detection limits of the ICP–OES instrument used are displayed in Table II.

Table II. Lower bounds of the ICP–OES detection of elements in ppm.

Ca	Al	Si	Hf	Sm	Cl
<0.1	<0.01	<0.01	<0.01	<0.005	<0.5

The chemical durability of the glass waste form was calculated according to the ASTM PCT-B [7]. Therefore the normalised release rate (*NR*) for each element *i* was calculated using:

$$NR_i = \frac{C_i(\text{sample})}{(f_i)\times\left(\frac{SA}{V}\right)\times(t)} \quad (1)$$

where $C_i(sample)$ is the concentration in grams per litre of element *i* in the leachate, *f* is the fraction (wt. %) of element *i* in the unleached wasteform, *SA* is the initial surface area of the glass in meters squared, *V* is the initial leachate volume in litres, *t*, is the duration of the test in days. Alternatively the units of time were omitted from the NR_i calculation to give the normalised release for element *i* (NL_i):

$$NL_i = \frac{C_i(\text{sample})}{(f_i)\times\left(\frac{SA}{V}\right)} \quad (2)$$

NR_i is expressed as grams of glass wasteform dissolved per m^2 per day (g m^{-2} d^{-1}). NL_i is expressed as grams of glass wasteform dissolved per m^2 (g m^{-2}).

RESULTS

Density

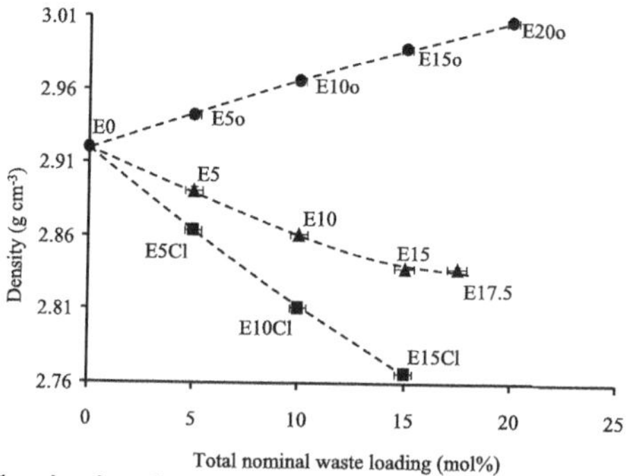

Figure 1. Glass density for glasses E#Cl (■), E#o (●), and E# (▲).Dashed lines drawn as a guide for the eye.

Energy Dispersive Spectroscopy

EDS results are reported as at. % for the elements Ca, Al, Si (Table III), Hf, Sm and chlorine (Figure 2). Oxygen content was calculated on a stoichiometric basis as the oxides in Table I.

Table III. Nominal glass composition (N.) and EDS analysed glass composition for Si, Al, and Ca.

Name	Glass components (at. %)					
	Si		Al		Ca	
	N.	EDS	N.	EDS	N.	EDS
E5Cl	15.5	15.0	5.4	6.5	19.2	18.7
E10Cl	15.2	14.9	5.3	6.4	18.8	18.3
E15Cl	14.9	14.6	5.2	6.0	18.5	18.6
E0	15.8	15.9	5.5	5.3	19.5	19.4
E5o	15.7	15.6	5.4	5.4	19.4	19.6
E10o	15.7	16.1	5.4	5.4	19.2	18.6
E15o	15.7	15.8	5.4	5.0	19.1	19.2
E5	15.5	15.2	5.4	6.1	19.1	18.7
E10	15.2	15.0	5.3	6.5	18.6	17.9
E15	14.9	14.9	5.2	6.2	18.1	17.6
E17.5	14.8	14.6	5.1	6.0	17.9	18.1

The results in Table III suggest that the glasses are of the expected (nominal) composition with a slight enrichment in alumina due to corrosion of the alumina crucibles during melting.

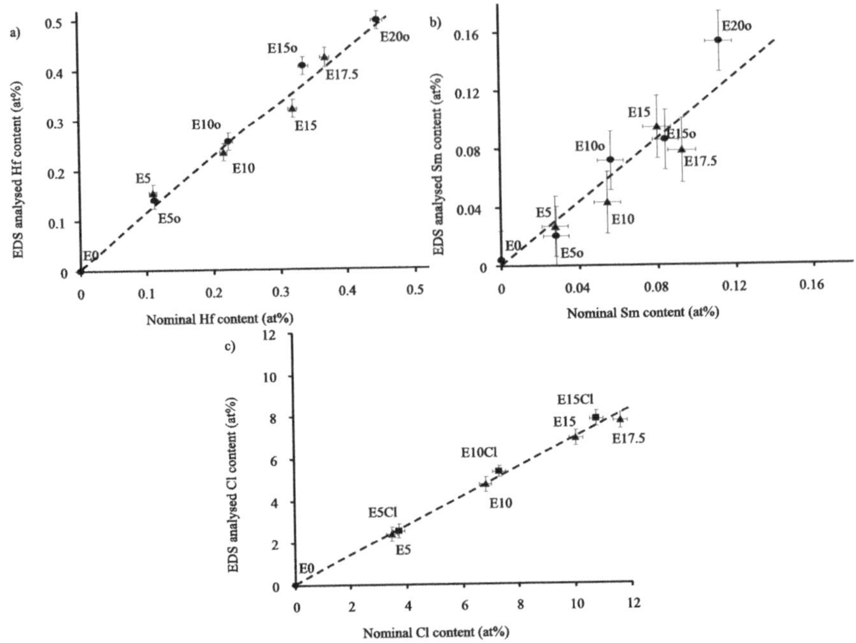

Figure 2. Analysed contents of a) Hf, b) Sm and c) Cl in the glasses, E#Cl (■), E#o (●), and E# (▲), plotted against nominal (batched) amounts in at. %. Dashed lines drawn as guide for the eye.

Dilatometry

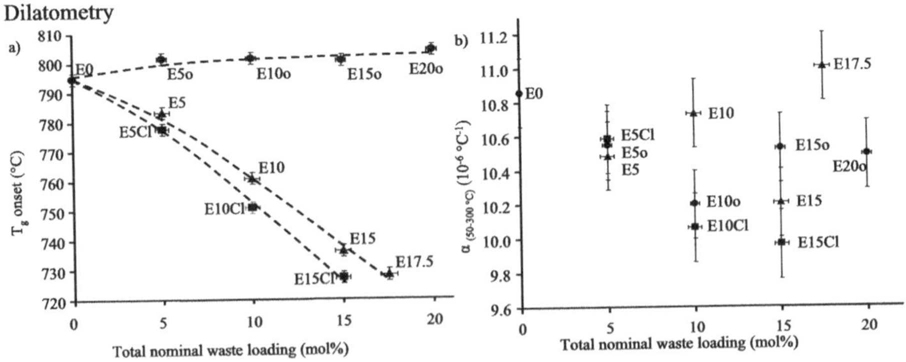

Figure 3. Total nominal waste loading (mol %) versus a) glass transition temperature and b) thermal expansion coefficient for glasses E#Cl (■), E#o (●), and E# (▲). Dashed lines drawn as a guide for the eye.

Dilatometry results indicate that both the glass transition temperature (T_g) (Figure 3 (a)) and thermal expansion coefficient ($\alpha_{50\text{-}300}$) (Figure 3 (b)) decrease with increasing Cl content of the glass. The addition of Hf and Sm to these glasses leads to a slight increase in T_g onset.

Fourier Transform Infra-Red Spectroscopy

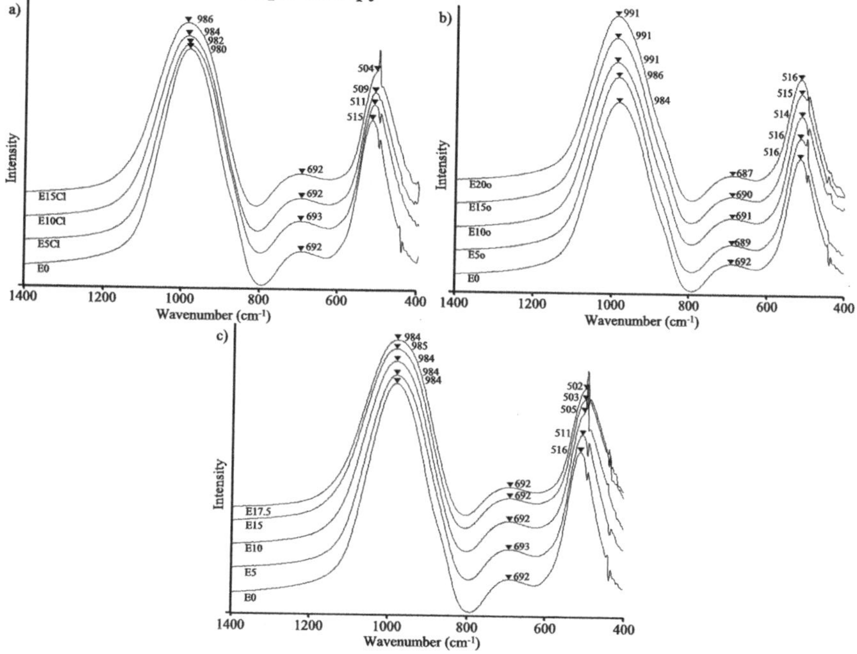

Figure 4. The absorbtion index of glasses E#Cl (a), E#o (b), and E# (c), measured using FTIR. Intensity in arbitary units versus wavenumber (cm^{-1}).

Durability

Modified PCT–B tests were performed on glasses E0, E5, E10 and E15. The results are presented in terms of the normalized release rate (NR_i) and the normalized release (NL_i).

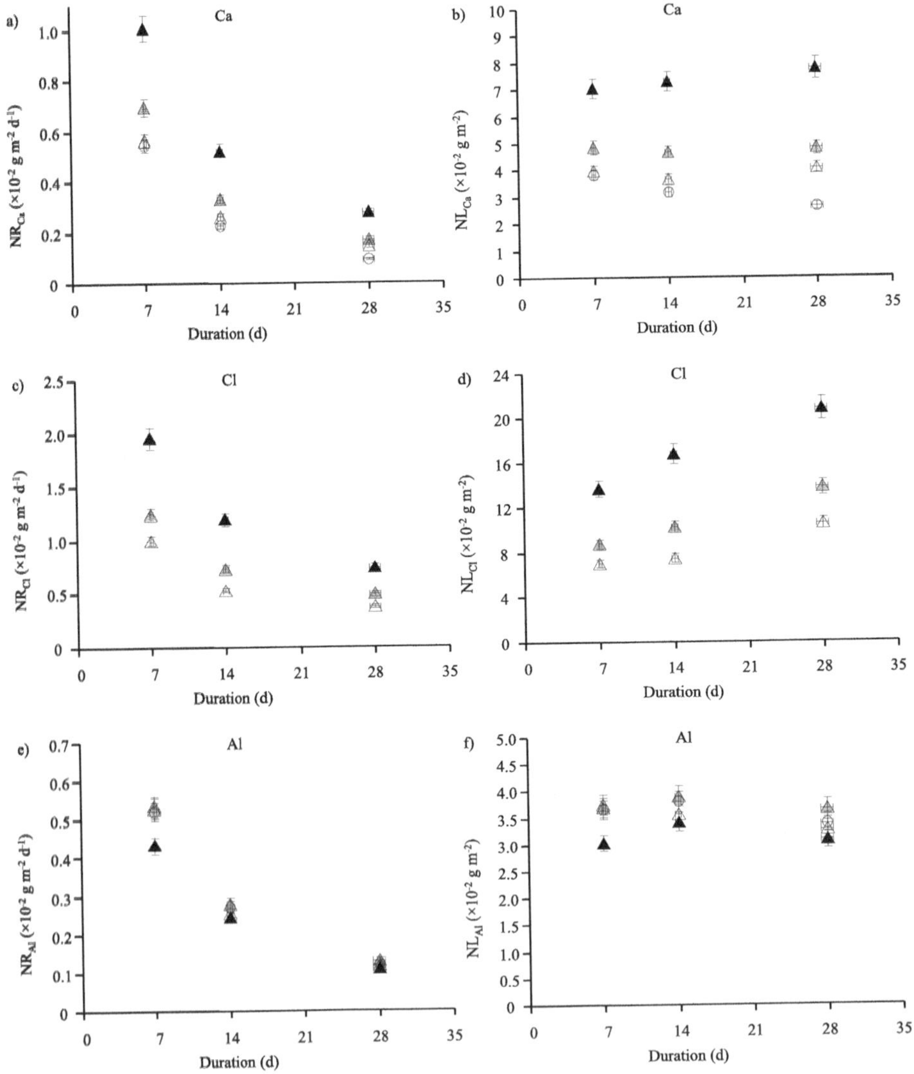

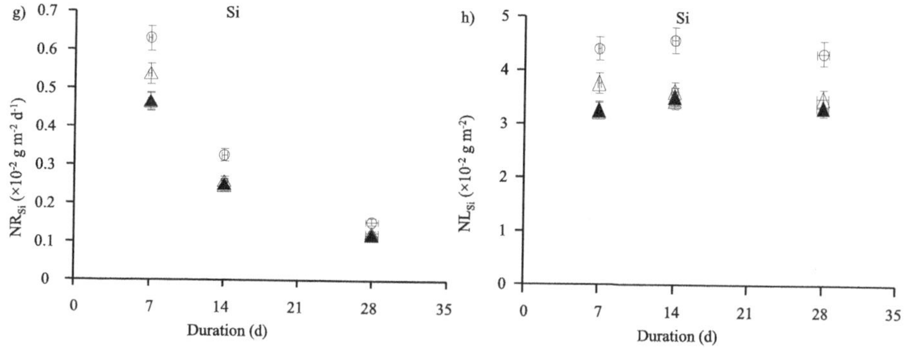

Figure 5. Wasteform durability of glasses E0(○), E5(△), E10(△) and E15(▲), normalised rate (NR) and normalised release (NL) of elements a,b) Ca, c,d) Cl, e,f) Al and g,h) Si, versus test duration in days.

DISCUSSION

Upon casting, it was observed that increased halide content appears to have reduced the viscosity of the glass, an effect which has been observed in other studies of halide containing silicate glasses [8]. Such behaviour should prove beneficial during the vitrification of these wastes as it would enable a lower processing temperature, provided that this temperature is sufficiently in excess of the liquidus temperature.

Results in Table III suggest that the glasses are of the expected (nominal) composition with differences being ± 0.5 at. % Si, ± 0.9 at. % Al and ± 0.7 at. % Ca. The difference in the alumina values suggests that some Al_2O_3 has been dissolved from the crucible wall. The EDS results in Figure 2 suggest that Hf (a) and Sm (b) are successfully incorporated into these calcium aluminosilicate glasses. There is a slight downturn in Sm (Figure 2 (b)) content in the E17.5 glass, in comparison to the trend, which could be due to the solubility of Sm being exceeded in these glasses. However, observations of the cast glass samples during SEM did not provide any evidence for this or any other inhomogeneities. The ratio of nominal Cl to EDS analysed Cl is ~ 0.7 (Figure 2 (c)).

According to Siwadamrongpong *et al.* [5] the maximum fraction of chlorine, X_c, that can potentially be incorporated by a base oxide glass can be related to the non-bridging oxygen (NBO) fraction. Figure 2 in reference 5 suggests that

$$X_c = 0.25\frac{[\mathrm{NBO}]}{[\mathrm{O}]} - 0.035 = 0.25\frac{2([\mathrm{CaO}]-[\mathrm{Al_2O_3}])}{[\mathrm{CaO}]+2[\mathrm{SiO_2}]+3[\mathrm{Al_2O_3}]} - 0.035 \quad (3)$$

where X_c is the critical solubility of chloride and [oxide] indicates mole fraction of that oxide.

The base glass has a NBO to O ratio of 0.568, and thus according to equation 3, it should be able to incorporate a maximum of 11 mol % $CaCl_2$. In glass E17.5 we have incorporated 10.7 mol % $CaCl_2$ successfully, with the rest of the $CaCl_2$ added having been converted to CaO with the loss of Cl [9]. Experiments were run with the same base glass with a 20 mol % $CaCl_2$ waste, but a gall layer was formed during casting. The presence of a gall layer indicates that the melt was supersaturated in $CaCl_2$ and had exceeded its Cl solubility limit. This result is in line with the prediction of equation 3. Kiprianov and Karpukhina [10] suggest that if the number of chlorine atoms exceeds the number of aluminium atoms then volatile $AlCl_3$ will be evolved during melting. With the E# glasses in going from the E17.5 to the E20 composition the number of chlorine atoms will have exceeded the number of aluminium atoms. Takeda *et al.* [9] suggest that Cl can be evolved during melting as either HCl or Cl_2 depending on the glass environment. Without thermal gravimetry-mass spectrometry (TG–MS) it is not possible to determine how Cl leaves the glasses under consideration in this study.

Glass density in this system appears to change in an approximately linear fashion with waste loading (Figure 1). When only oxide simulant was used (E#o), density increases with waste loading, whereas when chloride containing simulant wastes were used density decreased with waste loading, with the calcium chloride only simulant waste (E#Cl) showing a greater decrease in density than the calcium chloride and hafnium and samarium oxide simulant waste loading (E#). It has been suggested that chlorine and other halides depolymerise glass networks by replacing bridging oxygens [11, 12]. Chlorine has a lower ionic field strength than oxygen, so this coupled with the depolymerisation of the glass network could result in looser packing, thereby explaining the density decreases [11]. A more depolymerised network would also explain the observed reduction in glass transition temperature with increasing Cl content.

With an increase in Cl content, FTIR results indicate a reduction in the relative intensity of the 1200 - 810 cm^{-1} peak and a shift toward higher wavenumbers (Figure 4 (a, c)) whilst the 510 cm^{-1} peak reduces in relative intensity and shifts toward lower wavenumbers. Little change is observed in the glasses E#o (Figure 4 (b)). FTIR peaks around 710 cm^{-1} are believed to be associated with the stretching vibrations of the Al-O bonds in four fold coordination, Al(IV), [13] as the Raman spectra of mullite ($3Al_2O_3.2SiO_2$) displays a broad band at 700 cm^{-1} [14]. Huang *et al.* allocates the peak at 960 cm^{-1} toward a stretching vibration of an Si–O in the $Si(OAl/Ca)_2$ group *i.e.* Si–O tetrahedra with two corners shared with Al–O or Ca–O polyhedra [14]. Wang *et al.* [13] observed a broad peak between 810 and 1200 cm^{-1} which they subdivided into four components associated with vibrations arising from different Q species. Specifically they identified a peak at 980 cm^{-1} to the vibration of Si–O[NBO] in the glass network. The NBO content of these glasses is large ~ 0.568 which may explain why the peaks seen by Wang *et al.* are not seen here, as the Si–O[NBO] is the possible Si–O bond represented in the region 810 to 1200 cm^{-1}. Therefore we associate the ~984 cm^{-1} peak with Si–O[NBO] (Q3). It is also notable that the intensity of the ~984 cm^{-1} peak reduces as Cl content increases which indicates there is less oxygen available for formation of Si-O bonds in the glass network (Figure 4 (a, c)). In this case it Cl could be replacing the O in Si–O[NBO]. Huang and Behrman [14] indicate the FTIR peak, found in their study, at around 475 cm^{-1} could be attributed to motion of a bridging oxygen in the plane perpendicular to Si–O–Si(Al). They also observed this peak increases in intensity, narrows and shifts towards higher wavenumbers with increasing SiO_2/Al_2O_3 ratio. With this in mind we attribute the peak at 510 cm^{-1} to motion of a bridging oxygen in the plane perpendicular to Si–O–Si(Al). This peak is at a higher wavenumber than the 475 cm^{-1} peak as the glasses in this study have a relatively high SiO_2/Al_2O_3 ratio. The shift of the peak at 510 cm^{-1} toward lower frequencies and relative absorbance with the addition of Cl can be explained by a decrease in Si–O–Si bonds and a relatively small change in the number of Si–O–Al bonds, which have a smaller force constant (giving a less intense, lower wavenumber shifted peak) [14]. There also appears to be a decrease in the intensity of the shoulder on the low wavenumber side of the 980 cm^{-1} peak found at ~865 cm^{-1} with increasing waste loading (Figure 4 (a, c)).

Glass corrosion is usually thought to be initially dominated by ion exchange followed by dissolution of the network forming elements [15]. For corrosion under stagnant or semi-stagnant conditions the initial high rates of alteration are followed by a gradual reduction in alteration rate until a low, so–called final rate, is achieved. PCT–B is an aggressive, high surface area, test and thus the final rate can be approached following a relatively short test duration. In the tests conducted here it was found that the network modifier element Ca was leached more readily than Al and Si with NL_{Ca} remaining relatively consistent up to 28 days. With increasing waste loading of Cl, Hf, and Sm, the leach rate of Ca increases (Figure 5 (a, b)), whereas the leach rates of Al and Si both decrease (Figure 5 (e-h)). The dissolution rate of Cl in these glasses is the highest amongst the elements that could be detected in the leachates (Figure 5. (c, d)). Levels of Hf and Sm in the leachates were below the detection limits (Table II).

Without leach testing on a monolithic sample (ASTM C 1220 – 98) [16] it is impossible to determine whether there are any near-surface regions of the glass that are enriched or depleted in any specific element following the leach test. However it is notable that Cl and Ca leaching follow a similar NL_i pattern, increasing with increasing test duration (Figure 5 (b, d)) whereas NL_i for Al and Si first increases and then decreases with increasing test time (Figure 5 (f, h)) possibly suggesting the precipitation of an aluminosilicate phase with increasing test duration.

CONCLUSION

Calcium aluminosilicate glasses were prepared with up to 17.5 mol % of the $CaCl_2$ HfO_2 Sm_2O_3 containing wastes. These glasses have been reported as having favourable waste immobilisation properties and retain ~70 % of Cl of the initial Cl in the waste, whilst still retaining the Hf and Sm surrogate elements.

ACKNOWLEDGMENTS

JMS thanks AWE and Engineering and Physical Sciences Research Council (EPSRC, UK) for funding to undertake this work and The Armourers & Brasiers' Company for the award of a travel grant. PAB thanks EPSRC, UK for financial support.

REFERENCES

[1]I. W. Donald, B.L. Metcalfe, & R.N.J. Taylor, The immobilization of high level radioactive wastes using ceramics and glasses, *J. Mater. Sci.*, **32 [22]**, 5851-5887 (1997).

[2]B.L. Metcalfe, S.K. Fong, & I.W. Donald, Preparation, thermal characterisation and chemical durability of calcium iron phosphate glasses, *Glass Technol.*, **46 [2]**, 130-133 (2005).

[3]I. W. Donald, B. L. Metcalfe, S. K. Fong, L. A. Gerrard, D. M. Strachan, & R. D Scheele, A glass-encapsulated calcium phosphate wasteform for the immobilization of actinide-, fluoride-, and chloride-containing radioactive wastes from the Pyrochemical reprocessing of plutonium metal, *J. Nucl. Mat.*, **361**, 78-93 (2007).

[4]S. K. Fong, I. W. Donald, & B. L. Metcalfe, Development of a glass-encapsulated calcium phosphate wasteform for the immobilization of actinide and halide containing radioactive wastes from the Pyrochemical reprocessing of plutonium metal, *J. Alloys Compd.*, **444-445**, 424-428 (2007).

[5]S. Siwadamrongpong, M. Koide, & K. Matusite, Prediction of chloride solubility in CaO–Al_2O_3–SiO_2 glass systems, *J. Non-Cryst. Solids*, **347**, 114-120 (2004).

[6]J. E. Shelby, Formation and properties of calcium aluminosilicate glasses, *J. Am. Ceram. Soc.*, **68 [3]**, 155-158, (1985).

[7]American Society for Testing and Materials (ASTM), Standard Test Methods for Determining Chemical Durability of Nuclear, Hazardous, and Mixed Waste Glasses and Multiphase Glass Ceramics: The Product Consistency Test (PCT), *ASTM C1285-02*, West Conshohocken, PA, (2002).

[8]P. A. Bingham, & R. J. Hand, Sulphate incorporation and glass formation in phosphate systems for nuclear and toxic waste immobilization, *Mat. Res. Bullet.*, **43**, 1679-1693 (2008).

[9]D. Takeda, K. Matusita, & M. Koide, Thermal properties of CaO–Al_2O_3–SiO_2 glass containing chlorine, *J. Ceram. Soc. Jpn.*, **111 [5]**, 339-342 (2003).

[10]A.A. Kiprianov & N.G. Karpukhina, Oxyhalide silicate glasses, *Glass Phys. Chem.*, **32**, 1-27 (2006).

[11]G. El-Damrawi, $PbCl_2$ conducting glasses with missed glass formers, *J. Phys.: Condens. Matter*, **7**, 1557-1563 (1995).

[12]M. Zimova, & S. L. Webb, The combined effects of chlorine and fluorine on the viscosity of aluminosilicate melts, *Geochim. Cosmochim. Ac.*, **71**, 1553-1562 (2001).

[13]F. Wang, A. Stamboulis, D. Holland, S. Matsuya, & A. Takeuchi, Solid state MAS-NMR and FTIR study of barium containing alumino-silicate glasses, *Key. Eng. Mat.*, **361-363**, 825-828 (2008).

[14]C. Huang, & E. C. Behrman, Structure and properties of calcium aluminosilicate glasses, *J. Non-Cryst. Solids*, **128**, 310-321 (1991).

[15]United States Department of Energy, Office of Waste Management, High level waste borosilicate glass, A compendium of corrosion characteristics, *DOE-EM-0177*, **1**, (1994).

[16]American Society for Testing and Materials (ASTM), Standard Test Method for Static Leaching of Monolithic Waste Forms for Disposal of Radioactive Waste, *ASTM C1220-98*, West Conshohocken, PA, (1998).

INTERNATIONAL STUDIES OF ENHANCED WASTE LOADING AND IMPROVED MELT RATE FOR HIGH ALUMINA CONCENTRATION NUCLEAR WASTE GLASSES

Kevin M. Fox, David K. Peeler and James C. Marra
Savannah River National Laboratory
Aiken, South Carolina USA

Albert Aloy, Roman Soshnikov and Alexander V. Trofimenko
V.G. Khlopin Radium Institute
St. Petersburg, Russia

John D. Vienna, Brian J. Riley, Dong-Sang Kim and Jarrod V. Crum
Pacific Northwest National Laboratory
Richland, Washington USA

ABSTRACT

The goal of this study was to determine the impacts of glass compositions with high aluminum concentrations on melter performance, crystallization and chemical durability for Savannah River Site (SRS) and Hanford waste streams. Glass compositions for Hanford targeted both high aluminum concentrations in waste sludge and a high waste loading in the glass. Compositions for SRS targeted Sludge Batch 5, the next sludge batch to be processed in the Defense Waste Processing Facility (DWPF), which also has a relatively high aluminum concentration. Three frits were selected for combination with the SRS waste to evaluate their impact on melt rate. The glasses were melted in two small-scale test melters at the V. G. Khlopin Radium Institute. The results showed varying degrees of spinel formation in each of the glasses. Some improvements in melt rate were made by tailoring the frit composition for the SRS feeds. All of the Hanford and SRS compositions had acceptable chemical durability.

INTRODUCTION

The United States Department of Energy (DOE) is currently processing at the Savannah River Site (SRS), and planning to process at the Hanford Site, high-level waste (HLW) through Joule-heated melters. The process combines the HLW sludge with prefritted or mined mineral glass forming additives that are subsequently melted. The molten glass is poured into stainless steel canisters to create the final waste form. In preparation for the qualification and receipt of each sludge batch, development and definition of various tank blending and washing strategies have been initiated at SRS and are pending at Hanford. The various strategies are the result of efforts to meet critical site objectives or constraints that include tank volume space, transfer options, and settling issues. Although these objectives or constraints are critical, the ability to meet both process and product performance criteria associated with the final waste form must also be maintained. The product performance issues relate to the durability of the glass waste form. Process related issues (e.g., liquidus temperature, viscosity, electrical conductivity, and melting rate) ultimately dictate the efficiency and effectiveness of the melter operation.

Tank retrieval and blending strategies at both SRS and Hanford have identified high Al_2O_3 concentration waste streams that are scheduled to be processed through their respective HLW vitrification facilities. For example, the Liquid Waste Organization (LWO) at SRS provided compositional projections for the next two sludge batches (Sludge Batch 4 and Sludge

Batch 5) to be processed in the Defense Waste Processing Facility (DWPF). These streams have Al_2O_3 concentrations of approximately 25-40 wt % (on a calcined oxide basis). In addition, physical limitations in the Tank Farms and/or settling issues associated with the sludge coupled with the need to maintain feed for DWPF have prevented advanced washing which has resulted in relatively high Na_2O (approximately 22-26 wt %) concentrations. Current Hanford projections suggest that Al_2O_3 concentrations in sludge could be much greater than those currently projected for DWPF, with Al_2O_3 concentrations as high as 80 wt %.

While it is well known that the addition of small amounts of Al_2O_3 to borosilicate glass generally enhances the durability of the waste form (through creation of network-forming tetrahedral Na^+-$[AlO_{4/2}]^-$ pairs), nepheline ($NaAlSiO_4$) formation, which depends in part on the Al_2O_3 concentration, can result in a severe deterioration of the chemical durability of the glass through residual glass compositional changes. The primary driver for this reduction in durability is that nepheline removes three moles of glass forming oxides (Al_2O_3 and $2SiO_2$) from the continuous glass phase per each mole of Na_2O consumed. Nepheline formation produces an Al_2O_3 and SiO_2 deficient continuous glass matrix (relative to the same composition which is void of crystals), which reduces the durability of the final product. The magnitude of the reduction ultimately depends on the extent of crystallization and on the initial glass composition.

The formation of nepheline and other aluminum/silicon-containing crystals is a potential for both DWPF and Hanford due to the projected compositional views recently evaluated coupled with the frit development strategy (e.g., higher alkali frits have lead to enhanced melt rates at DWPF for those sludge batches). Although durability is obviously a critical constraint that the HLW glass must meet, other process-related issues must also be considered. Additionally, the inclusion of higher concentrations of Al_2O_3 will generally increase the liquidus temperature of the melt and decrease the processing rate.

The objective of this task was to develop glass formulations for DOE waste streams with high aluminum concentrations to avoid nepheline formation while meeting waste loading and waste throughput expectations as well as satisfying critical process and product performance related constraints. At SRS, frit compositions must be developed for high aluminum concentration wastes that will avoid detrimental crystallization while enhancing melt rate. At Hanford, high aluminum concentrations have necessitated melter operation with a small volume of spinel crystallization in the melt pool. Both of these systems are evaluated in this study in partnership with the V.G. Khlopin Radium Institute (KRI) in St. Petersburg. Small-scale test melters at KRI were used to evaluate test glass compositions for both SRS and Hanford.

EXPERIMENTAL PROCEDURE

The Steklo Metallicheskie Konstruktsii (SMK) Melter System

The SMK melter system at KRI is a batch melter intended for experiments on the rate of melting (by monitoring cold cap consumption) and crystal accumulation (through sectioning of the melt crucible) of simulated HLW glass. The components of the SMK melter system make it possible to provide for a dry or liquid feeding, off-gas treatment, air bubbling, glass product pouring, and vitrification temperature monitoring, with a continuous stirring of the feed. Fig. 1 shows a general layout of the SMK melter system.

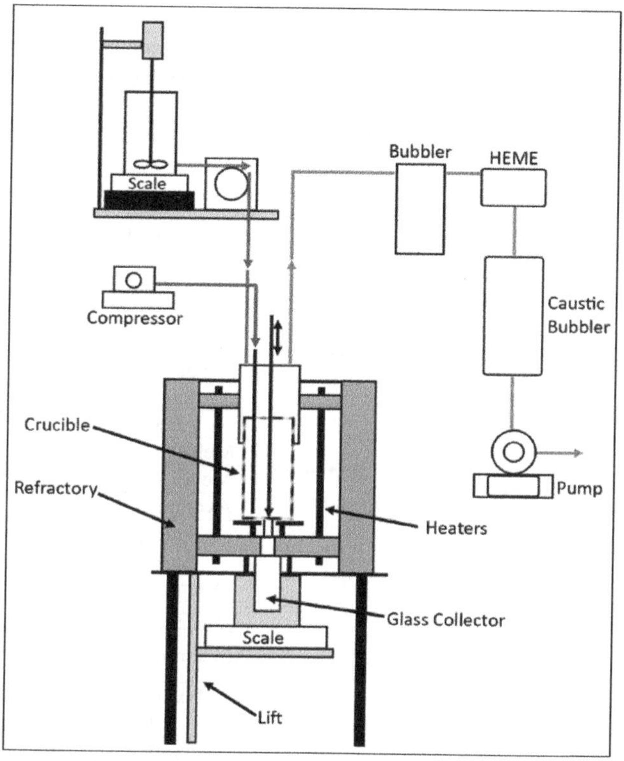

Figure 1. General Layout of the SMK Melter System.

Prior to an experiment, a quantity of melt startup product estimated to generate 500 g of glass was placed into the crucible. The composition of the melt startup product was the same as the glass composition to be evaluated. After the melter was heated and the melt startup product reached the target temperature (1150 °C for these experiments), the bubbler was turned on 0.6 L/min air) followed by the start of liquid feeding. Note that bubbling was not used for the experiments with SRS feeds since the DWPF melter is not currently configured for bubbling. The feed tube was water cooled to minimize the possibility of plugging. At the completion of feeding and when the cold cap had been consumed, the bubbler was turned off. A portion of the melt was then poured into an insulated can and later characterized. After the final pour of an experiment, the crucible was removed from the melter for quenching the remaining glass to support an evaluation of crystallization within the melt pool.

The following parameters are monitored during the experiments: feeding rate, melt temperature, bubbling rate, voltage, and current. Visual observations are also performed. The melt temperature was measured by using a thermocouple submerged into the melt next to the crucible bottom, 3 cm away from the side wall.

The Electricheskaya Pech-5 (EP-5) Melter System

The EP-5 melter at KRI is configured for longer-term, continuous feeding and batch pouring. It is equipped with systems and mechanisms that provide for the following: dry or liquid feeding, off-gas treatment, air bubbling of the melt, pouring of the glass product, and temperature monitoring of the melting process. A schematic of the EP-5 melter system and the operational parameters of the melter are shown in Fig. 2.

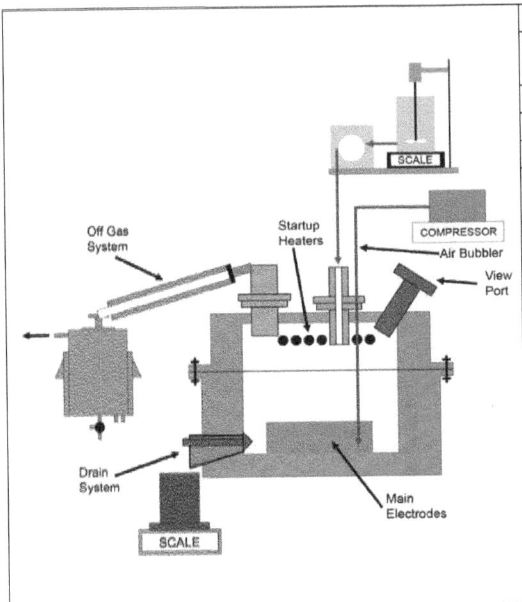

Parameter	Value
Maximum Operating Temperature	1200 °C
Melter Cavity Length	260 mm
Melter Cavity Width	130 mm
Melter Cavity Depth	210 mm
Total Glass Pool Surface Area	340 cm^2
Feed Area	229 cm^2
Melt Production Area	106.5 cm^2
Melter Cavity Height	285 mm
Average Glass Depth	110 mm
Average Glass Volume	3.7 L
Average Glass Mass	9.3 kg
Refractory Material	Bacor-33
Electrode Dimensions	200 mm x 75 mm
Maximum Current Density	2 A/cm^2
Maximum Current	300 A
Maximum Power	22 kW

Figure 2. Schematic and Operational Parameters of the EP-5 Melter System.

Prior to an experiment, a melt startup product consisting of crushed glass remaining from the SMK experiments was placed into the melter. The chemical composition of the startup product matched the targeted glass composition. After the melter was heated and the startup product was melted at 1150 °C, the bubbler was turned on (for the Hanford glass only) followed by the start of liquid feeding. After a sufficient quantity of glass was generated, the melt was poured without interrupting the feeding.

Feed Selection and Preparation

KRI prepared the sludge simulants in accordance with recommendations provided by Savannah River National Laboratory (SRNL) and Pacific Northwest National Laboratory (PNNL). Details of the sludge stimulant preparation methodology are available elsewhere.[1, 2] Table I provides the Hanford simulated (i.e., no radioactive components) sludge, frit and glass compositions selected for this study. The glass composition is referred to as HAL-17 and represents an oxide waste loading of 48.6 wt % sludge in glass. The HAL-17 composition was selected based on its high waste loading, high alumina concentration and acceptable physical and chemical characteristics of the final product.

Table I. Composition of the HAL-17 Glass and Its Components (wt % calcined oxides).

Oxide	Sludge	Frit	Glass composition (48.6 wt % WL)
Al_2O_3	53.27	–	25.87
B_2O_3	0.42	31.0	16.14
BaO	0.12	–	0.06
Bi_2O_3	2.54	–	1.24
CaO	2.39	12.0	7.33
CdO	0.054	–	0.03
Cr_2O_3	1.16	–	0.56
F	1.48	–	0.72
Fe_2O_3	13.11	–	6.37
K_2O	0.31	5.0	2.72
Li_2O	0.38	7.42	4.0
MgO	0.26	–	0.13
Na_2O	7.96	4.28	6.07
NiO	0.89	–	0.43
P_2O_5	2.34	–	1.14
PbO	0.91	–	0.44
SO_3	0.44	–	0.22
SiO_2	10.88	40.30	26.0
TiO_2	0.022	–	0.01
ZnO	0.18	–	0.09
ZrO_2	0.88	–	0.43

KRI used two types of simulated sludge for the SRS feeds: Sludge 1 which represents a projected Sludge Batch 5 (SB5) without the implementation of the Al-dissolution process and Sludge 2 which represents a projected SB5 composition after the completion of Al-dissolution. Their compositions are given in Table II.

Table II. SRS Target Sludge Compositions (wt % calcined oxides).

Oxide	Sludge 1 (without Al-dissolution)	Sludge 2 (with Al-dissolution)
Al_2O_3	33.22	16.60
BaO	0.11	0.15
CaO	2.09	2.90
Cr_2O_3	0.20	0.28
CuO	0.07	0.11
Fe_2O_3	26.40	36.77
K_2O	0.16	0.22
MgO	1.41	1.96
MnO_2	6.33	8.87
Na_2O	24.40	24.36
NiO	2.31	3.23
P_2O_5	0.12	0.11
PbO	0.10	0.14
SO_3	0.96	1.35
SiO_2	1.82	2.53
ZnO	0.07	0.10
ZrO_2	0.23	0.32

Three frits were evaluated for combination with the simulated SRS sludges to evaluate the impact of frit composition on melt rate. The frit compositions are given in Table III. The process used to select these frit compositions is documented elsewhere.[3] The frits batched and then melted at 1250 °C, with the melt being held at this temperature for 30 minutes. The frits were ground and sieved to a particle size of -80/+200 mesh (-177/+74 μm, respectively). Melter feed targeted a waste loading of 35 wt %.

Table III. Target Frit Compositions (as wt % oxides) for the SRS Feeds.

Frit	B_2O_3	CaO	Li_2O	Na_2O	SiO_2
503	14	0	8	4	74
517	17	0	10	3	70
520	8	1	10	4	77

Durability Testing

The Product Consistency Test (PCT), Method A[4] was performed in triplicate on samples of the glasses resulting from the melter runs to assess their chemical durability, which is a critical HLW glass performance criterion for acceptance into the federal repository. Included in the experimental test matrix was the Environmental Assessment (EA) glass, which is used for comparison to assure acceptable durability.[5] Glass samples were ground, washed, and prepared according to the ASTM procedure. At the completion of the 7-day test, the solutions were sampled (filtered and acidified) and analyzed by Inductively Coupled Plasma – Atomic Emission Spectroscopy (ICP-AES). Samples of a multi-element standard solution were included in the analyses as a check on the calibration of the ICP-AES instrument. Normalized release rates were calculated based on the target or measured (when available) glass compositions.

RESULTS AND DISCUSSION

SMK Experiments with Hanford Feeds

KRI performed three runs using the Hanford HAL-17 composition in the SMK melter. The feeding took 3 hours in the first run, 7 hours in the second run, and 11.2 hours in the third run, all at a temperature of 1150 °C. The poured glass distributed evenly in the can, with a sink hole forming on the surface upon cooling. KRI sampled the glass from both the can and the crucible upon completion of the runs. The samples were taken from upper, middle, and bottom part of the can. The glass samples were analyzed using scanning electron microscopy (SEM) for identification and volume fraction determination (using image analysis) of the crystalline phases.

A summary of the feeding rates and the amount of glass produced during each melter run is provided in Table IV. The feeding rate was reduced in order to achieve longer feeding times since the first run showed that the melter would not be able to handle evaporation if operated at higher feeding rates for an extended time. The glasses appeared heterogeneous when poured.

Table IV. Summary of SMK Melter Runs for Hanford Feeds

Run Number	Feeding Rate (L/hour)	Feeding Time (hours)	Number of Pours	Mass of Glass Poured (g)
1 (1150 °C)	1.23	3	1	1415
2 (1150 °C)	0.53	7	2	1420
3 (1150 °C)	0.48	11	4	2035

SEM analyses of the glasses from each of the three runs showed that spinels were present in all areas of the poured glass as well as in the glass remaining in the crucible after the final pour at concentrations of about 1-4 vol %. An example of the measured spinel distribution for Run 3 is given in Table V. Increasing the residence time in the melter appeared to have some influence on reducing the spinel content in the poured glass. There may also have been an effect of the startup material on crystallization of the glass.

Table V. Spinel Distribution in HAL-17 Glass after 11 hours of Feeding

Sampling Location	Spinel Content (vol %)				
	Pour 1	Pour 2	Pour 3	Pour 4	Quenched Glass in the Crucible
Top	1.6	1.1	1.1	1.3	1.2
Middle	1.8	1.1	1.1	1.2	1.5
Bottom	1.6	1.2	1.2	1.2	1.4

Fig. 3 shows a representative micrograph of the spinels identified in the HAL-17 glass sampled from the can (poured glass) after melting in the SMK for 11 hours. Similar crystallization was seen in all regions of the poured glass regardless of feeding rate or feeding time. Energy dispersive spectroscopy (EDS) analysis indicated that the spinels consist mainly of Fe, Cr and Mn.

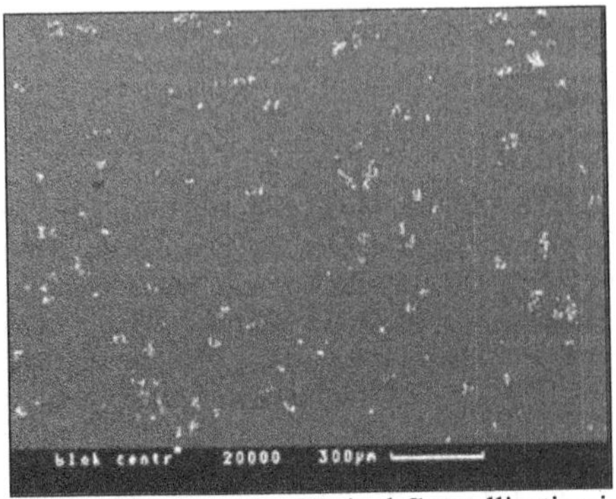

Figure 3. Representative Micrograph of Spinel Crystallization in HAL-17 Glass.

EP-5 Experiments with Hanford Feeds

KRI processed 41 L of the HAL-17 composition in the EP-5 melter. The feeding lasted 39 hours, and the overall duration of the EP-5 melter experiment was 50 hours. Approximately 20 kg of glass were produced during this experiment, including about 16 kg of glass generated from the feed. The remaining quantity of glass was generated from the startup material. Table VI summarizes the EP-5 operation with the HAL-17 feed.

Table VI. Quantities of Processed Sludge and Produced Glass for Hanford Waste Test

Pour No.	Feeding rate (L/hour)	Specific productivity ($L/dm^2 \cdot hr$)	Feeding Time (hours)	Mass of Glass Poured (g)
1	0.8	0.35	5	1580
2	0.78	0.34	5.5	1640
3	1.0	0.43	4	1584
4	0.8	0.35	4.25	1390
5	1.0	0.43	3.4	1300
6	1.3	0.57	5	2600
7	1.2	0.52	5	2875
8	1.2	0.52	3.8	1470
9	1.2	0.52	3	1490
Total			39	15929

The glass from all of the pours appeared heterogeneous. The spinel content in the poured glass as measured using SEM varied from 2-5 vol %. There was no clear relationship between the volume fraction of spinel crystallization and melter feeding time. Spinel crystals were distributed throughout the glass, with some agglomerates of smaller crystals appearing as striations.

SMK Experiments with SRS Feeds

The results of the SMK runs with the SRS feeds are summarized in Table VII.

Table VII. Results of the SMK Testing with the SRS Feeds

Frit Type	503		517		520	
Sludge Type	Sludge 1	Sludge 2	Sludge 1	Sludge 2	Sludge 1	Sludge 2
Feeding Time (minutes)	195	320	215	260	200	130
Average Feeding Rate (L/hour)	1.03	0.55	0.90	0.67	1.03	0.95
Mass of Glass Poured (g)	1490	1400	1460	1382	1470	1120
Pouring Time (minutes)	35.00	20.00	15.75	11.58	29.00	20.00

Visual observations of the melter experiments provide some insight into the behavior of the frit-sludge systems. When Sludges 1 and 2 mixed with Frits 503 and 517 were fed onto the melt surface, the cold cap distributed unevenly, mostly accumulating in the area where the feed entered the system. When the cold cap coverage reached the target 80-95%, mounding of the feed was observed on the cold cap. Melting of Sludge 1 mixed with Frit 503 resulted in a calcinate accumulation in the pour closure rod due to spills of the sludge. Melting of Sludge 1 mixed with Frit 517 resulted in a calcinate accumulation on the crucible wall. The calcinate accumulation did not make it possible to accurately measure the time required for consumption of the cold cap. No calcinate formations were associated with Frit 520. The feed material and the cold cap distributed evenly on the melt surface and the melting process continued as intended. Partial loss of the feed during feeding of Sludge 2 mixed with Frit 520 was experienced due to a damage of a hose in the peristaltic pump.

SEM analysis showed that the Sludge 1 glass formulated with Frit 503 glass had heterogeneous areas in the upper part of the poured glass. These areas were identified as spinel crystals via EDS. No spinel was found in the glass samples taken from the middle and lower part of the poured glass. The middle part of the Sludge 1 glass formulated with Frit 517 contained 14 vol % spinel crystals. No spinel was found in the samples taken from the upper and lower parts of the poured glass. The Sludge 1 glass formulated with Frit 520 contained spinel in the upper (0.1 vol %) and middle parts (0.5 vol %) of the poured glass, but not in the lower part.

No crystallization was visible via SEM in the Sludge 2 glasses formulated with Frits 503 and 520. The glass formulated with Sludge 2 and Frit 517 contained approximately 0.1 vol. % spinel crystals

The melt rate provided by the three frits was evaluated by comparing the average feeding rates for each experiment. In general, the melt rate for the Sludge 1 glasses was higher than that for the Sludge 2 glasses. For the Frit 503 and Frit 517 glasses, the melt rates differ significantly between the two sludges, while for Frit 520 glass the melt rates for the two sludges differ insignificantly. Frit 520 provided a fairly high melt rate for processing both sludges. Frit 520 also produced a reasonably homogenous glass when combined with either sludge. Frit 520 was therefore selected for processing Sludges 1 and 2 in the EP-5 Melter.

EP-5 Experiments with SRS Feeds

The goal of the EP-5 experiments with the SRS feeds was to identify the optimal feeding rate and determine whether a difference in melt rate could be observed between the sludges combined with Frit 520. The evaluation was made by monitoring the cold cap growth associated with a gradual increase of the feeding rate. The initial feeding was performed at a given rate for 30 minutes. If, during this time, the melt surface was not 80-90% covered with the cold cap, the feeding was stopped until the cold cap was consumed. The feeding rate was then increased and the process repeated. The feeding rate was subsequently reduced if the melt surface was 100% covered with the cold cap within 30 minutes. At the completion of feeding and cold cap consumption the melted glass was held in the melter for one hour to be followed by complete pouring into a collector can.

The Sludge 1 glass was poured once (6840 g). The feeding rate ranged from 1.6-2.0 L/hr, with the optimal feeding rate being 1.7 L/hr. The Sludge 2 glass was poured twice (805 g and 4156 g). The feeding rate ranged from 1.4-1.8 L/hr. The optimal feeding rate was 1.6 L/hr. There was no obvious difference in melt rate between the two sludge types.

Table VIII shows data on the spinel content in the glass from the two EP-5 experiments. The poured glass from the Sludge 1 experiment had a zoned distribution of spinel. The middle part of the glass contained a larger amount of spinel than the upper and bottom parts, which may relate to cooling rate. No spinel was observed in the glass from the first pour with Sludge 2. The spinel content in the glass from the second pour of Sludge 2 ranged from 0.05 to 0.8 vol %.

Table VIII. Spinel Content in the SRS Glasses Melted in the EP-5 (averages of 3 measurements)

Sludge 1		Sludge 2	
Sample Location	Spinel Content (vol %)	Sample Location	Spinel Content (vol %)
Top	0.4	Top	0.5
Middle	1.6	Middle	0.1
Bottom	0.2	Bottom	0.3

Chemical Durability Analysis

Representative samples of each of the glass systems were evaluated for chemical durability using the PCT. A select set of the results of the PCT for some of the study glasses in terms of the normalized leachate releases for boron (NL [B]) is given in Table IX. All of the glasses had NL [B] values that are well below that of the EA reference glass (16.695 g/L).[5] This indicates that these glasses would be acceptable for disposal in the federal repository. The PCT results also indicate that the crystallization (spinels) identified in the glasses does not have a significant impact on their chemical durability.

Table IX. Product Consistency Test Results for the Study Glasses

Melter	SMK	SMK	SMK	EP-5	EP-5
Sludge	2	2	2	1	HAL-17
Frit	503	517	520	520	
NL [B] (g/L)	0.84	1.15	0.95	1.08	0.22

CONCLUSIONS

The goal of this study was to determine the impacts of simulated high level waste glass compositions with high aluminum concentrations on melter performance, crystallization and chemical durability for SRS and Hanford waste streams. Glass compositions for Hanford targeted both high aluminum concentrations in waste and a high waste loading in the glass. Compositions for SRS targeted the next sludge batch to be processed in the DWPF, which also has a relatively high aluminum concentration. Three frits were selected for combination with the SRS waste to evaluate their impact on melt rate. The simulated sludges and frits were mixed and then melted in two small-scale test melters at KRI. The SMK melter allows for a determination of optimum feeding rate using a liquid feed over relatively short time periods. The EP-5 melter allows for longer feeding times between pours to evaluate longer term melter operation and glass crystallization.

The results showed varying degrees of spinel formation in all of the glass compositions using both of the melters. Spinels are known to have little influence on chemical durability and therefore were not of significant concern. Spinel accumulation within the melter did not appear to be an issue for the Hanford feeds, even after melter feeding times of more than 30 hours. Melt rates were estimated based on the feeding rate that provided optimum cold cap coverage on the melt pool. Some improvements in melt rate were made by tailoring the frit composition for the SRS feeds. Differences in cold cap behavior (flowing of the feed material) were also noted. All of the Hanford and SRS glass compositions had measured chemical durability performance that was acceptable for disposal in the federal repository. The highest measured NL [B] value was 1.15 g/L, which is significantly lower than that of the EA benchmark glass (16.695 g/L). Future studies will continue to assess the ability to tailor frits for improved melt rate through the control of boron and alkali additives, as well as other additives that may influence reactions in the cold cap.

ACKNOWLEDGEMENTS

The authors would like to thank Irene Reamer, Phyllis Workman, James Gillam Jr., David Best, Pat Toole, and Whitney Riley for their assistance in fabricating and characterizing test glass compositions at SRNL, Tanya Albert for expert facilitation of communication between the US and Russian laboratories, and Kurt Gerdes and the U.S. Department of Energy Office of Environmental Management for funding this task through the EM-21 International Program.

REFERENCES

[1]Peeler, D. K., "Savannah River Site Simulant Sludge Compositions and Preparation Instructions for SMK Melter Testing – Sludge Batch 5 with Aluminum-Dissolution," *U.S. Department of Energy Report SRNL-PSE-2007-00139,* Washington Savannah River Company, Aiken, SC (2007).

[2]Peeler, D. K., "Savannah River Site Simulant Sludge Compositions and Preparation Instructions for SMK Melter Testing – Sludge Batch 5," *U.S. Department of Energy Report SRNL-PSE-2007-00141,* Washington Savannah River Company, Aiken, SC (2007).

[3]Gillam Jr., J. H., K. M. Fox, T. B. Edwards and D. K. Peeler, "Frit Selection to Support Steklo Metallicheskie Konstruktsii (SMK) Melter Testing with SRNL Feeds," *U.S. Department of Energy Report WSRC-STI-2007-00363, Revision 0,* Washington Savannah River Company, Aiken, SC (2007).

[4]Herman, C. C., T. B. Edwards, D. R. Best, D. M. Marsh and R. J. Workman, "Reduction of Constraints: Phase 2 Experimental Assessment for Sludge-Only Processing," *U.S. Department of Energy Report WSRC-TR-2002-00482, Revision 0,* Westinghouse Savannah River Company, Aiken, SC (2002).

[5]Jantzen, C. M., N. E. Bibler, D. C. Beam, C. L. Crawford and M. A. Pickett, "Characterization of the Defense Waste Processing Facility (DWPF) Environmental Assessment (EA) Glass Standard Reference Material," *U.S. Department of Energy Report WSRC-TR-92-346, Revision 1,* Westinghouse Savannah River Company, Aiken, SC (1993).

LEACH TESTING APPLIED TO THE INVESTIGATION OF LONG-TERM BEHAVIOR OF HIGH-LEVEL WASTE GLASS: FRENCH EXPERIENCE

S. Gin, P. Frugier
CEA Marcoule, DTCD/SECM/LCLT,
BP 17171, 30207 Bagnols-sur-Cèze Cedex, France

ABSTRACT

Experimental methods and protocols applied to HLW glass dissolution are closely related to the understanding of the leaching mechanisms and the way these mechanisms are taken into account in models. The model developed in recent years for French nuclear glass performance calculations in a geological repository is described, the main techniques for characterizing altered glass samples are discussed, and the main routine leaching methods are detailed. The methods are classified according to two factors having a significant impact on the alteration kinetics: solution renewal, and the effects of the groundwater composition and environmental materials. The key glass alteration mechanisms are identified together with the conditions under which they predominate.

INTRODUCTION

Under the provisions of the French radioactive waste management act of 30 December 1991, ANDRA (French agency for nuclear waste management) has performed safety analyses concerning a geological repository for vitrified high-level waste [1]. For R7T7-type glass (incorporating fission products and minor actinides arising from reprocessing the spent fuel from 58 pressurized water reactors) the simulations are performed using a radionuclide release model known as $V_0 \rightarrow V_r$, based on a simplified, conservative, but relatively realistic phenomenology of glass degradation in a clay repository environment [2]. This model was developed and configured on the basis of research carried out mainly in the CEA laboratories at Marcoule since the mid-1980s, involving not only empirical but also more theoretical developments (for the general methodology, see [3]). Leach tests were carried out extensively to obtain detailed data and understanding, and to refine several generations of models. This article concerns only aqueous leaching phenomena affecting vitrified wasteforms, as it is generally acknowledged that in a geological repository water will be both the principal source of waste package alteration and the radionuclide migration vector. After describing the main characteristics of different types of leach tests, their scope of application and their limitations are examined.

THE $V_0 \rightarrow V_r$ MODEL

The $V_0 \rightarrow V_r$ model applied to R7T7 glass is based on the following main hypotheses. As long as the surrounding materials that are reactive with respect to glass (at least the canister corrosion products and possibly the products formed by corrosion of the overpack) are not saturated with silica, the glass is supposed to be leached at its initial alteration rate r_0. The interactions between the glass and corrosion products are modeled by the silicon sorption capacity of corrosion products, as determined by laboratory measurements of model compounds. The initial rate r_0 depends on the glass composition, temperature and pH.

During silica sorption on the corrosion products, the quantity of altered glass per unit time is the product of the initial alteration rate by the effective surface area, which in turn is equal to the product of the geometric surface area glass block by the effective cracking factor τ_0:

$$QAG = r_0\, S_0\, \tau_0$$

As soon as the corrosion products are saturated, the glass alteration rate drops as a result of the formation of a protective gel until residual rate conditions are reached [4]. The quantity of glass altered during the rate drop regime, because of its short duration, is considered negligible compared with the quantities of glass altered during the initial and residual rate phases.

The residual rate, r_r, depends only on the glass composition and the temperature [2]. The temperature-dependence implies time-dependence (the temperature diminishes over time due to radioactive decay). Under residual rate conditions, the quantity of altered glass per unit time is the product of the residual alteration rate by an effective surface area equal to the geometric surface area multiplied by the effective cracking factor τ_r:

$$QAG = r_r\, S_0\, \tau_r$$

The model is applicable within the following range:

R7T7 glass
$7 < pH < 10$; $25°C < T < 100°C$
Limited groundwater water flow

Although radionuclide retention in the glass alteration products is not taken into account and the overall glass source term is therefore overestimated, the model provides a generally realistic description of glass matrix alteration.

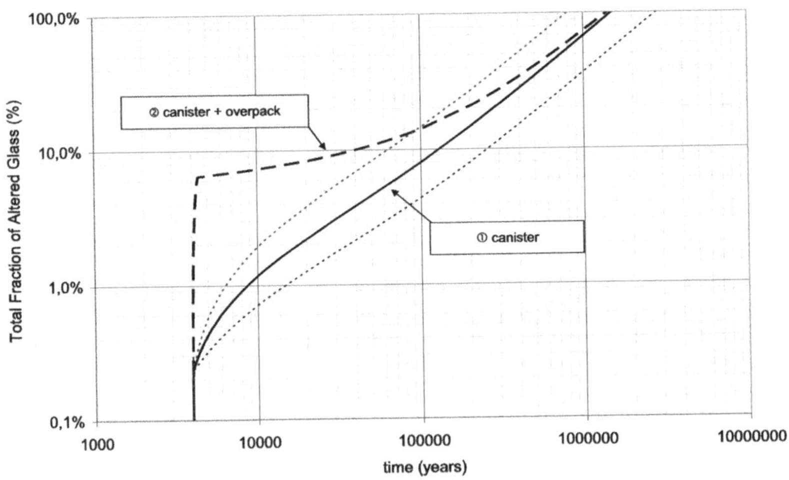

Figure 1. Total altered glass fraction: ① taking into account the saturation of the corrosion products of the canister alone, ② taking into account saturation of the corrosion products of the canister and the overpack. In both cases alteration begins after 4000 years. The corresponding uncertainties are indicated in broken lines, especially in the reference case ①.

Figure 1 is a typical performance estimate with the $V_0 \rightarrow V_r$ model assuming first that only the primary canister corrosion products must be saturated by silicon released during glass leaching.

In this case about 0.25% of the package is altered at the initial rate in order to saturate the metal corrosion products before the transition to the residual rate, which persists until complete alteration of the glass package. Under normal repository conditions, the altered glass fraction is about 1% after 10 000 years and the package lifetime is between 750 000 and 2.5 million years, given the uncertainties on the model parameter values. Taking into account in the calculation all the corrosion products from the canister and the overpack will drastically increase the amount of altered glass in the first alteration regime but will not significantly modify the life time package.

TESTS FOR IMPROVING OUR UNDERSTANDING OF THE MECHANISMS INVOLVED AND FOR CONFIGURING MODELS

The $V_0 \rightarrow V_r$ model is a simple and conservative representation of the complex phenomenology that has already been described in detail [4]. The cognitive approach applied in France uses glass leach tests to accurately measure the glass dissolution kinetics as a function of the environmental parameters and characterize the chemical and structural properties of the altered solid, with the objective of establishing correlations between them that can be substantiated at a theoretical level. Table I lists the main characterization tests carried out either on R7T7 glass or on simplified glasses (used to avoid some phenomena particular to nuclear glass), and indicates some of their applications and limitations.

Table I. Main techniques used to characterize glass alteration during aqueous leaching. Note that the references cited here are not exhaustive.

METHOD	SCOPE OF APPLICATION	LIMITATIONS	RESULTS OBTAINED
SEM	Morphological and semi-quantitative chemical analysis of the alteration film	Limited spatial resolution (1 μm). Light elements cannot be determined.	Revealed that leaching is a constant-volume process. 2D characterization of crack network in a full-scale glass block [18].
TEM	Morphological, structural and chemical examination of the alteration film at scales ranging from 1 nm to 10 μm. Allows qualitative assessment of porosity.	Very small analysis zone. As for SEM, the sample must be submitted to extensive treatment.	Demonstration of pore closure in the gel [19]. Microstructure characterization of the alteration film at nanoscopic level.
SIMS	Profilometric chemical analysis of alteration films.	The spatial resolution of the mixing zone may reach 100 nm.	Local composition as a function of the depth [5].
ToF-SIMS	Profilometric chemical analysis of alteration films.	The spatial resolution of the mixing zone may reach 5 nm.	Demonstration and quantification of diffusion phenomena in the gel [6].
NMR	Structural examination of pristine glass and gel	The two materials (glass and gel) must be studied separately. Only a few elements can be probed. Applicable to simple glass compositions.	Role of elements as Si, Al, Ca or Zr on the gel structure [7-9].

METHOD	SCOPE OF APPLICATION	LIMITATIONS	RESULTS OBTAINED
EXAFS & XANES	Structural examination of pristine glass and gel		Consequence of alteration on the coordination number of elements such as Fe, Zr, La, Ce, U [10,11].
SAXS	Porous gel structure	Measurement quality diminishes with the altered glass fraction	Quantification of gel porosity and pore sizes [12].

Laboratory tests performed to understand and rank glass alteration mechanisms can be classified into several categories. Two main test parameters can have a significant impact on the alteration kinetics and are therefore the subject of considerable attention: the effect of leaching solution renewal, and the effect of the solution composition and the nature of the environmental materials near the glass. The principal types of leach tests can be arranged in a double-entry table (Table II).

Table II. Organization of leach tests for long-term nuclear glass behavior studies

HYDRODYNAMIC ENVIRONMENT	INITIALLY PURE WATER	SPECIFIED SOLUTION COMPOSITION	PRESENCE OF SOLIDS
Open system: High solution renewal rate	A1	B1	C1
Partially open system: Low solution renewal rate	A2	B2	C2
Closed system: No solution renewal	A3	B3	C3

The objective of leach tests becomes clearer in the light of a few key leaching mechanisms. Consider the *alteration drivers* (any physicochemical phenomenon that diminishes the activity of chemical species subject to condensation in the amorphous gel —especially silicon — can be an alteration driver):

- solution renewal affected by molecular diffusion and/or convective water transport into the nearfield;
- supply of a chemical complexing agent;
- supply of element that can lead to the precipitation of crystallized secondary mineral (involving a "gel former" element);
- high affinity of materials surface with respect to silicon.

In a closed system, reactions associated with diffusive transport of aqueous species through the pristine glass, the hydrated glass, and the gel (depending on the authors) must also be taken into account. This *reactive diffusion* arises from the residual chemical potential difference between the glass and the Si-saturated solution. This reaction simply reflects the fact that glass is not the most thermodynamically stable solid at low temperatures. The release of mobile elements during this reaction can induce chemical modifications in the system and requires that complex formation and precipitation phenomena be taken into account.

We can now consider the scope, characteristics, and limitations of the test configurations defined in Table II.

A: Leaching in initially pure water

A1. HIGH SOLUTION RENEWAL RATE	A2. LOW SOLUTION RENEWAL RATE	A3. CLOSED SYSTEM
Scope	Scope	Scope
Qualify the existence of an initial rate by observing the dissolution congruence: i.e. check that hydrolysis of the silicate network controls and limits solid dissolution. For example, elements that enter solution more rapidly than silicon can reveal the existence of a highly soluble percolating subnetwork. Measure the maximum initial alteration rate in pure water. Measure the activation energy of the hydrolysis reaction	Qualify the reduction of the alteration rate versus the flow rate per unit area. Qualify the rate drop according to the concentrations of elements, especially silicon, in solution: determine a rate law.	Qualify the glass alteration rate drop over time and measure the residual rate (or indicate a value "less than" the determination limit) Qualify the possible long-term resumption of alteration (especially at high glass-surface-area-to-solution-volume ratios, when glass dissolution results in very alkaline pH values). Establish a relation between the dissolved elements concentrations and the solubility of the mineral phases in the alteration film when steady-state conditions are reached. Identify and quantify reactive diffusion of mobile elements in the gel (by varying the glass-surface-area-to-solution-volume ratio).
Characteristics	Characteristics	Characteristics
The flow rate per unit area Q/S (in m/d) must exceed the critical flow rate above which elements released from the glass no longer affect the dissolution kinetics (generally above 1 m/d). Typical devices and normative references: Soxhlet [13], modified Soxhlet, column test [14].	The flow rate per unit area Q/S (in m/d) is less than the critical flow rate above which elements released from the glass no longer effect the dissolution kinetics (between 10^{-1} and 10^{-8} m/d). In most cases, the solution is stirred and the diffusive flow of dissolved elements is negligible compared with the convective flow controlled by the solvent. Typical devices and normative references: flowing solution reactors, flowing autoclaves.	Powder samples of different particle size fractions are used to obtain glass-surface-area-to-solution-volume (S/V) ratios ranging over several orders of magnitude between 10^{-2} and 10^{+5} cm^{-1}. The predominant mechanisms thus also vary: with a large water volume the glass must release a large quantity of silicon before the dissolution rate drops. Conversely, with a small water volume (S/V > 10 cm^{-1}) a few nanometers of alteration are sufficient to saturate the solution with respect to silicon: under these conditions, the predominant mechanism is reactive diffusion among concentrations of dissolved mobile elements [4].

A1. HIGH SOLUTION RENEWAL RATE	A2. LOW SOLUTION RENEWAL RATE	A3. CLOSED SYSTEM
Limitations	Limitations	Limitations
The maximum glass alteration rate is not representative of the rate expected in a geological repository. Moreover, the effect of the glass composition is not the same for the initial rate as for a more realistic rate. It is therefore important not to specify a glass composition based on this measurement alone even in a case of a conservative approach.	Diffusion phenomena in a passivating reactive layer are no longer negligible over short time periods at the very low flow rates per unit area that yield the lowest dissolution rates ($Q/S < 10^{-4}$ m/d). The rate law for the material is always determined in a solution with a stoichiometry identical or very closes to that of the initial solid.	The very thin altered glass thickness, especially under residual rate conditions, make it difficult to characterize the alteration film.

B: Specified solution composition

B1. HIGH SOLUTION RENEWAL RATE	B2. PARTIALLY OPEN SYSTEM	B3. CLOSED SYSTEM
Scope	Scope	Scope
Qualify the catalytic/inhibiting effect of a dissolved ion on the hydrolysis kinetics: for example to establish the pH-dependence of the initial rate.	Refine the rate law by determining the role of a dissolved ion in solution in different proportions than the other constituent ions of the material (contrary to A2).	Qualify the effect of elements supplied by solution on the rate drop and reactive diffusion, and identify and quantify the underlying mechanisms (complex formation, precipitation, condensation).
Characteristics	Characteristics	Characteristics
Same as A1	Same as A2	Same as A3
Limitations	Limitations	Limitations
The effect of dissolved ions on the initial rate has been demonstrated to be generally much less significant that the effects of the pH and temperature [15].	The wide range of possible solution compositions and solution flow rates per unit area can significantly multiply the number of experiments. As a result, the study is often limited to the effect of the solution chemistry imposed by the environment selected for the leaching scenario to validate a rate law based on experiments in initially pure water (A2).	The effects of complex solutions are assessed by separately examining the effects of the individual constituents. Some elements may exhibit negligible effects only because they are sparingly soluble and are therefore found only in small concentrations in solution: the effect on glass alteration of a dissolved element that precipitates to form a crystallized secondary phase consuming silicon will be limited by the initial quantity available, which depends on the solubility of its oxide or hydroxide form.

C: Presence of solids

C1. HIGH SOLUTION RENEWAL RATE.

This situation is of little practical interest: rapid solution renewal is necessary to measure the initial rate, on which the effect of solids is negligible.

C2. PARTIALLY OPEN SYSTEM	C3. CLOSED SYSTEM
Scope	Scope
Could satisfactorily represents the scenario in a geological repository, but is complex because of the presence of potentially several alteration drivers: dilution effect due to solution renewal; effect of the leaching solution chemistry; consumption of elements in solution by precipitation of secondary phases; solution resupply through dissolution of solid materials; sorption on these solids of elements dissolved from the glass; diffusive transport through the solids. Used to validate a modeling approach based on simpler experiments designed to investigate each mechanism separately.	Representative of a highly confined system. Compared with B3, it is used to qualify the effect of a low-solubility element that is consumed very quickly but remains at a nonzero concentration due to continuous dissolution of the added solid. This is useful for qualifying interactions between materials at the interfaces. This approach can avoid transport problems by thorough mixing of the glass and the added solid; conversely, it can be used to examine how transport can limit the reactivity of the solids simply by moving them farther away from the glass.
Characteristics	Characteristics
Percolation experiments with or without direct contact between the glass and the environmental solid; *in situ* leaching experiments; etc.	Thorough blending of powder test material and environmental solid or, on the contrary, separation of the solids in a specified geometry to assess the drop in reactivity due to element transport effects.
Limitations	Limitations
These complex experiments require prior knowledge of the hydrodynamic properties of the reactor, the properties of the solids (permeability, porosity) and their respective chemical reactivity. Attempting to impose experimental conditions too near the actual repository conditions is not necessarily scientifically relevant. The time factor cannot be taken into account in the laboratory, and the chemical and hydrodynamic phenomena observed in the laboratory could well be merely transient phenomena on a repository time scale.	Requires prior knowledge of the chemical reactivity of the solids in the system. For example, configurations can be imagined in which the dissolution kinetics of the added solid control by the glass dissolution kinetics.

All these tests allow direct measurements of the parameters used in the model. The number of model parameters to be measured depends on the complexity of the model. For the $V_0 \rightarrow V_r$ model, for example, type A2 tests are used to determine $V_0 = f(T, \mathrm{pH})$, and type A3 tests to measure $V_r = f(T)$. The C3 configuration is used to validate the time to saturation with respect to the corrosion products. The most complex scientific models require a description of the speciation of each element in solution and its distribution between the solution and the amorphous or crystallized solid phases of the alteration film. All the experimental configurations from A1 to C3 could thus potentially be used either to configure or to validate the models.

Validation is a key step in the French modeling approach. In addition to the tests described above, validation also requires *in situ* tests and specific examinations of natural and archaeological analogs. Under the terms of the waste management act of 28 June 2006 that should lead to the qualification of a clay repository site, *in situ* tests are planned to validate the transient rate drop and long-term residual rate regimes. These tests are scheduled to begin in early 2010 and will be conducted on nonradioactive glass, but under realistic chemical conditions; the instrumentation will allow solution and solid analysis during the experiment.

Natural and archaeological analogs are studied to provide long-term validation of the key mechanisms taken into account in the models. Studies in recent years on basaltic glass samples have demonstrated their behavioral similarity with nuclear glasses [16]. The characterization of basaltic glass samples leached in clay media over very long time periods corroborates the expected orders of magnitude for the lifetime of nuclear glasses in a geological repository [17]. Initial results on coupling of chemistry and transport phenomena were recently obtained for Roman archaeological glass altered in seawater [18]. This work will continue over the coming years.

CONCLUSION

This paper provides an overview of leach tests as applied in the French approach for evaluating nuclear glass performance in a geological repository. This assessment is based on a model, and the robustness of the model is due primarily to the choice of parameter values and their uncertainties. Some model parameters (the initial rate r_0 and, to a lesser extent, the residual rate r_r) are inferred from the results of a very large number of experiments, and are thus known with good precision; the uncertainty limits incorporate all the variations due to environmental parameters that are not explicitly taken into account in order to simplify the model (solution composition, radiolysis, pressure, etc.). The other model parameters (effective cracking factors under r_0 and r_r conditions) are based on a limited number of experiments; the proposed values are based on proven knowledge at this time, or include a safety margin.

REFERENCES

[1] Jacq, F., 2005. The research program of Andra on HLLLW disposal in clay rocks. Main results and perspectives. International meeting "Clays in Natural & Engineered Barriers for Radioactive Waste Confinement", March 14-18, 2005, Tours (France).

[2] Gin S., Ribet I., Frugier P., Chave T., Angeli F., Lartigue J.E., de Combarieu G. and Godon N., 2005. Assessment of nuclear glass behavior in geological disposal conditions: Current state of knowledge and recent advances. European Nuclear Conference proceeding. Versailles, France.

[3] European Standard ENV 12920. Characterisation of waste—methodology guideline for the determination of the leaching behavior of waste under specified conditions, 1996.

[4] Frugier P., Gin S., Minet Y., Chave T., Bonin B., Godon N., Lartigue J.E., Jollivet P. Ayral A., De Windt L. and Santarini G., 2008. SON68 Nuclear Glass Dissolution Kinetics: Current State of Knowledge and Basis of the New GRAAL Model. Journal of Nuclear Materials, under press.

[5] Valle, N., Sterpenich, J., and Libourel, G., 2001. ^{29}Si and ^{18}O tracing of borosilicate glass alteration. Implications on the mechanisms of formation of the altered layer. Geochimica and Cosmochimica Acta, 1-40.

[6] Chave, T., Frugier, P., Ayral, A., and Gin, S., 2007. Solid state diffusion during nuclear glass residual alteration in solution. Journal of Nuclear Materials, 362: 466-473.

[7] Angeli F., Charpentier T., Gaillard M. and Jollivet P., 2008. Influence of zirconium on the structure of pristine and leached soda-lime borosilicate glasses: Towards a quantitative approach by O-17 MQMAS NMR. Journal of non-crystalline solids, 354:3713-3722.

[8] Angeli F., Gaillard M., Charpentier T., and Jollivet P., 2007. Contribution of Ca-43 MAS NMR for probing the structural configuration of calcium in glass. Chemical Physics Letters, 440: 324-328.
[9] Angeli F., Gaillard M., Jollivet P., and Charpentier T., 2006. Influence of glass composition and alteration solution on leached silicate glass structure: A solid-state NMR investigation. Geochimica et Cosmochimica Acta, 70: 2577-2590.
[10] Calas G., Cormier L, Galoisy L. and Jollivet P., 2002. Structure-property relationships in multicomponent oxide glasses. Comptes Rendus Chimie, 5 (12): 831-843.
[11] Jollivet P., Den Auwer C., Delaye J.M. and Simoni E., 2007. Evolution of the local environment of lanthanum during simplified SON68 glass leaching. Journal of Non-Crystalline Solids, 353 (4): 344-353.
[12] Spalla, O., Barboux, P., Sicard, L., Lyonnard, S., and Bley, F., 2004. Influence of insoluble elements on the nanostructure of water altered glasses. Journal of Non-Crystalline Solids, 347(1-3): 56-68.
[13] ISO 16797, 2004. Chemical durability test – Soxhlet-mode chemical durability test – Application to vitrified matrixes for high-level radioactive waste, 2004.
[14] French Standard XP X30-450. Déchets - Vitrifiats silicatés - Essai de détermination en fonction du temps de la vitesse de dissolution au contact d'une solution aqueuse de composition spécifiée suivant un scénario et renouvelée régulièrement à un débit supérieur au débit critique - Détermination de la vitesse initiale de dissolution, 2002.
[15] Gin S., 1994. Etude expérimentale de l'influence d'espèces aqueuses sur la cinétique de dissolution du verre nucléaire R7T7. PhD thesis, University of Poitiers, France.
[16] Techer I., Advocat T., Lancelot J. and Liotard J.M., 2001. Dissolution kinetics of basaltic glasses: control by solution chemistry and protective effect of the alteration film. Chemical Geology, 76 (1-4): 235-263.
[17] Techer I., Lancelot J., Clauer N., Liotard J.M. and Advocat T., 2001. Alteration of a basaltic glass in an argillaceous medium: The Salagou dike of the Lodève Permian Basin (France). Analogy with an underground nuclear waste repository. Geochimica et Cosmochimica Acta, 65: 1071-1086.
[18] Verney-Carron A., Gin S. and Libourel G., 2008. A fractured Roman glass block altered 1 800 years in seawater: analogy with nuclear waste glass in deep geological repository. Geochimica et Cosmochimica Acta, under press.
[19] Cailleteau C., Angeli F., Devreux F., Gin S., Jestin J., Jollivet P. and Spalla O., 2008. Insight into silicate glass corrosion mechanisms. Nature Materials, submitted.

SCK•CEN R&D ON THE INTERACTION BETWEEN NUCLEAR WASTE GLASS AND CLAY NEAR- AND FAR- FIELD MATERIALS

Pierre Van Iseghem, Karel Lemmens, Elie Valcke and Marc Aertsens
SCK•CEN, Mol (Belgium)

ABSTRACT

A large R&D programme has been performed in Belgium to evaluate the chemical durability of candidate high-level waste glasses in clay environments. This paper discusses the methodology followed and the main results obtained over the past years. The methodology involves laboratory test set-ups focusing on the interaction with solid clay, modelling, and tests in the underground research laboratory in clay. The results discussed include dissolution rates, mobile radionuclide concentrations in the interacting solutions, analytical modelling and results from an *in situ* test. The results show that HLW glass will act as a substantial barrier in such an environment. The paper also discusses the main dissolution processes involved and the remaining uncertainties.

INTRODUCTION

Belgium is studying a Boom Clay formation as a host rock for the final disposal of the vitrified Belgian high-level waste. Presently various other countries (Switzerland, France, Japan, Spain, *etc*) are considering clay host rocks as well. Since the mid 90's a reference geological disposal concept "SAFIR2" was proposed by NIRAS/ONDRAF (the Belgian management agency for radioactive waste and fissile materials)[1]. As part of that, SCK•CEN has been performing extensive R&D on the chemical durability of HLW glass in clay, as in the SAFIR2 disposal concept clay is both the candidate host rock (Boom Clay) and the major constituent of the candidate near field. More recently, the supercontainer concept has been selected as the new reference in Belgium [22]. In this concept the near field is dominated by concrete instead of clay. The R&D on glass obviously has to be reconsidered.

The Boom Clay beneath the Mol-Dessel nuclear zone is the reference site for methodological R&D, without making any presumption on the site where a repository would possibly be implemented.

Our approach consisted in performing laboratory experiments coupled with modelling (geochemical, analytical) so as to identify and understand the major interaction processes occurring between glass and clay [2]. The second objective was to determine the overall dissolution behaviour of glass in clay, in order to determine the lifetime of HLW glass upon disposal in clay. We also performed *in situ* tests in the underground laboratory "HADES" in clay, to evaluate the results from the laboratory tests in realistic conditions.

The results of this R&D are implemented in the Safety Assessment [3] or Safety Case [4] of the geological disposal. As one of the input parameters, the R&D should provide best estimates of the long-term glass dissolution rate and associated lifetime. These informations will also lead to identify the safety function of the HLW glass in the Safety Case that is being elaborated by NIRAS/ONDRAF.

METHODOLOGY

The glass-clay interaction has been studied since many years by various laboratories [5,6, 18]. The earlier programme focused on a parametric approach, investigating the effect of the temperature, the solid (clay) to solution ratio's contacting the glass, redox potential values, presence of γ-radiation, *etc.* [2]. More recently we have developed experimental approaches where we considered more realistic temperatures and contacting clay media [7,8]. We considered two clay media: Boom Clay * (which is the candidate far field), and FoCa ** clay based materials (FoCa clay has been a candidate backfill material), in agreement with the SAFIR2 reference disposal concept[1]. Certain tests were performed to derive parameter values for simplified subsystems, whereas other tests were performed to study the coupled subsystems and for parameter validation.

Through our continued research, we developed various experimental set-ups, such as:

- Closed vessels (PTFE, polycarbonate, stainless steel,...), filled with glass (coupons or powdered), solution and (in many tests) solids (clay, corrosion products, cement) – see Figure 1. Currently one vessel per test duration is used. This configuration enables to sample sufficient amounts of solution for analysis of the leached glass constituents. If solids (clays, *etc*) are present, these leachates are centrifuged and (ultra-)filtered before analysis.
- Percolation or flow-through cells, filled with a solid (*e.g.*, clay), together with other materials (*e.g.* glass coupon, radioactive tracers, container corrosion products – see Figure 2. This set-up is reproducing closely conditions prevailing in real disposal situations, and can be used under static or dynamic conditions. It is also suited for studying coupled processes, such as the combination of glass dissolution and diffusion through clay, but it is more difficult for sampling of the solution.

For both types of experimental approaches, we spent large attention to controlling and monitoring the environmental conditions of importance. A very important parameter is the pH. The p_{CO_2} is important because dissolved carbonate affects pH and can complex certain radionuclides. E_h, and especially the related p_{O_2} is important because it affects the speciation of certain radionuclides, and because the oxidation of Boom Clay changes its properties (a.o. pH decrease). Especially for radionuclide studies, we also took much care to separate the leaching media into a mobile (< 5nm size), soluble (0.45 μm filter) and solid fraction. Another point of attention was quality assurance, involving amongst other the validation of the experimental data and parameters (such as E_h).

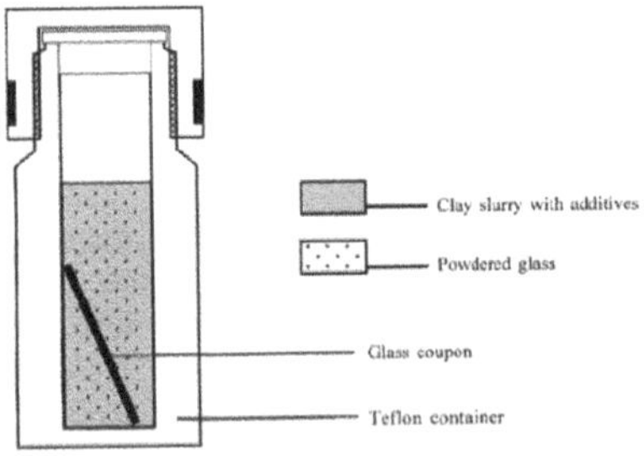

Figure 1. Experimental set-up used to study the interaction between glass samples and pure solutions in static conditions

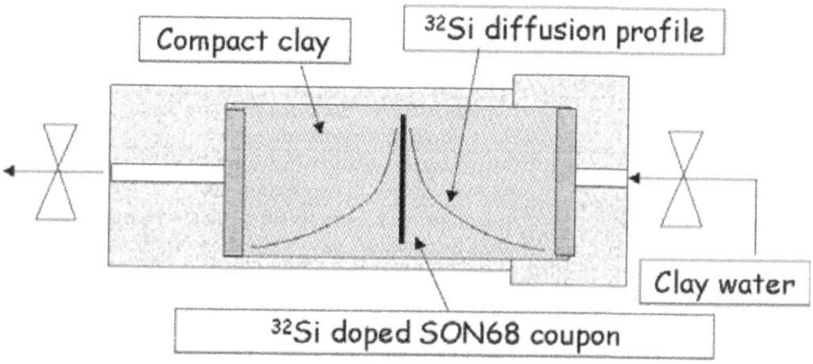

Figure 2. Experimental set-up to measure both glass dissolution and silica diffusion in one single experiment

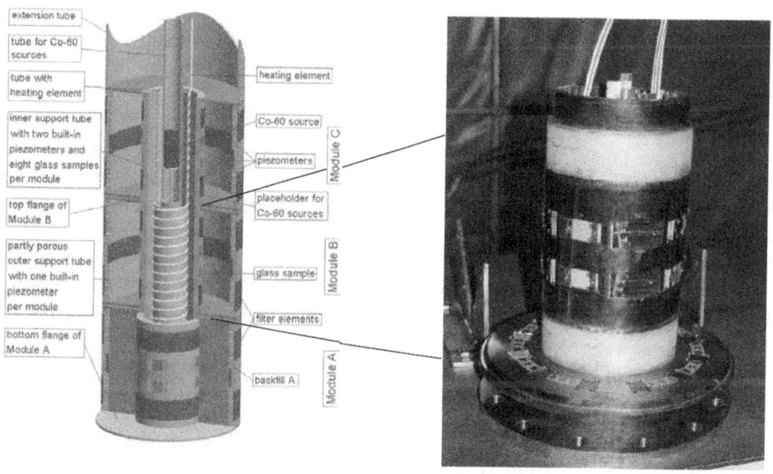

Figure 3. (left) Three-dimensional cut-away view of a test tube with ^{60}Co sources. (right) Inner support tube, bottom flange, and connection flange of one module. The photo shows the position of the samples (SON68 and other glass types, and container materials) on the inner support tube and on the flange, and the filter screens and connecting tubes of the inner piezometers (in grey).

Table I summarizes the different laboratory experiments performed in the past 10-years period. Within this paper we will discuss the results of few of them. More detailed reports can be found in [7-10].

Table I: Type of experiments on HLW glass durability performed during the 1998-2004 programme. Further experimental details are given in [7-10].

Description	Main output
Influence of silica release on clay characteristics	Determination of the silica sorption parameters on clay at high Si concentrations
Si sorption characteristics of Boom Clay and FoCa clay (25°C)	Determination of the silica sorption parameters of the clays at low Si concentrations
Combined leaching - diffusion test with ^{32}Si doped glass in Boom Clay (30°C, 90°C)	Simultaneous measurement of glass dissolution and silica diffusion in clay. Validation of processes and parameters for coupled systems
Dissolution in clay at in situ density (30°C, 90°C)	Measurement of dissolution of glass in contact with compact Boom Clay or candidate backfill (FoCa clay mixed with glass frit). Validation of processes and parameters for more realistic conditions
Dissolution in clay slurries (40°C); the solid fraction consists of either Boom Clay or FoCa clay, mixed with glass frit.	Measurement of glass dissolution in presaturated clay (complementary to the tests at *in situ* density
Dissolution of glass doped with ^{99}Tc and ^{237}Np in a FoCa clay – synthetic clay water slurry (40°C)	Determination of radionuclide concentration and speciation in reducing near field conditions , and determination of the main influencing mechanisms
Dissolution of glass doped with Sn, Se, Pd, Zr in clay slurry (40°C); the solid fraction consists of a mixture of FoCa clay, Boom clay, sand, graphite, stainless steel powder	Measurement of the concentration of these critical nuclides in reducing near field conditions, and comparison with background concentrations.

Whereas our R&D on non-radioactive systems was considering both far- and near-field clay environments, our R&D on radioactive glasses was focused on their interaction with candidate near-field (backfilling) clay media. This is because the radioactive glass will eventually be in contact with an environment that will be for long time (thousands to hundreds of thousands of years) dominated by these clay backfilling materials. We decided to investigate the release of ^{237}Np, ^{99}Tc, ^{79}Se, ^{126}Sn, ^{93}Zr and ^{107}Pd from HLW glass, as the Safety Assessment studies concluded that they were amongst the most critical radionuclides for the long-term safety of a disposal in Boom Clay [1,3].

The rationale for choosing the test temperatures is that 30°C corresponds to the temperature that is expected at the interface between HLW glass and the near field after the initial thermal phase [1]; 90°C is another reference temperature in many R&D programmes on HLW glass [13].

Figure 3 shows the design of the *in situ* experiment CORALUS ***, that provides a demonstration in realistic disposal conditions of the results and conclusions of laboratory work. The CORALUS test is conceived as an integrated test, as it involves various coupled processes (glass dissolution/migration through clay; effects of radiation, radiolysis, thermal gradient). The test includes highly active α-doped SON68 glass samples, clay-based backfill materials exerting a high swelling pressure, contact with the clay host formation, controlled temperature, and presence of γ-radiation. Each module includes glass samples doped with ~0.85 weight% of one of the radionuclides selected ($^{237}NpO_2$, $^{239}PuO_2$, or $^{241}Am_2O_3$) and inactive glass samples. Three modules per CORALUS tube are loaded with different backfill materials: module A contains dried Boom Clay; module B contains Ca-bentonite mixed with sand and graphite (60/35/5 wt%), this is a Belgian candidate backfill); module C contains Ca-bentonite mixed with powdered SON68 glass frit (95/5 wt%), this is a French candidate backfill. The addition of the powdered glass frit aims at establishing high solution concentrations of the main glass constituents (Si, Al,

B, Na, Ca, Zr, Li, and Zn), to reduce the chemical potential difference of these elements between glass and interstitial solution, thus decreasing the glass dissolution rate

The CORALUS *in situ* tests comprise four modular test tubes, placed in the Boom Clay in the underground research facility (URF) HADES (Mol, Belgium), for durations up to 10 years. Two tubes are operated at 30°C and do not include γ-sources, and are running for 3.3 and 10 years, respectively. They simulate the long-term during geological disposal, after the thermal phase, and advanced corrosion of the overpack and canister. The two other tubes are operated at 90°C and include γ-sources, and run for 1.2 and 6 years, respectively. They simulate an early failure scenario, ~50 years after disposal. Full details are given in [16].

The Belgian programme considers various HLW glasses, resulting from different sources of high-level reprocessing waste, and vitrification processes [1]. The compositions are listed in Table II. An important distinction is to be made between the glasses with low alumina content (SON68 and SM513) and those with high alumina content (SM527 and SM539).

Table II. Main constituents of the HLW glasses considered in Belgium. SON68 is the COGEMA R7T7 reference glass, the other glasses are DWK/PAMELA reference glasses.

(wt%)	SON68	SM513	SM527	SM539
SiO_2	45.48	52.15	38.75	35.28
B_2O_3	14.02	13.08	21.70	25.58
Na_2O	9.86	9.05	8.64	9.21
Li_2O	1.98	4.18	3.1	3.49
Al_2O_3	4.91	3.59	19.96	19.83
CaO	4.04	4.54	3.87	5.05
TiO_2	/	4.54	1.55	0.003
FP oxides	13.38	4.46	1.58	0.96
Balance	6.33	4.41	0.85	0.6

DISSOLUTION OF HIGH- LEVEL WASTE GLASS IN CLAY

We will discuss some of the experimental results obtained upon interaction between glass and solid clay at 30°C. Figure 4 shows the results in terms of mass losses in function of the duration, upon interaction with Boom clay (left Figure) or with a mixture of FoCa clay with 5% SON68 glass frit (right Figure). This 95%-5% mixture was considered as a potential candidate backfill. Experimental data till more than six years duration are available.

In compact Boom Clay (Figure 4, left), the initial rate is close to the forward rate for the considered pH (8.5) and the rate decreases with time, be it very slowly. Based on a reproduction of the curve with the Pescatore model [11], a dissolution rate of 0.0042 g/m²day is found after 5 years, compared to a short term rate of 0.014 g/m²day [8]. This is a decrease with a factor 3. The short term rate is close to the forward rate in pure water for the considered pH (8.5) [8]. The right plot in Figure 4 shows much smaller mass loss rates in the order 10^{-3}–10^{-4} g/m² day, but with large uncertainties. These lower dissolution rates will probably be achieved in the test in pure Boom Clay as well, after much longer durations. Additional analyses – such as surface analyses - will be performed to provide a more accurate value of this dissolution rate. Although rates in the order 10^{-3}–10^{-4} g/m² day may seem very low, the combination with a relatively large effective glass surface area (due to the glass cracking) still leads to a life time of a few hundreds of thousands of years at most.

The impact of the addition of the (silica rich) glass frit to clay in reducing the glass dissolution appears to be very strong, and agrees with earlier findings [12]. This will contribute to the saturation of the sorption sites of the clay, and will therefore simulate the long-term situation when the HLW glass has sufficiently dissolved. We can interpret this very low dissolution rate in the FoCa-glass frit medium therefore as a residual rate, which is currently accepted to occur in pure solutions after silica saturation [13]. Nevertheless, the alteration processes in the tests with compact clay and glass frit after six years or less may not be completely identical to the long term processes on geological time scales. The presaturation of the clay with silica effectively slows down the dissolution of Si from the clay, resulting in thin gel layers, but the effect of the potential slow conversion of the gel into secondary phases on the long term cannot be accelerated by addition of glass frit. Because this process is so slow at 30°C, it should not cause a corrosion resumption, though.

We assume that the glass dissolution in compact clay (here, Boom Clay) can be reproduced with a first order rate-law equation [14], but including a constant residual rate [13]. A validation of this assumption is difficult to achieve for glass in contact with compact clay, though. When clay is involved, there are problems to define correctly the silica saturation concentration at which the residual rate takes over. If this concentration is influenced by the gel composition, it should be influenced also by the clay, because the clay alters the gel. Moreover, the Si concentration in the clay pore water is assumed to show a gradient (in absence of glass frit), so the average Si concentration over the clay plugs in our tests (< 10 mg/L at 30°C for SON68 in contact with Boom Clay) will be different from the concentration at the glass/clay interface. The combined leaching-diffusion test allow to reconstruct the Si concentration profiles in the porewater, but the uncertainty on the results remains high. Moreover, glass is not the only dissolving phase when clay is present. The measured Si concentrations are the net result of dissolution and reprecipitation of the glass and of the instable minerals in the clay. We also experienced that Si concentrations in the interstitial solution and the Si sorption on the clay are still evolving with time, even in the FoCa – glass frit medium. There is also evidence of Si precipitation, at least at higher temperatures. At 30°C, the evidence for Si precipitation in compact clay is rather weak so far.

The complex Si geochemistry is further interpreted. The silica saturation value used in our modelling calculations is mainly a fitting parameter, not yet based on a full understanding.

Noteworthy is that the different glass compositions present quite similar mass losses.

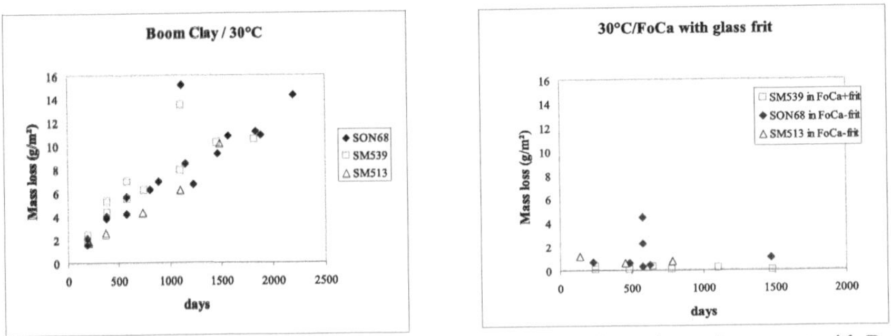

Figure 4. Mass losses measured during the dissolution of various glasses in contact with Boom Clay (left) or a mixture of FoCa clay with 5% glass frit (right).

Figure 5 shows the experimental and fitted ^{32}Si-profile in the clay after 887 days at 30°C, for the combined dissolution – diffusion test. The mass loss data are in agreement with those from the test in pure Boom Clay shown in Figure 4 (left). But by doping the glass with ^{32}Si we were able to record in addition ^{32}Si diffusion profiles through the clay [8]. The interest of this test is mainly that we were able to interpret the coupled process of glass dissolution and Si migration through clay with existing models for the "simple" systems [15]. An analytical model was elaborated by considering a constant, congruent glass dissolution rate and silica diffusion through and sorption onto the clay. The following parameter values were used: a forward glass dissolution rate (at zero silica concentration) of 0.028 $g.m^{-2}day^{-1}$, an apparent silica diffusion coefficient in the clay of 1.4 10^{-12} m^2sec^{-1}, and a distribution coefficient for silica on Boom Clay between 0.010 and 0.075 m^3kg^{-1}. These parameter values are close to the range found in literature. The silica saturation concentration predicted by the analytical model would be 15-21 mg/L, however the measured average silica concentrations were max 10 mg/L. More details can be found in [15].In the meantime, the results of longer term tests are available and an update of the interpretation is ongoing.

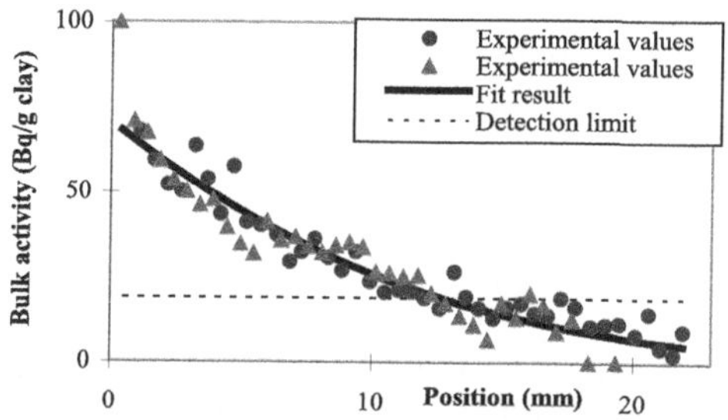

Figure 5. Experimental and fitted ^{32}Si diffusion profile in the clay after 887 days at 30°C. The experimental values of the right and left profile are similar.

RADIONUCLIDE LEACHING FROM HLW GLASS IN CLAY

The leaching behaviour of Np and Tc in a FoCa clay slurry (simulating a candidate backfill) was particularly sensitive to the redox condition of the medium. When appropriate reducing conditions were achieved (E_h about -200mV, similar as in real Boom Clay) Np and Tc reduced from V, resp VII valence (in the glass) to IV (both Np and Tc) valence state. The Np and Tc concentrations in solution (ultrafiltered through 100,000 MWCO filter) were nearly constant during the two years leach test, and were extremely low [9]. Their average values are given in Table III (see "near field"). One can observe that these values are much lower than the solubilities obtained from percolation migration tests ("far-field" values) [1] and the solubility as calculated from the existing thermodynamic database (fourth column). In the last column it can be seen that performance assessment is very conservative in selecting radionuclide solubilities from the R&D programmes, keeping large safety margins. Some of the questions are whether the RN concentrations measured in our tests correspond to the thermodynamically stable values, or whether the solubility limiting phase predicted is really the right one. The constant concentrations below the expected solubility of the pure Np and Tc phases suggest that their solubility near the glass/clay interface may be suppressed by coprecipitation, related to the dissolution/reprecipitation of glass and clay in these systems out of equilibrium.

Table III: Comparison of the near-field concentrations obtained from our leach tests with the far-field concentrations

(Mol/l)	[RN] near-field	[RN] far-field	Solubility limiting solid phase	Best estimate PA (values)
^{237}Np	$1.5x10^{-10}$	$6.7x10^{-10}$	$Np(OH)_4$ $(5x10^{-9}M)$	10^{-6}
^{99}Tc	$5.6x10^{-10}$	10^{-8}	$TcO(OH)_2$ $(10^{-8}M)$	$3x10^{-8}$

The leaching behaviour of ^{79}Se, ^{126}Sn, ^{93}Zr and ^{107}Pd from glass in near-field clay media was investigated by melting inactive isotopes in the glass. Their concentrations in the leachate were recorded by ICP-MS, after subtracting their concentration in the blank. The results for Se are shown in Figure 6 (complete results, see [10]). The main result is that the release of Se, Zr, Pd and Sn from the reference HLW glasses SON68 and SM539 (inactive isotopes simulating the radionuclides Se-79, Zr-93, Pd-107 and Sn-126, respectively) did not increase their element concentration in the contacting solution beyond the natural background in the clay medium. Yet, we can't conclude for sure if their leaching from the glass is limited by solubility, and whether the concentration of their stable isotopes in the clay medium will affect the radionuclide leaching behaviour, because we only considered the concentration in the aqueous phase of the clay medium. We can expect a similar behaviour in other (clay-based) near field environments with similar characteristics. The solubility limitation would imply that the diffusion of the radioisotopes through the near field will be retarded by isotopic dilution with the inactive, naturally occurring isotopes.

The most interesting results were obtained for the Se-leaching. The initial hypothesis was that Se would be in its oxidized form (VI) in the glass, and that the concentration of Se in the aqueous phase of the clay medium would not be solubility limited. However, we rather believe that Se is present as Se(IV) and Se (0) in the glass, and that its concentration in the aqueous phase might be solubility limited [10]. We assume that FeSe or Se (cr) are the solid phases, and selenide or selenite, and possibly traces of selenate are the dissolved species.

Like for Se, also Zr, Pd and Sn show rather steady state concentrations in the aqueous phase, which are interpreted as solubility controlled. These concentrations are much smaller than the corresponding radionuclide solubilities used in the safety assessment study [19].

Table IV: Comparison of the near-field concentrations obtained from our leach tests with the far-field concentrations

Element	**[element] (near-field)**	**Solubility limiting Solid phase**	**[RN] (far-field)**	**[element] (in natural porewater)**	**[RN] (best estimate PA values)**
	(mol/l)	(mol/l)	(mol/l)	(mol/l)	(mol/l)
Se	$4x10^{-8}$ - $2.5x10^{-7}$	FeSe 1.9 10^{-9} Se (cr) 2.9 10^{-7}	$1 - 0.3\ x10^{-8}$	$2.4\ x10^{-8}$	- $5x10^{-8}$(Se 0, -II) - No solubility limitation (Se VI)
Zr	$5x10^{-9}$ - $1x10^{-7}$	ZrO_2, 6.7 10^{-10} $ZrSiO_4$ 1.3 10^{-13}	/ (no data)	$6.4 - 34\ x10^{-7}$	$1x10^{-6}$
Pd	$3x10^{-9}$ - $3x10^{-8}$	Pd poorly soluble	$1.6–1.9\ x10^{-8}$	/	$1x10^{-7}$
Sn	$3x10^{-9}$ - $1x10^{-8}$	SnO_2 2.7 10^{-8}	/ (no data)	/	$5.5x10^{-7}$

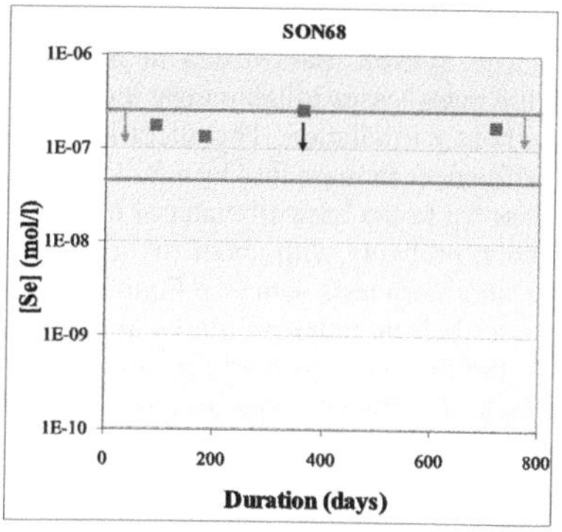

Figure 6. Average concentrations of Se in the leachates after ultrafiltration for the SON68 glass. The pink dashed lines show the background concentrations in the medium. Concentrations reported as 'lower than' (for the leach tests and for the blank tests) are indicated with a downward arrow.

COMPARISON OF LABORATORY WITH *IN SITU* TEST DATA

The four CORALUS tubes have been installed by the latest at the end of 2003. The interstitial solutions of the three backfill materials have been regularly analyzed chemically (anions, cations, organic constituents, dissolved gas, pH-E_h). One tube at each temperature has been stopped, and the glass and clay (radionuclide migration, mineralogy) samples analyzed. Analysis and interpretation is still in progress. We show some radionuclide migration profiles through the clay in Figure 7. More details are given in [16,17].

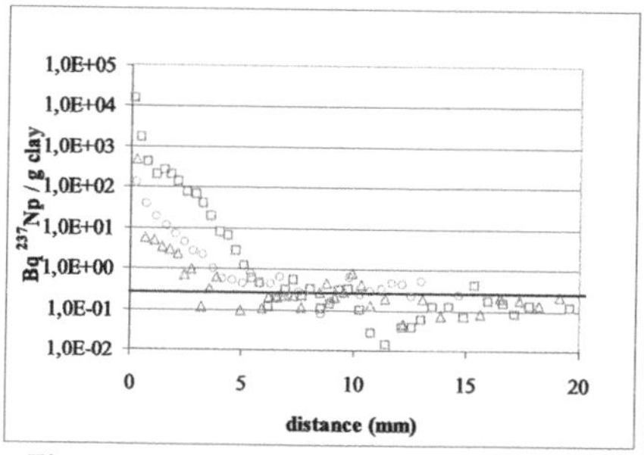

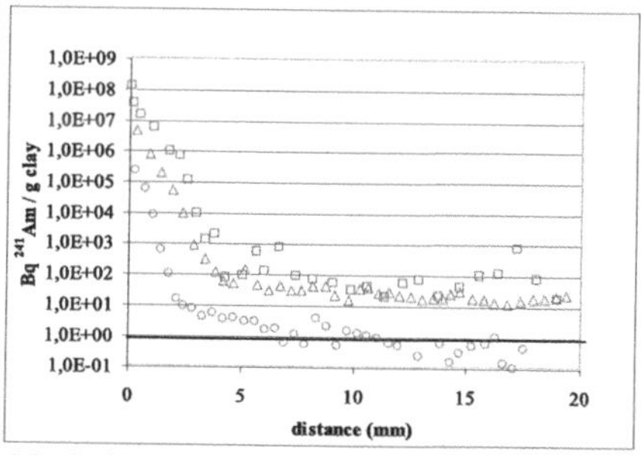

Figure 7. Radionuclide migration profiles for Np (left) and Am (right) for the CORALUS test at 90°C, 15 months, including ^{60}Co sources. (□) dried Boom Clay, (Δ) Ca-bentonite mixed with sand and graphite, (○) Ca-bentonite mixed with powdered SON68 glass frit

Many of the glass samples were broken after retrieval and analysis. We believe that this is due to the design of the sample holder, rather than due to intrinsic glass properties. The main conclusions on the glass dissolution and radionuclide migration so far are:

1. The results compare fairly well with the surface laboratory test results for the same conditions in undisturbed Boom Clay. The SON68 mass losses in module A (with dried Boom Clay backfill) compare well with the mass losses in laboratory test: 0.01 $g/m^2/day$ at 30°C, and 0.3–0.5 $g/m^2/day$ at 90°C, but without γ irradiation. The alteration at 90 °C is considerably higher than at 30 °C. We measure differences in mass loss by a factor of 20.
2. The addition of powdered glass frit to the backfill material has a beneficial effect: it diminishes the glass alteration considerably, probably with about two orders of magnitude, or even more. This is in agreement with the laboratory tests data (see Figure 4).
3. The α-doped and inactive glasses behave rather similarly. We observe slightly thicker alteration layers for the active samples, the thickness seems to increase with increasing α-β-γ-activity of the samples. Because of the lack of sufficient mass loss data, we cannot conclude that this goes along with an increased mass loss.
4. The order of radionuclide activity at the interface glass/clay is: Dried Boom Clay > Bentonite with sand and graphite > Bentonite with glass frit, in agreement with the different degree of alteration of the glass in the different backfill materials. The radioactivity migrated in the backfills drops very sharply over the first few mm of clay, demonstrating the strong radionuclide retardation.
5. The migration distance of Am is longer than for Np (and Pu). We attribute this to colloid migration in case of Am, as Am is forming soluble complexes with the humic acids present in Boom Clay. Further interpretation of the migration profiles based on migration processes in undisturbed clay is impossible, because in CORALUS processes are too complex (different radionuclide source, clay was partially disturbed, *etc*).

CONCLUSIONS

SCK•CEN has been using several methodologies to investigate the complex interaction processes between HLW glass and clay materials (considered either as the candidate far field, or as candidate near-field materials). Some are focusing on specific process parameters, other on the entire, coupled interaction processes. They allowed to identify the major processes (matrix dissolution, sorption on clay, dissolution decrease due to saturation of the clay), and to provide estimations for the very low long-term dissolution rates. Resulting lifetimes of HLW glass packages would be of the order of $\sim 10^5$ years and more. The release of specific radionuclides critical for the long-term safety (^{99}Tc, ^{237}Np, ^{79}Se, ^{126}Sn, ^{107}Pd, ^{93}Zr) in the clay media is characterized by very low mobile concentrations in the soluble phase, which also tend to constant values. The composition of the glasses considered in Belgium (main difference is the Al_2O_3 concentration) does not have a measurable effect on the dissolution behaviour in the clay media considered.

The paper further identified main remaining uncertainties, deserving further research:

- The Si behaviour at the interface between glass and clay, where various processes (sorption, precipitation) may affect its solubility,
- The long-term dissolution behaviour of the HLW glasses in clay media; long-term data are in particular needed, amongst other to estimate and explain the long-term "residual" dissolution rate,

Another important remaining uncertainty, which has had little attention in the SCK•CEN programme concerns the increase of the effective surface area of glass packages in long-term disposal situations; this will have a serious impact on the predicted lifetime of the glass packages.

HLW glass can be considered as a significant barrier upon geological disposal in the Belgian Boom Clay only if we can predict lifetimes of ~10^6 years. With the current status of R&D, this is not (yet) possible. This clearly sets the challenge for further work on glass-near field interaction.

ACKNOWLEDGEMENTS

The authors gratefully acknowledge financial support through various projects sponsored by NIRAS/ONDRAF and the European Commission. They also acknowledge the contributions of Dr. P. De Cannière and Dr. N. Maes (Si sorption on clay), Dr. D. Jacques (geochemical modelling) and Dr. K. Ferrand (basic glass studies on the glass durability).

FOOTNOTES

* Boom Clay consists for ~60% of clay minerals (illite, smectite, kaolinite, chlorite, other), ~20% of quartz, and smaller amounts of feldspars-K, carbonates, pyrite and organic carbon [20].
** FoCa : Fourges-Cahaignes bentonite, which consists for (average values) 84% of clay minerals (kaolinite, smectite), 6% of quartz, 6% of goethite, and minor amounts of carbonates-calcite, hematite, gypsum and organic carbon [21]
** CORALUS : Corrosion of α-active glass in underground storage conditions

REFERENCES

[1]SAFIR 2. Safety Assessment and Feasibility Interim Report 2. NIROND 2001-06 E, December 2001.
[2]K. Lemmens, M. Aertsens, V. Pirlet, H. Serra, E. Valcke, P. De Cannière, P. Van Iseghem, "Measurement of glass corrosion in Boom Clay disposal conditions". Proceedings of the 8th International Conference on Radioactive waste management and environmental remediation (2001)
[3]D. Mallants, J. Marivoet, X. Sillen, "Performance assessment of the disposal of vitrified high-level waste in a clay layer". Journal of Nuclear Materials Vol 298 (2001), 125-135
[4]L. Johnson and P. Zuidema, "Evaluating R&D progress in the context of the Safety Case". Mat. Res. Soc. Symp. Proc. Vol 932, 2006, 201-210
[5]S. Gin, P. Jollivet, J. Mestre, M. Julien, C. Pozo, "French SON 68 nuclear glass alteration mechanisms in contact with clay media". Applied Geochemistry 16 (2001), 861-881,
[6]K. Lemmens, "The effect of clay on the dissolution of nuclear waste glass". Journal of nuclear Materials Vol 298 (2001), , 11-18
[7]K. Lemmens, M. Aertsens, V. Pirlet, N. Maes, H. Moors, P. Van Iseghem, "Measurement of glass corrosion in Boom Clay disposal conditions : first results of the experimental programme 2000-2003 of SCK•CEN". ICEM'03, 9th International conference on environmental Remediation and Radioactive Waste Management, Sept 2003, Oxford (U.K.) Proceedings on CD-Rom, published by ASME
[8]K. Lemmens, C. Cachoir, E. Valcke, K. Ferrand, M. Aertsens, T. Mennecart, The strategy of the Belgian nuclear Research Centre in the area of high-level waste form compatibility research. ICEM'07, 11th International conference on environmental Remediation and Radioactive Waste Management, Sept 2007, Bruges (Belgium). Proceedings on CD-Rom, published by ASME
[9]V. Pirlet, K. Lemmens, P Van Iseghem, "Influence of the near-field conditions on the mobile concentrations of Np and Tc leached out from vitrified HLW". Scientific Basis for Nuclear Waste Management XXVIII, Mat. Res. Soc. Symp. Proc. Vol 824, 2004, 385-390.

[10]P. Van Iseghem, K. Lemmens, V. Pirlet, The Leaching of Se, Sn, Zr, Pd from Vitrified High-Level Waste in Clay Slurries. Int. Workshop on Mobile fission and Activation products in nuclear waste disposal (La Baule, January 2007). Proceedings to be published.
[11]C. Pescatore, "The dependence of waste form leaching on migration parameters in the host medium". Radiochimica Acta Vol 66/67 (1994), 439-444
[12]N. Godon, Z. Andriambolonona, E. Vernaz, "Effect of a siliceous additive in a clay engineered barrier on aqueous corrosion of R7T7 nuclear waste glass". Mat. Res. Soc. Symp. Proc. Vol 257 (1992), 135-141
[13]P. Van Iseghem, M. Aertsens, S. Gin, D. Deneele, B. Grambow, D. Strachan, P. McGrail, G. Wicks, GLAMOR – a critical evaluation of the dissolution mechanisms of high level waste glasses in conditions of relevance for geological disposal. Final report to the EC, EUR 23097, 2007.
[14]B. Grambow, "A general rate equation for nuclear waste glass corrosion", Mat. Res. Soc. Symp. Proc. Vol. 44, 15-27 (1985)
[15]K. Lemmens and M. Aertsens, Validation of glass dissolution and Si diffusion parameters with a combined glass dissolution-diffusion experiment in Boom Clay. Scientific Basis for Nuclear Waste Management XXIX, Mat. Res. Soc. Symp. Proc. Vol 932, 2006, 329-336
[16]Valcke E., Smets S., Labat S., Peeters J., Boven P., Thomas P., Dupuis E., Van Iseghem P., Godon N., Jollivet P., Parisot G., Jockwer N. and Wieczorek K. Design and operation of an integrated in situ corrosion test on alpha-active HLW glass (CORALUS), Scientific Basis for Nuclear Waste Management XXIX, Mat. Res. Soc. Symp. Proc. Vol 932 (2006), 67-78
[17]Valcke E., Van Iseghem P., Gysemans M., Van Bree P., Moors H., Godon N. and Jollivet P. Leaching and migration of Np, Pu and Am from α-doped SON68 HLW glass in contact with dense clay, Scientific Basis for Nuclear Waste Management XXIX, Mat. Res. Soc. Symp. Proc. Vol 932 (2006), 999-1006
[18]P. Van Iseghem, K. Lemmens, V. Pirlet, "Assessment of the compatibility of vitrified high-level waste with geological disposal in a Boom Clay formation: remaining uncertainties". In Radioactive Waste Products – RADWAP 2002, Proceedings of an International Seminar, Eds R. Odoj, J. Baier, P. Brennecke and K. Kühn, 229-235 (2003).
[19] J Marivoet, E Weetjens, "The importance of mobile fission products for long-term safety in the case of disposal of vitrified high-level waste and spent fuel in a clay formation". Int. Workshop on Mobile fission and Activation products in nuclear waste disposal (La Baule, January 2007). Proceedings to be published.
[20] B. Baeyens, A. Maes, A. Cremers, P.N. Henrion, "*In situ* physico-chemical characterization of Boom Clay", Radioactive Waste Management and the Nuclear Fuel Cycle, Vol 6 (3-4), 391-408 (1985)
[21] A. Lajudie, J. Raynal ,J.C. Petit, P. Toulhoat, "Clay-based materials for engineered barriers: a Review", Scientific Basis for Nuclear Waste Management XVIII, Mat. Res. Soc. Symp. Proc.Vol 353, Part I, 221-230 (1995)
[22] J.Bel, S Wickham, R Gens, "Development of the *Supercontainer* Design for Deep Geological Disposal of High-Level Heat Emitting Radioactive Waste in Belgium", Scientific Basis for Nuclear Waste Management XXIX, Mat. Res. Symp Proc. Vol. 932, pp.23-32 (2006)

GLAMOR – OR HOW WE ACHIEVED A COMMON UNDERSTANDING ON THE DECREASE OF GLASS DISSOLUTION KINETICS

P. Van Iseghem[1], M. Aertsens[1], S. Gin[2], D. Deneele[3], B. Grambow[4], D. Strachan[5], P. McGrail[5], G. Wicks[6]
[1] SCK•CEN, Mol (Belgium)
[2] CEA, Bagnols-sur-Cèze (France)
[3] LCPC, Bouguenais (France)
[4] SUBATECH, Nantes (France)
[5] Pacific Northwest National Laboratory, Richland, WA (USA)
[6] Savannah River National Laboratory, Aiken, SC (USA)

ABSTRACT

The objective of the EC funded GLAMOR project was to achieve a common understanding of the processes that control the decrease of the dissolution rate of high-level waste glass in water when silica becomes saturated. Is affinity controll or a protective layer dominating? The following steps were taken: (1) review of the literature, (2) selection of an experimental dataset, and selection of the models r(t) and GM2003, and (3) application by the GLAMOR partners of the models to the datasets. The main focus has been on dissolution tests in pure water at different values of surface-to-volume and pH. Some of the main conclusions were: (1) both affinity and protective layer concepts must be considered in the interpretation of the rate decreasing stage, (2) the residual dissolution rate observed beyond the silica saturation stage is far more important in view of the long-term performance of the glass, and deserves more attention in future R&D. In the GLAMOR final report, we also discussed in detail the modeling parameters such as the silica saturation concentration, the silica diffusion coefficient, the silica retention factor in the reaction layer, and the water diffusion coefficient.

INTRODUCTION

From the early 1980s, large experimental programmes have been performed worldwide, to investigate the dissolution behaviour of candidate high-level waste glasses in pure water or candidate repository waters. We refer to some excellent review documents [1-3]. It was soon clear that the dissolution rate of the glasses decreases with progressing duration. Two kinds of interpretation of the rate decrease were proposed then, and later on these were further elaborated (see below). Noteworthy is one of these review documents in which it was concluded that the glass dissolution would stop completely at a certain stage [1].

During the past decade a consensus has grown that the dissolution rate of vitrified high-level waste decreases with prolonged (months to years) contact with geological disposal media. Very low values of the dissolution rate are achieved under solution saturation conditions, as expected under most geological environments [4-8]. This decrease results in significantly increased lifetimes of the vitrified waste, and contributes to a higher safety of the geological disposal in the long-term. However, the understanding of the long-term dissolution rate and its precise value are poorly documented or known. These findings and conclusions were summarized in the Proceedings of a dedicated international Workshop [9].

The Concerted Action, GLAMOR*, was conceived to achieve a common interpretation of the decrease in dissolution rate of nuclear waste glasses observed in most experimental programmes. Two explanations for the decrease in dissolution rate have been considered. The decrease could arise as the solution concentration of H_4SiO_4 approaches a solubility limit thereby decreasing the driving force for the dissolution. In the second explanation, the layers that develop on the surface of the glass act as transport barriers. Both interpretations refer to concepts that were proposed as early as the 80's [10,11].

To achieve a common interpretation, an international group of experts was brought together to select from the many available data a set that could be used to reach a common understanding of the mechanism responsible for the rate drop. Selected analytical models were then planned to be applied to these data by the different participating laboratories. Therefore, from GLAMOR no new experimental data or models would be generated. Instead, a joint interpretation of the same experimental data would be made with the same models. In particular, GLAMOR was expected to elucidate in more detail the relative importance of the processes that control the decrease in dissolution rate: saturation effects in the solution or the diffusion limitation due to the surface "alteration" layer formed on the glass surface. Another goal of GLAMOR was to gain a better understanding of the uncertainties of the various parameter values used in the analytical models, and of the models themselves therefore.

Within GLAMOR we limited the discussion to glass alteration data in solutions, *i.e.* in absence of near-field or host rock solids.

GLAMOR consists of the following actions:

1. Overview of the status of the R&D on the dissolution mechanisms of HLW glasses.
2. Selection of experimental data to be used as input to the analytical models and selection of these analytical models.
3. Application of the models to the data in a consistent manner.

GLAMOR ran for 3 years between 2001 and 2004. A full report of the results of the project can be found in [12].

BACKGROUND

The decrease in the dissolution rate of HLW glass in confined systems is currently explained in two ways:

- the affinity concept, in which the thermodynamic driving force for glass dissolution decreases as the saturation of silica in solution is approached with respect to an alkali depleted glass surface region [10].
- the protective layer concept in which the formation of a dense surface layer at the interface between glass and solution is responsible for the decrease in glass dissolution rate [13,14].

Both concepts are currently coupled to describe the corrosion of complex glasses, since neither process alone can explain complex glass dissolution. The approach to low affinity as the solution approaches saturation conditions may be retarded by silica mass transfer in the surface layers. The protective layer argument is mathematically related to a saturation concentration at the interface between the unreacted glass and the surface layer. The scientific debate on the glass dissolution processes controlling the long-term dissolution has given different weights to both concepts for about two decades. Within GLAMOR we performed an analysis of the pros and cons of both processes. The coupling of only affinity and protective layer concepts appears to be insufficient to describe long-term dissolution of glass beyond the time at which saturation with respect to silica occurs. Other key processes such as water diffusion, ion exchange (inter-diffusion) and precipitation of secondary minerals (i.e., clays, zeolites, *etc*) may be important in the long-term and may occur in parallel to glass matrix corrosion. However, their impact on the long-term dissolution was poorly understood thus far.

The starting point of the GLAMOR project was the general scheme of glass dissolution in confined solution, as proposed by CEA (Figure 1) [15].

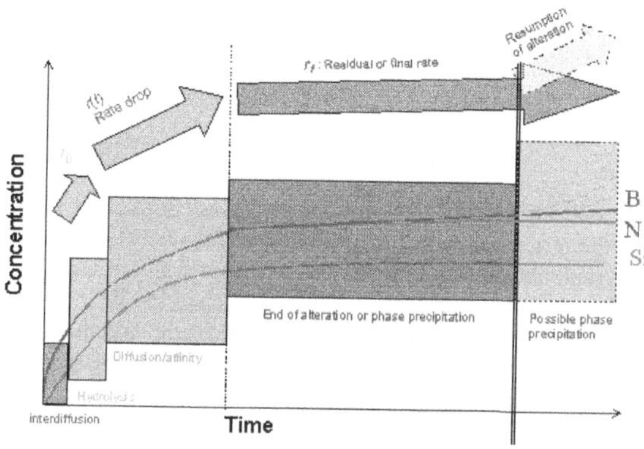

Figure 1: Schematic representation of the predominant mechanisms and the resulting kinetics as they affect the concentration in solution of the glass constituents Si, B, and Na.

MAIN RESULTS

State-of-the-art on glass dissolution based on a literature review

We focused our attention on experiments showing dissolution rate decrease, silica saturation and subsequent phenomena. Such conditions are typically achieved at high reaction progress** values, approximately $10^3 - 10^4$ g·m^{-3} with powdered glass. These results at high S/V values are known to enhance the accumulation of dissolved species in solution [4]. These studies were performed over the past ~20 years in various laboratories around the world. They were conceived for diverse reasons: to enlarge the database of glass dissolution (i.e. product qualification), to investigate the dissolution behaviour in the long-term, or to increase the understanding of the dissolution mechanism. Most of the datasets did not include results from solutions loaded with solids, such as container corrosion products or clay. The review showed that:

- The rate decreasing stage before silica saturation is now currently interpreted as a combined effect of affinity-limited dissolution and kinetically-reduced dissolution from the alteration layer barrier [16,14].
- There is an agreement that beyond the silica saturation stage dissolution continues, but there is no unanimous interpretation of the results [12]. Sometimes a dissolution rate decrease proportional with $\sqrt{t}$ is considered, sometimes the rate is interpreted as constant with time (Figure 2). These "final" or "residual" dissolution rates may be ~5000 times lower than the initial dissolution rate in case of the French R7T7 glass SON68*** (~10^{-4} g.m^{-2}.d^{-1}) [12]. Different hypotheses are proposed for the final dissolution rate (e.g. [13, 23,25]), such as water diffusion in glass, precipitation of secondary phases, and network hydrolysis, but there is insufficient evidence at present to conclude on "the" mechanism.

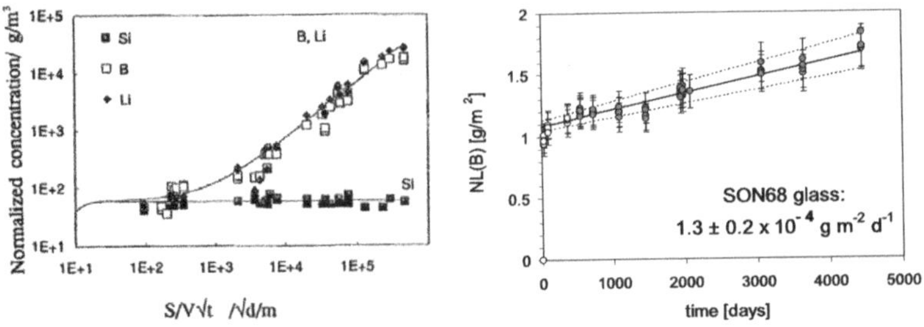

Figure 2 : Results from long-term dissolution tests on glass SON68 at 90°C in pure solutions with glass powder: the left graph shows normalized losses obtained in brine at different S/V values [4], the right graph shows normalized losses in pure water at a single S/V condition [5].

An important conclusion, however, is the identification and interpretation of this final dissolution rate as more important than the elucidation of the relative impact of the processes leading to the decrease of the dissolution rate before silica saturation, because that strongly affects the lifetime of the HLW glass package. In addition, the proponents of the affinity and kinetics controlled decrease of the dissolution rate before silica saturation mainly disagree about which of these processes dominates, not about the fact that both contribute [14,16].

Selection of experimental data and models

The GM2003 [16] and r(t) [17] analytical models were selected. Both models consider a first order rate law and Si diffusion in the "gel" (i.e. main alteration layer), although in r(t) diffusion through the alteration layer is considered more important than in GM2003. The mathematics of both models is quite similar, although the significance given to the various parameters (modeling strategy) is different. Table I summarizes the fitting parameters of both models. The forward and the final dissolution rate are parameters in both models, but they are not considered as fitting parameters. Environmental parameters to consider are the solution renewal rate, S/V, temperature, p_{CO_2}, and pH. From a mechanistic point of view, the main difference between the two models is that GM2003 also considers diffusion/sorption of H_2O, B, and Li in a diffusion layer next to the unreacted glass. The r(t) model considers as well the pH evolution of the solution, a parameter that is constant in the version of GM2003 used in this comparison. The fact that both models contain similar mathematics makes it possible to use both models with the same set of parameters to check whether they lead to the same results. If the pH is kept constant, one can use the GM model with the choice $D_{H2O} = 0$ to reproduce the r(t) model.

The datasets selected for GLAMOR covered S/V of 50 to 200 000 m^{-1}, solution free or fixed pH values from 2.5 to 11.5, glass composition (SON68, a high Al_2O_3 glass, and simple glass), solution composition (high purity water as reference; various chemicals added), temperature (90°C, 50°C, and 150°C), presence/absence of solids (clay), flow rate (static as reference; 14.4 mL/d). Up to 50 experimental datasets were selected.

Table I. Different fitting parameters in the two glass dissolution models used in the GLAMOR project

r(t) model	*GM2003 model*
α (Si retention factor in the gel: the fraction of silica retained in the gel is $1 - \exp(-\alpha C)$ with C the silica concentration in solution)	α (Si retention factor in the gel: the fraction of silica retained in the gel is $1 - \exp(-\alpha C)$ with C the silica concentration in solution) or fpc (Si retention factor in the gel: a constant fraction fpc of silica is retained in the gel)
D_g (effective Si diffusion coefficient through the gel)	D_{Si} (T) (pore Si diffusion coefficient through the gel)
C* (Si concentration at the alteration front at which the glass-to-gel transformation ceases)	C* (T, pH) (Si saturation concentration in solution with respect to an alkali depleted surface region, T, pH)
	D_{H2O} (water diffusion coefficient in the glass and/or diffusion layer)
	Kd_B, Kd_{Li} (apparent sorption constants for boron and lithium to account for the slightly higher mobility of water molecules as compared to B and Li in the diffusion layer)
r_{fin} (residual dissolution rate)	r_{fin} (residual dissolution rate)

Results of the joint modeling of the experimental results

The EU project partners applied both models r(t) and GM2003 to all of the datasets [12]. The discussion was divided into the effect of the experimental parameters S/V, solution pH, saturation of the solution, and glass composition. In this paper, we used the S/V parameter as an example, since it is the most informative. Here, most effort was spent on the tests performed at 90°C in deionized water with SON68 glass. We only modeled the Si and B concentrations, as Si is the most important element in HLW glass dissolution and B is a soluble element from which the dissolution of the glass matrix can be inferred.

We first discuss an experiment at a moderate S/V of 50 m^{-1}. This is approximately the same as the 10 m^{-1} used in the ASTM standard leach test [18]. The results display a quite "simple" dissolution sequence (Figures 3 and 4): initial dissolution, rate decrease with the approach to silica saturation, and a continued "residual" dissolution afterwards. Figure 3 shows the results of modeling with r(t). The left graph shows the fit to the first two stages; the right graphs the later stage. Figure 4 shows the best fit with both r(t) and GM2003 models from the result of the test. The parameter values used in the different plots are shown in Table II.

From these modelling results, the following observations can be made:

- The parameter values for the two models are nearly equal, except for the use of D_{H20} and Kd_B in GM2003 (Table II).
- The decrease in the dissolution rate before Si saturation is modeled by optimizing the D_{Si} and C* values [12] (Figure 3). The best fit is shown.
- The residual rate is needed in both models – compare Figures 3 and 4 (note: the value for r_{fin} used here is much higher than the value reported above (see section "State-of-the-Art on Glass Dissolution"), because the high reaction of ~10^{-4} g/(m^2·d) progress was not reached in this test. We also observe that GM2003 cannot account for the continued Si release beyond Si saturation.
- This example shows that the key parameters used in the models are C* and D_{Si} (to calculate Si release), α (to relate B with Si) and the final rate (r_{fin}). Some of the modeling parameters are correlated with each other: D_{Si} with α, and D_{H20} with K_{dB} [12].

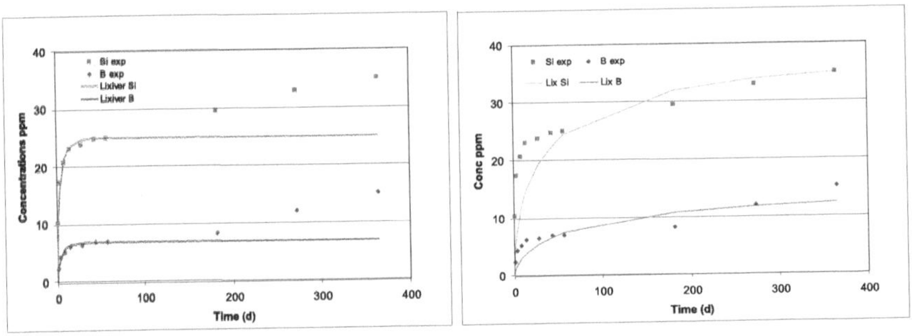

Figure 3: Model fits to results from tests performed at 50 m^{-1} and 90°C in deionized water. The graph on the left shows the fit to the dissolution rate decrease and on the right the final rate. The parameter values are listed in Table II.

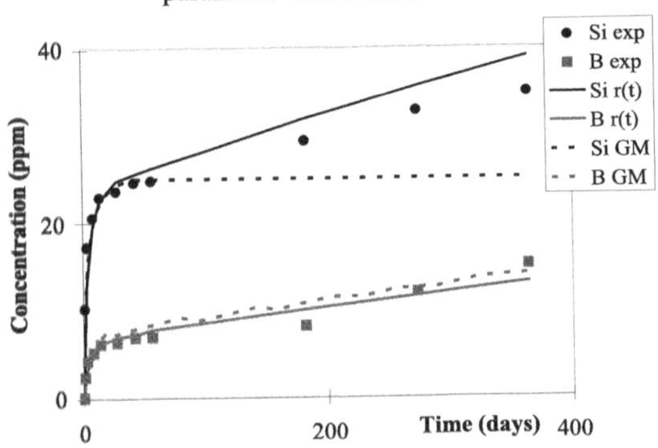

Figure 4: Fit obtained with the r(t) and GM2003 models for the same test as in Figure 3, but now fitting the entire test data. The parameter values are listed in Table II.

Table II Summary of r(t) and GM2003 parameters used in the plots in Figures 3 and 4.

	FIG 3, left r(t) (CEA)	**FIG 3, right r(t) (CEA)**	**FIG 4, (r(t) SCK•CEN**	**FIG 4 GM 2003 SCK•CEN**
Forward rate ($g\ m^{-2}\ day^{-1}$)	r_o= *f*(pH) ~0.95	r_o= *f*(pH) ~1.5	r_o= *f*(pH) ~1.33	0.7
Silica saturation ($mg.L^{-1}$)	C^*=16 , C^*_{tot}~25	C^*=35 , C^*_{tot}~66	C^*= 16 , C^*_{tot}~25	K_{SiO2} =25.05
Si-diffusion coefficient ($m^2\ s^{-1}$)	D_g=5.0 10^{-14}	D_g=7.0 10^{-16}	D_g=5.0 10^{-14}	D_{si}=2.6 10^{-13}
Si-retention α (ppm^{-1})	0.023	0.031	0.022	0.022
Final rate, r_{fin} ($g\ m^{-2}\ day^{-1}$)	1.0 10^{-8}	1.0 10^{-8}	8 10^{-3}	8.0 10^{-3}
Water diffusion coefficient D_{H2O} ($m^2\ s^{-1}$)	Not considered	Not considered	Not considered	5.0 10^{-21}
Kd_B ($kg\ m^{-3}$)	Not considered	Not considered	Not considered	100

The second dataset discussed in this paper is from an experiment at very high S/V, performed at the same conditions but where B and Na concentrations in solution reached values as high as 3000 mg/L and 4000 mg/L, respectively. Here we explore the limits of the models, because the experimental data suggest resumption of glass dissolution after rate decrease, and total Si concentrations as high as ~350 mg/L are measured for a solution pH of ~9.4. At moderate S/V values as in the first example Si concentrations remain in the range between 20-40 mg/L. Table III lists the parameter values obtained from the different model fits. Figures 5 and 6 show the corresponding model fits.

We observe the following:

- Both r(t) and GM2003 models are able to reproduce the decrease in rate (see B-data up to one month duration in Figures 5 and 6, except those with r(t) in Figure 6), but the models cannot reproduce the resumption in dissolution that is observed for B after some 400 days (both Figures 5 and 6). On the other hand, the data obtained for B with r(t) in Figure 6 correspond with an optimisation of the curve fit over the entire test duration. It is obvious that this does not account for the mechanistic interpretation of the test, e.g. initial dissolution, rate decrease, resumption of the dissolution, as shown in Figure 1. Note, this interpretation of the experimental data is supported by much evidence [24]. This example clearly shows the limits of both approaches. If one needs to elaborate a conceptual model of experimental data, it is of course necessary to account for the dissolution mechanisms, and, therefore, we should discard the curve fitting approach in our discussion.
- The Si data can be fit satisfactorily by selecting a C^*_{tot}**** value of ~300 mg/L (Figure 6), which is much higher than the theoretical solubility of silica with respect to amorphous silica of ~100 mg/L at pH ~9.4 [19]. This classical interpretation of solubility should be questioned, however. Recent calculations showed that in this high S/V test the Si_{tot} increase is from $NaHSiO_3$ complexes [20] while the H_4SiO_4 concentration remains close to the value for equilibrium with amorphous SiO_2. This evolving C* could in this test be associated with the dissolution resumption (see B data in Figures 5-6). The main difference between the fit of the Si data with GM2003 in Figure 6 and both plots in Figures 5 is the choice of C* and D_{Si}..

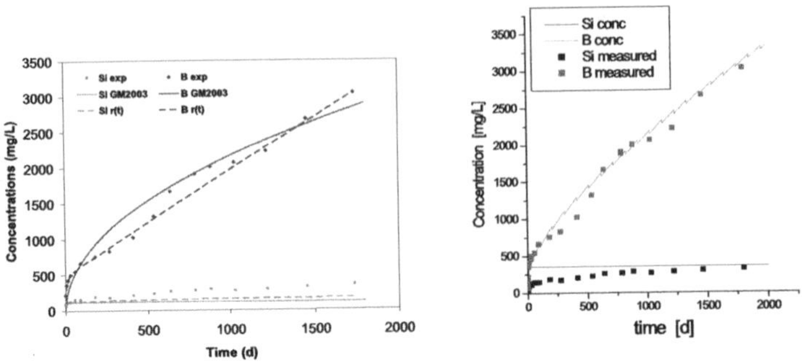

Figure 5: Fits obtained with both models for results from the test at high S/V (200 000 m^{-1}) in deionized water at 90°C at CEA (left graph) and SUBATECH (right graph). The parameter values are listed in Table III.

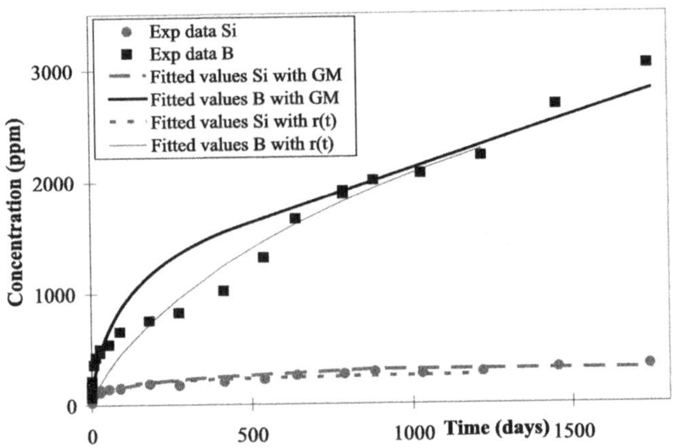

Figure 6: Two more fits of the same experimental dataset as in Figure 5 are shown: a best fit with GM2003, based on the best choice of parameter values, and a best fit with r(t), with optimized curve fitting. The parameter values are listed in Table III.

Table III: Summary of parameters used with r(t) and GM2003 models in Figures 5-6
C^*_{tot} is the total silica concentration in solution at saturation.

	r(t) (CEA) Figure 5, left	**GM2003 (CEA) Figure 5, left**	**GM 2003 (SUBATECH) Figure 5, right**	**r(t) (CEA) (curve fitting) Figure 6**	**GM 2003 (SCK•CEN) Figure 6**
Forward rate ($g.m^{-2}.d^{-1}$)	r_o=f(pH) ~3.2	0.7	0.7	1.5	0.69
Silica saturation concentration ($mg.L^{-1}$)	C^*=32 C^*_{tot}=115	K_{SiO2} =115	K_{SiO2}=350	C*=284	K_{SiO2}= 307
Si-diffusion coefficient ($m^2\ s^{-1}$)	D_g =1.0 10^{-19}	D_{Si}= 5.2 10^{-14}	D_{Si}= 5.2 10^{-14}	Dg=2.5 10^{-21}	D_{Si}= 5.2 10^{-20}
Si-retention α (ppm^{-1})	0.038	0.004	0.025	0.020	fx=0.97
Final rate, r_{fin} ($g\ m^{-2}\ day^{-1}$)	1.5 10^{-4}	7.0 10^{-11}	8 10^{-5}	0	1.0 10^{-4}
Water diffusion coefficient D_{H2O} ($m^2\ s^{-1}$)	Not considered	5.0 10^{-23}	2.5 10^{-23}	Not considered	7 10^{-20}
K_{dB} ($kg\ m^{-3}$)	Not considered	70	100	Not considered	10^{+5}
Ratio D_{H2O}/K_{dB}	Not considered	7 10^{-25}	2 10^{-25}	Not considered	7 10^{-25}

We can further comment the modeling attempts with both datasets. Whereas the two models consider the same processes – with GM2003 considering water diffusion in addition – the fit strategy used by the model developers is different. The strategy used in r(t) is to choose first the appropriate value for C*, next choose the right value for D_{Si} to fit the rate decrease, finally choose the right value for α to fit the B data relative to the Si data. With GM2003 however, much less credit is given to D_{Si} , and therefore a constant, high value for D_{Si} is choosen. As for the dissolution beyond the Si saturation, the stategy can focus on r_{fin} or D_{H2O} as controlling parameter.

These two strategies worked quite well at moderate conditions of S/V (first test), but much less in the very large reaction progress situation achieved in the second test. This problem is obvious in case of the C* (both models) and D_{Si} (GM2003 model). A better fit can be obtained by choosing other values for these main parameters (see Figure 6, GM2003 plot).

This discussion shows very clearly that a better knowledge ànd understanding of the modeling parameters is needed, in order to reduce the uncertainties in modeling. This is typically true for D_{Si} , where the database of experimental data is limited to the range larger than 10^{-14} m^2/s [12].

The datasets discussed here did not allow us to evaluate the role of the D_{H2O} parameter. The GLAMOR team agreed that water diffusion is occurring, and yet this parameter is well documented [21,22]. Both D_{H2O} and r_{fin} did impact the rate decrease in the silica saturation stage and during the residual dissolution. Their relative importance will determine the time dependence in that stage.

CONCLUSION

We have shown that both models can produce reasonable fits to the rate decreasing stage for various types of tests, two examples have been shown in the above discussion. This is also true for the larger database (different values of S/V and pH) [12]. The modeling was done with a number of known or proposed parameter values and the fit strategy of the two models. However, beyond the saturation stage, which is currently assumed to occur beyond the rate decreasing stage (Figures 3 and 4), the agreement between the modeling predictions and experimental data diverge. The main deviations are (i) that with the GM2003 model Si concentration after saturation does not increase any longer, contrary to the experimental observations that the total Si saturation concentration can change, and (ii) resumption of dissolution beyond the saturation stage. As discussed above, the first deviation may be overcome by properly choosing C*. This also leaves the question open if additional changes in C* occur at still higher reaction progress than we achieved so far in the experiments. The second deviation (resumption of dissolution) could be handled by the models if this resumption would fit with a single final dissolution rate beyond silica saturation. This is hardly possible at very high S/V (Figures 5 and 6).

In addition, even if predictions with both models might be satisfactory for a specific experimental dataset, the agreement between both models for longer term data might become worse. This is because of the different mathematics. With r(t) a constant residual rate is used, whereas in GM the long term prediction depends on the relative weights given to water diffusion and residual rate. In the end, the residual rate controls the long-term release in the GM model, as well.

The methodology chosen in GLAMOR to have various external experts work with both models and datasets was very fruitful. It enabled us to achieve a common understanding about some issues that have been the subject of long-time debate, such as:

- It is agreed that both affinity and protective layers concepts are valid, and both must be taken into account. However, the total silica saturation concentration (C*) is not necessarily constant with time as it may change with the reaction progress. And, whereas it is clear that dissolved Si diffusion through a surface layer occurs, it is less clear under which conditions this process is rate limiting. The role of D_{Si} is also poorly documented [12].
- There is a consensus that a residual ("final" but not necessarily constant with time) dissolution rate exists beyond the rate decreasing stage, at least in pure solutions and the closed systems considered in GLAMOR. This residual rate significantly controls the long-term release from the glass, and it might become an input parameter in the safety assessment calculations for a geological disposal site. Various processes have been proposed to account for this residual rate: water diffusion, ion-ion-exchange reactions, network hydrolysis, precipitation of secondary phases. More efforts should be devoted to document ànd understand the processes behind this residual rate under standard conditions (pure water), but as well in repository waters loaded with near- or far-field solids.

The way models (analytical or other) are applied also must be done with caution. When the interpretation of the experimental data and further extrapolation to longer timescales is to be based on the understanding of the dissolution processes (as is done here), the models should be used in a conceptual way. This should include a consistent documentation of the main model parameters that result in reducing the uncertainties of the model calculations. The "optimized curve fitting" methodology is inappropriate if one wishes to perform conceptual modeling.

ACKNOWLEDGEMENTS

The GLAMOR project was sponsored through the EC contract FIKW-CT-2001-00140. The authors wish in particular to thank Mr Thomas McMenamin of the EC, who strongly supported this project.

FOOTNOTES

* GLAMOR = A Critical Evaluation of the Dissolution **M**echanisms of High Level Nuclear Waste **Gla**sses in conditions **o**f **R**elevance for Geological Disposal

** Reaction progress is the amount of glass dissolved per volume of solution (in $g \cdot m^{-3}$)

*** SON68 (full code: SON 68 18 17 L1C2A2Z1) is a reference glass with a composition simulating the nominal composition of the COGEMA R7T7 glass.

**** C* corresponds with the dissolved silica concentration ($Si(OH)_4$), C^*_{tot} with the sum of the concentration of all silica species in solution.

REFERENCES

[1]Mendel, J.E. 1984 "Final Report of the Defense High-Level Waste Leaching Mechanisms Program," PNL-5157, Pacific Northwest National Laboratory, Richland, WA.

[2]Lutze W. 1988 "Silicate glasses," in Radioactive Waste Forms for the Future. ed. W. Lutze and R. Ewing, North Holland, Amsterdam

[3]Cunnane J.C Editor, 1994 "High-Level Waste Borosilicate Glass. A Compendium of Corrosion Characteristics". US-DOE-EM-0177 (3 volumes). See also: Ebert W. L.and Mazer, J.J. Mat. Res. Soc. Symp. Proc. Vol 333, 1994, 27-40.

[4]Grambow, B., Lutze, W., Müller, R., 1992, "Empirical Dissolution Rate Law for the Glass R7T7 contacting Halite- and Silica-saturated Brines". Mat. Res. Soc. Symp. Proc. Vol 257, 143-150

[5]Curti, E. 2003, "Glass dissolution parameters: Update for "Entsorgungsnachweis"". PSI Bericht Nr 03-18

[6]Gin, S. and Frugier, P. 2003, "SON68 Glass Dissolution Kinetics at high Reaction Progress: Experimental Evidence of the Residual Rate". Mat. Res. Soc. Symp. Proc. Vol. 757, 175-182

[7]Ebert, W.L., and Mazer, J.J. 1994, "Laboratory Testing of Waste Glass aqueous Corrosion; Effects of Experimental Parameters". Mat. Res. Symp. Proc. Vol 333, 1994, 27-40

[8]Lemmens, K., Aertsens, M., Pirlet, V., Maes, N., Moors, H., Van Iseghem, P.,2003, "Measurement of Glass Corrosion in Boom Clay Disposal Conditions : First Results of the Experimental Programme 2000-2003 of SCK•CEN". ICEM'03 – 9th international conference on environmental remediation and radioactive waste management. Proceedings on CD-Rom, paper Icem03-4774.

[9]Glass in its Disposal Environment. Journal of Nuclear Materials Vol 298 (2001), Nos 1,2.

[10]Grambow, B. "A general rate equation for nuclear waste glass corrosion", Mat. Res. Soc. Symp. Proc. Vol. 44, 15-27 (1985).

[11]Barkatt, A. B., Macedo, P.B., Gibson, B.C., Monrose, C.J., 1985, "Modeling of Waste Form Performance and System Release" Mat. Res. Soc. Symp. Proc. Vol 44, 3-13

[12] Van Iseghem, P., Aertsens, M., Gin, S., Deneele, D., Grambow, B., Strachan, D., McGrail, P., Wicks, G., 2007, GLAMOR – A Critical Evaluation of the Dissolution Mechanisms of High Level Waste Glasses in Conditions of Relevance for Geological Disposal. Final report to the EC, EUR 23097.

[13]Gin, S., 2000, "Protective Effect of the Alteration Gel : A key Mechanism in the Long Term Behavior of Nuclear Waste Glass", Mat. Res. Soc. Symp. Proc. Vol 663, 207-215

[14]Vernaz, E., Gin, S., Jégou, C., Ribet, I., 2001, "Present Understanding of R7T7 Glass Alteration Kinetics and their Impact on Long-Term Behaviour Modeling", Journal of Nuclear Materials 298, 27-36

[15]Godon N., *et al.*, 2002, "Reference Document: Long Term Behaviour of Nuclear Waste Glasses". CEA Rapport Technique DIEC/2002/02

[16]Grambow, B.and Müller, R., 2001, "First-order Dissolution Rate Law and the Role of Surface Layers in Glass Performance Assessment". *J. of Nuclear Materials*, 298, 112-124.

[17]Ribet, I.,Gin, S., Minet, Y., Vernaz, E., Chaix, P., Do Quang R., 2001 "Long-Term Behaviour of Nuclear Glass: The r(t) Operational Model". GLOBAL-International conference on: "*Back-End of the fuel cycle : from research to solution*",9th-13th September, Paris.

[18]"Standard Test Method for Static Leaching of Monolithic Waste Forms for Disposal of Radioactive Waste". ASTM C 1220 – 98, 2004

[19]Stumm W., and Morgan, J.J., "Aquatic Chemistry, an Introduction emphasizing Chemical Equilibria in Natural Waters". Second edition. Publ. Wiley-Interscience. John Wiley & Sons.

[20]Frugier, P., Gin, S., Lartigue, J.E., Deloule, E., 2006, "SON68 Glass dissolution Kinetics at high Reaction Progress: Mechanisms accounting for the Residual Alteration Rate". Mat. Res. Soc. Symp. Proc. Vol 932, 305-312

[21]Ferrand, K., Abdelouas, A., Grambow, B., Crovisier, J.L., 2004, "The Role of Water Diffusion in the Corrosion of the French Nuclear Waste Glass SON68 under Solution Saturation Conditions". Mat. Res. Soc; Symp. Proc. Vol. 807, 193-198

[22]Yanagisawa N., Fujimoto K., Nakashima S., Kurata Y., Sanada N., 1997, "Micro FT-IR Study of the Hydration Layer during Dissolution of Silica Glass", Geochimica et Cosmochimica Acta, Vol 61, n° 6, pp 1165-1170

[23]Grambow, B., Lutze, W., Müller, R. 1992 "Empirical Dissolution Rate Law for the Glass R7T7 contacting Halite- and Silica- saturated Brines", Mat. Res. Soc. Symp. Proc. Vol 257, 39-46

[24]Vernaz, E., Gin, S., 2001, "Apparent Solubility Limit of Nuclear Glass", Mat. Res. Soc. Symp. Proc. Vol 663, 217-225

[25]McGrail, B.P. *et al*, 2001, "Near-field Performance Assessment for a Low-Activity Waste Glass Disposal System: Laboratory Testing to Modeling Results". Journal of Nuclear Materials Vol 298, 95-111.

CHARACTERIZATION OF VITRIFIED SAVANNAH RIVER SITE SB4 WASTE SURROGATE PRODUCED IN COLD CRUCIBLE INDUCTION MELTER

S.V. Stefanovsky
SIA Radon, Moscow, Russia

J.C. Marra
Savannah River National Laboratory, Aiken, SC, USA

A.A. Akatov
Institute of Technology, St-Petersburg, Russia

ABSTRACT

Savannah River Site (SRS) sludge batch 4 (SB4) waste surrogate with high aluminum and iron content was vitrified using commercially available Frit 503-R4 (8 wt % Li_2O, 16 wt % B_2O_3, 76 wt % SiO_2) by cold crucible inductive melting using lab- (56 mm inner diameter), bench- (236 mm) and large-scale (418 mm) cold crucible induction melters. The waste loading ranged between 40 and 60 wt %. The vitrified products obtained in the lab-scale cold crucible were primarily amorphous, had traces of unreacted quartz in the product at a 40 wt % waste loading and contained traces of a spinel phase in the product at a 50 wt % waste loading. The glassy products obtained from the bench-scale cold crucible were primarily vitreous with minor iron-rich spinel-type phases whose content at ~60 wt % waste loading was ~10 vol %. The vitrified waste obtained in the large-scale cold crucible was also composed of major vitreous and minor spinel structure phases. No nepheline phase was found. The average degree of crystallinity was estimated to be ~12 vol %. Infrared spectroscopy studies showed that the anionic motif of the network in these glasses was built from short metasilicate chains and boron-oxygen constituents based on boron-oxygen triangular units.

INTRODUCTION

Currently borosilicate glass is considered a high level waste (HLW) form in most countries developing nuclear technologies. In the United States, borosilicate glass is used as the HLW form at Savannah River Site (SRS) Defense Waste Processing Facility (DWPF) [1]. At the DWPF, the HLW vitrification process is performed in a Joule Heated Ceramic Melter (JHCM). A cold crucible induction melter, initially developed for production of fusible and high-purity single crystals, polycrystalline ceramics and glasses, is considered as an alternative to the JHCM. Major advantages of the cold crucible over the JHCM are higher productivity, higher temperature, longer lifetime, smaller dimensions, and attaining higher waste loadings while maintaining the same product quality. The active hydrodynamic regime in the cold crucible also prevents the segregation of crystalline phases normally formed under stationary conditions in a resistive heated furnace.

Vitrification of high-ferrous sludge batch 2 (SB2) HLW surrogate yielded a high-quality glass that contained a minor amount of spinel-type crystalline phases. The presence of the spinel-type crystals (highly substituted) did not negatively influence the chemical durability of the material [2]. In the present work, the phase composition, structure and chemical durability of vitrified SB4 HLW surrogate with high iron and aluminum content were investigated in detail.

EXPERIMENTAL

To identify a suitable frit for follow-on CCIM testing, crucible-scale testing was conducted at the Savannah River National Laboratory (SRNL). Glass samples were fabricated using the SB4 waste composition (Table I) and combinations of glass forming additives. The crystallinity of the glasses was determined using a Bruker D8 Advance X-Ray Diffraction (XRD) System. The chemical composition of the glasses was measured by dissolving the glasses and analyzing the digested glass using a Varian Vista AX Inductively Coupled Plasma – Atomic Emission Spectrometer (ICP-AES). The relative durability of glasses from the crucible-scale tests was determined using the Product Consistency Test – Method A (PCT-A) [3].

The vitrification tests of the SB4 waste surrogate were performed in three separate melting units:

- a 56 mm inner diameter, copper cold crucible lab-scale unit with a melt surface area of 24.6 cm^2 energized from a 5.28 MHz/10 kW generator [4];
- a 236 mm inner diameter, bench-scale cold crucible unit manufactured from stainless steel pipes and energized by a 1.76 MHz/60 kW generator [5]; and
- the Radon full-scale vitrification plant with a 418 mm inner diameter stainless steel cold crucible energized by a 1.76 MHz/160 kW generator [6].

The external surfaces of each of the crucibles were coated with a ground fireclay-based putty. Additional information about the vitrification runs are described in further detail in the cited references [4-6].

The products from the melter tests were analyzed by X-ray fluorescence (XRF) spectroscopy using a PW-2400 Philips Analytical BV unit equipped with quantitative analytical Philips SuperQuantitative & IQ-2001 Software and Inductively Coupled Plasma – Atomic Emission Spectroscopy (ICP-AES) at SRNL. The vitrified products were examined by X-ray diffraction using a DRON-4 diffractometer, scanning electron microscopy with energy dispersive spectroscopy (SEM/EDS) using a JSM-5300 + Link ISIS unit (metals, oxides and fluorides were used as standards). Infra-red spectra were recorded using an IKS-29 spectrophotometer (compaction of powdered glass in pellets with KBr). Chemical durability of the glassy products was determined using the PCT-A procedure.

RESULTS AND DISCUSSION

Crucible-scale Glasses

Frit 503-R4 which has a chemical composition of 8 Li_2O, 16 B_2O_3, 76 SiO_2 (in wt %) was found to be a suitable frit based on crucible-scale testing at the Savannah River National Laboratory (SRNL). When combined with the SB4 waste, a nepheline-free glass containing minor spinel phases was formed [7]. Lab-scale glass melting in a resistance heated furnace showed that uniform glasses could be produced at 1300° C. At lower temperatures, the melt was too viscous to pour and the resulting glass was inhomogeneous.

The normalized release values for B, Li, Na and Si as determined by PCT for the crucible-scale glasses are shown in Table II. Data are also included for glass samples subjected to the Canister Centerline Cooling Profile (CCC) [8]. The glasses at 45 and 50 wt % waste loading were quite chemically durable but the glasses at 55 wt % waste loading showed a slight decrease in durability under CCC conditions (Table II).

Table I. Target and Actual Chemical Compositions (wt %) of the SB4 Waste, Vitrified Waste Products Sampled from Canisters and the "Dead Volume" (DV) of the Cold Crucible

Oxides	SB4 waste	Target at WL			Bench-scale at WL			Full-scale, canisters #					
		40	50	60	40[1]	50[1]	60[2]	6,[2]	12,[2]	16,[2]	16,[1]	20,[2]	DV[2]
Li_2O	-	4.80	4.00	3.20	4.40	3.16	3.00	4.00	4.00	4.00	3.53	4.00	4.00
B_2O_3	-	9.60	8.00	6.40	11.50	8.72	6.70[3]	6.70	5.50	6.00	7.17	5.50	4.98
Na_2O	18.71	7.48	9.35	11.23	10.20	13.60	12.17	9.27	10.05	9.63	9.76	10.04	9.38
MgO	2.77	1.11	1.39	1.66	0.85	1.20	1.59	1.49	1.65	1.67	1.44	1.71	1.65
Al_2O_3	25.49	10.20	12.75	15.29	9.96	12.40	16.00	13.96	14.11	13.47	13.90	14.11	14.02
SiO_2	2.71	46.68	39.36	32.03	45.50	37.35	33.13	41.16	41.33	39.19	43.00	40.09	41.91
P_2O_5	-	-	-	-	-	-	-	0.09	0.12	0.08	0.08	0.12	0.12
SO_3	0.87	0.35	0.43	0.52	0.20	0.29	0.35	0.15	0.18	0.13	0.06	0.23	0.21
Cl	-	-	-	-	-	-	-	0.32	0.41	0.27	0.30	0.41	0.22
K_2O	0.07	0.03	0.03	0.04	0.04	0.04	0.04	0.08	0.08	0.08	0.10	0.09	0.07
CaO	2.77	1.11	1.38	1.66	2.05	3.36	2.55	1.64	1.58	1.76	1.52	1.49	1.41
TiO_2	0.04	0.01	0.02	0.02	0.02	0.02	0.01	0.02	0.04	0.01	0.04	0.02	0.03
Cr_2O_3	0.20	0.08	0.10	0.12	0.10	0.11	0.13	0.12	0.12	0.12	0.11	0.12	0.13
MnO	5.78	2.31	2.89	3.47	2.02	2.60	3.48	2.47	2.47	3.04	2.85	2.55	2.69
Fe_2O_3	28.99	11.60	14.49	17.39	13.40	15.70	18.50	12.86	12.98	13.79	10.85	13.61	13.32
NiO	1.66	0.66	0.83	1.00	0.62	0.78	1.00	2.36	2.11	2.56	2.19	2.14	2.09
CuO	0.05	0.02	0.03	0.03	0.03	0.02	0.03	0.04	0.03	0.03	0.03	0.02	0.05
ZnO	0.05	0.02	0.02	0.03	0.04	0.03	0.03	2.58	2.44	2.84	2.42	2.42	2.26
ZrO_2	0.09	0.04	0.05	0.54	0.07	0.04	0.06	0.05	0.07	0.07	0.07	0.05	0.07
Cs_2O[4]	<0.01	-	0.22	-	-	-	0.24	0.24	0.25	0.32	0.18	0.29	0.29
BaO	0.07	0.03	0.03	0.04	0.02	0.01	0.04	0.03	0.02	0.03	0.10	0.02	0.10
La_2O_3	0.03	0.01	0.02	0.02	<0.01	<0.01	0.02	0.03	0.02	0.03	0.01	0.05	0.07
Ce_2O_3	0.21	0.09	0.11	0.13	0.08	0.10	0.14	0.14	0.16	0.16	0.01	0.14	0.16
PbO	0.38	0.15	0.19	0.23	0.13	0.15	0.20	0.15	0.14	0.20	0.16	0.16	0.16
ThO_2[5]	0.03	0.01	0.02	0.02	-	-	-	-	-	-	-	-	-
U_3O_8[5]	9.03	3.61	4.52	5.42	-	-	-	-	-	-	-	-	-
Total	100.00	100.00	100.00	100.00	101.23	99.68	99.41	99.95	99.86	99.48	99.89	99.38	99.39

[1] Determined by ICP-AES at SRNL, [2] Determined by XRF at Radon, [3] Determined by potentiometric titration, [4] Introduced over 100 wt %,[5] ThO_2 and U_3O_8 were introduced only in the feed at the lab-scale tests.

Table II. PCT-A [7] Data for Frit 503-R4 with Various SB4 Sludge Loadings

Glass ID	Waste Loading, wt %	Normalized Release (g/L)			
		B	Li	Na	Si
503-R4	40	0.50	0.63	0.35	0.33
	45	0.41	0.52	0.35	0.32
	50	0.55	0.68	0.53	0.34
	55	0.42	0.56	0.53	0.34
503-R4 CCC	40	0.45	0.63	0.35	0.34
	45	0.41	0.52	0.35	0.32
	50	0.33	0.48	0.34	0.28
	55	0.59	0.75	0.49	0.32
SB4-40 (bench-scale)	40	0.28	0.48	0.36	0.26
SB4-50 (bench-scale)	50	0.48	0.60	0.65	0.28
SB4-60 (bench-scale)	60	0,66	0,71	0,57	0,32
SB4-50 (large-scale)	50	1,00	0,90	0,60	0,34
Environmental Assessment (EA) Standard		18.57	9.91	13.73	3.92
Approved Reference Material (ARM) Standard		0.48	0.56	0.48	0.26

CCIM Testing

In order to evaluate the effect of waste loading on the glass structure and elemental leaching, the waste loading in glass was varied from 40 to 60 wt % in the CCIM tests.

In the lab-scale CCIM tests, vitrification was performed by a rising melt level method where feed was periodically added after previously added feed was consumed into the melt and the glass specimens were sampled from the breakage removed from the cold crucible after it was cooled to room temperature. In the bench- and full-scale tests, molten glass was periodically poured into 10 L canisters and annealed to relax mechanical stress. After cooling, glass in some canisters was fractured and the specimens were sampled from the broken glass.

In the full-scale test, 30 canisters with vitrified waste surrogate were obtained and specimens from canisters numbers 6, 8, 10, 12, 14, 16, 18, 20, 21, 22 and "dead volume" (DV) of the cold crucible were sampled. All the specimens appeared glassy, dark-colored, opaque and visually uniform.

Chemical compositions of samples obtained from the bench-scale tests and from the products sampled from canisters numbers 6, 12, 16, 20 and the DV from the full-scale test were determined and compared to the target composition (Table I). Measured compositions were close to targeted with the exception of a few components. Measured concentrations for B_2O_3, SO_3, Cl, Cs_2O, PbO, and, to a less extent Na_2O, were lower than targeted in the glasses produced in the large-scale cold crucible likely due to their volatilization at high temperatures in the CCIM process, whereas excesses of NiO and ZnO may be explained by their impurities in some chemicals used. Overall, all the analytical sums were close to 100 wt %.

X-Ray Diffraction Results

The material with 40 wt % waste loading produced in the lab-scale cold crucible was found to be completely amorphous (Figure 1, *1*). Occurrence of quartz (major reflection at 3.33-3.34 Å) was due likely to traces of incompletely dissolved quartz (Figure 1, *2,3*). XRD patterns showed the presence of spinel-type phase in the glassy products at ≥50 wt % waste loading (Figure 1, *4-7*). Traces of spinel phase may have been present in the glassy material at 40 wt % waste loading produced in the bench-scale cold crucible and cooled relatively slowly in a heat-insulated box (Figure 1, *2*).

SEM/EDS Results

SEM/EDS study of the materials with ~50 wt % waste loading produced in the lab-scale cold crucible and in the bench-scale cold crucible showed similarities in microstructure. Primarily individual spinel-type crystals were distributed in a matrix vitreous phase (Figure 2, *a-c*). Minor inclusions of unreacted quartz were observed (Figure 2, *d*). No nepheline phase was found at even 60 wt % waste loading in glass produced in the bench-scale cold crucible and annealed in the heat-insulated box.

As followed from EDS data, partitioning of waste elements among vitreous and spinel type phases was as expected (Table III): alkali (Na, K, Cs), alkaline earth (Ca, Ba), aluminum, silicon, sulfur, lead, and uranium elements were concentrated in the vitreous phase; transition elements (Cr, Mn, Fe, Ni, Cu, Zn) were predominantly in the spinel phase. Some elements such as magnesium and cerium were distributed between both the phases.

Two glasses produced in the large-scale cold crucible were analyzed using optical microscopy and three of the glasses were examined by SEM/EDS to study the microstructures in

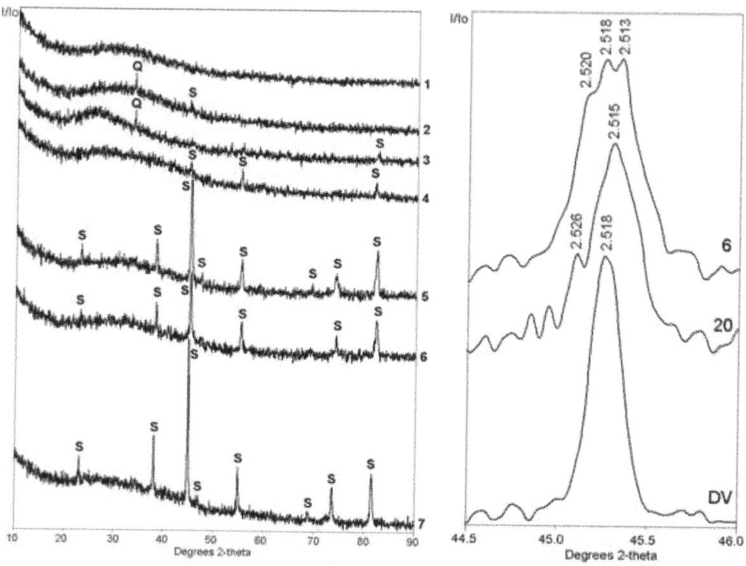

Figure 1. XRD patterns of glassy materials at waste loadings of 40 (1,2), 50 (3-6) and 60 wt % (7). Produced in the lab- (1,3), bench- (2,4,7), and large-scale (5,6) cold crucibles and major reflection due to spinel in the products sampled from canisters #6 and #20 and DV. Q – quartz, S – spinel.

more detail. In transmitted light, both the glasses were transparent, clear and brown colored in some areas. Spinel crystals appeared black. Skeleton-type spinel crystals were observed in the bulk (Figure 3*a*). In reflected light, the glass was gray-colored whereas the spinel was white or light-gray (Figure 3*b*).

SEM images showed non-uniform distribution of the spinel crystals in the bulk of the samples (Figure 3*e*). The bulk consisted of areas with different types of spinel crystals (Figure 3*f*). The bulk was primarily composed of vitreous phase with the skeleton-type crystals up to several microns in size distributed within. These areas were separated by layers composed of matrix glass containing individual spinel crystals up to 20-30 μm in size (Figure 4 *f-h*).

The occurrence of two different zones in the bulk glass may be due to the non-uniform cooling rate of glass in canisters especially when two pours were performed in the same canister. In this case, interaction of the hot melt with the solidified melt created temperature gradient zones with varying cooling rates. Fast cooling resulted in crystallization of fine skeleton-type crystals whereas larger individual crystals were formed in areas with slower cooling rates.

The EDS data showed that the chemical compositions of the vitreous phase and spinel in zones with different cooling rates varied especially in the contents of micro-components (Table III). The major constituents of the spinel were transition metal oxides (Fe, Mn, Ni, Cr, Zn) as well as alumina and magnesia. Elements such as Na, Si, K, Ca, Zr, Pb, and some Al and Mg were captured by the electron probe from the surrounding glass.

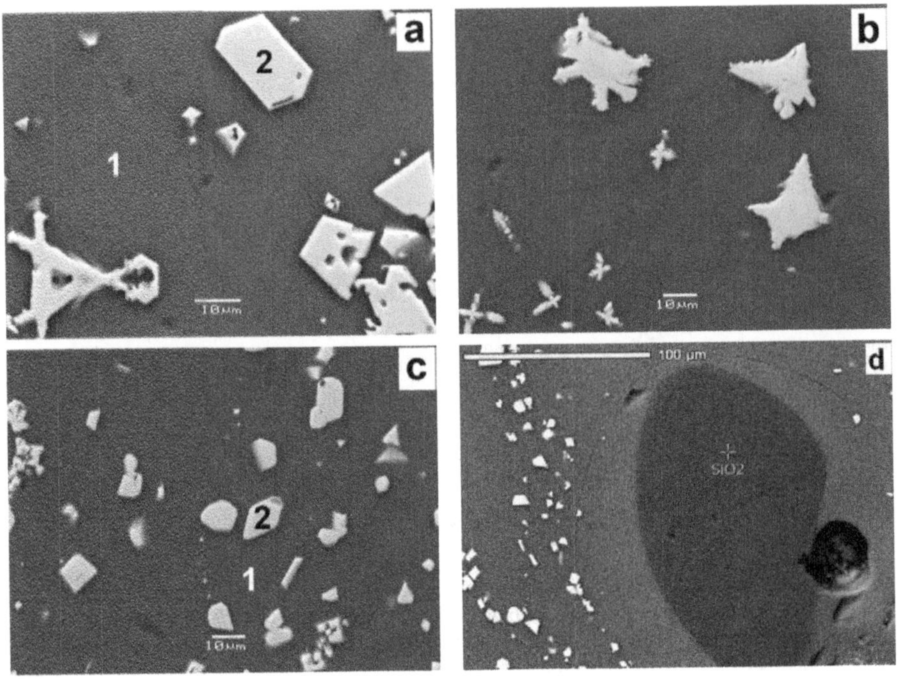

Figure 2. SEM images of the glassy materials with 50 (a, b) and 60 wt % (c, d) waste loading produced in the lab- (a) and bench-scale (b-d) cold crucibles.1 – vitreous phase, 2 – spinel.

The spinel structure phase segregated as the skeleton-type crystals was close to magnetite in chemical composition and its average formula may be represented as $(Mg_{0.03}Mn_{0.07}Fe_{0.05}Ni_{0.76}Zn_{0.09})^{2+}(Fe_{1.76}Al_{0.19}Cr_{0.05})^{3+}O_4$ (Spinel-1). In the zone of individual crystals, one of the varieties had a chemical composition similar to that in the zone of the skeleton-type crystals and formula $(Mg_{0.04}Mn_{0.07}Fe_{0.07}Ni_{0.73}Zn_{0.09})^{2+}(Fe_{1.81}Al_{0.16}Cr_{0.04})^{3+}O_4$ (Spinel-1). The second variety was enriched with chromium and, to some extent, with zinc and had a formula $(Mg_{0.02}Mn_{0.08}Fe_{0.11}Ni_{0.62}Zn_{0.17})^{2+}(Fe_{1.31}Al_{0.17}Cr_{0.52})^{3+}O_4$ (Spinel-2). Some grains had intermediate composition: $(Mg_{0.02}Mn_{0.07}Fe_{0.08}Ni_{0.69}Zn_{0.14})^{2+}(Fe_{1.58}Al_{0.16}Cr_{0.26})^{3+}O_4$ (Spinel-3). Based on SEM/EDS data the Spinel-1 variety may be identified as a trevorite (JCPDS-ICDD 44-1485 or 23-1119) and the two others – as solid solutions between trevorite and zincochromite, $ZnCr_2O_4$ (JCPDS-ICDD 22-1107).

Table III. Chemical Compositions (wt %) of Co-Existing Phases in the Vitrified SB4 Waste

Oxides	Lab-scale		Bench-scale		Large-scale							
					Canister #6				Canister #20			
	Glass on Fig. 2*a*	Spinel on Fig. 2*a*	Glass on Fig. 2*c*	Spinel on Fig. 2*c*	Zone with skeleton-type crystals		Border zone with individual crystals		Zone of skeleton-type crystals		Zone of individual crystals	
					VP	Spinel	VP	Spinel	VP	Spinel	VP	Spinel
Na_2O	8.44	0.54	13.04	0.43	10.16	1.91	9.58	1.24	8.96	1.59	9.32	1.24
MgO	1.06	1.09	1.65	1.66	1.40	0.81	1.44	0.68	1.66	1.07	1.52	0.90
Al_2O_3	12.42	2.54	15.39	2.71	14.22	6.71	14.26	6.61	14.07	6.01	14.23	6.41
SiO_2	42.16	0.51	39.75	0.35	43.92	5.51	43.30	3.04	43.28	4.94	43.85	3.48
SO_3	0.43	0.11	0.18	0.13	-	0.00	0.20	0.05	0.00	0.05	0.07	0.00
K_2O	-	-	-	-	0.22	0.01	0.20	0.02	0.16	0.05	0.18	0.04
CaO	1.27	0.10	3.38	0.10	1.64	0.17	1.71	0.13	1.80	0.12	1.67	0.11
TiO_2	-	-	-	-	0.09	0.27	0.00	0.11	0.04	0.04	0.06	0.16
Cr_2O_3	0.08	3.47	0.05	1.80	0.12	1.47	0.07	8.01	0.13	2.51	0.05	3.65
MnO	3.20	5.52	2.86	5.43	2.63	1.80	2.79	2.09	2.75	1.69	2.95	1.86
Fe_2O_3	12.58	71.03	8.07	73.92	6.88	56.66	7.05	52.05	6.74	56.13	6.37	57.36
NiO	0.40	13.91	0.07	13.63	1.04	21.71	0.98	19.93	0.79	21.98	0.91	20.88
CuO	<0.01	0.02	<0.01	0.03	0.08	0.08	0.05	0.07	0.19	0.07	0.04	0.19
ZnO	<0.01	0.32	<0.01	0.14	3.38	3.36	3.37	4.61	4.04	3.00	3.83	3.75
ZrO_2	-	-	-	-	-	0.05	0.20	0.00	0.19	0.27	0.14	0.35
Cs_2O	0.15	<0.01	0.44	<0.01	0.29	-	0.39	-	0.43	0.00	0.40	0.00
BaO	<0.01	0.01	0.02	<0.01	-	0.06	0.00	0.00	0.17	0.00	0.00	0.00
La_2O_3	-	-	-	-	-	0.21	0.24	0.22	0.06	0.00	0.00	0.06
Ce_2O_3	0.08	0.32	0.22	0.27	0.12	0.19	0.25	0.26	0.05	0.22	0.12	0.00
PbO	<0.01	<0.01	0.15	<0.01	0.38	0.21	0.25	0.00	0.22	0.11	0.15	0.09
U_3O_8	4.83	<0.01	-	-	-	-	-	-	-	-	-	-
Total*	87.12	99.48	85.25	100.60	86.53	101.20	86.32	99.13	85.74	99.84	85.86	100.53

Note: oxide sums for analyses of vitreous phase are less than 100 wt % because Li and B are not determined by EDS, VP – vitreous phase.

The glassy products from Canister #20 and DV of the cold crucible had a texture similar to that observed in the product from Canister #6. Both samples had zones with the skeleton-type crystal aggregates distributed in the glassy matrix and with larger-sized individual crystals (Figure 4). Chemical compositions of the phases co-existing in the zones with the skeleton-type and individual crystals in the glassy product from Canister #20 are given in Table III.

Similar to the sample from Canister #6, there was good agreement between the target glass composition and composition of the vitreous phase in the samples from Canister #20. In both zones with the skeleton-type and individual crystals, analytical sums of the vitreous phase constituents were about 84-87 wt %. Taking into account that the target sum of Li_2O and B_2O_3 was 12 wt % (4 wt % Li_2O + 8 wt % B_2O_3) the total is about 96-99 wt %. These results were very near the target considering that some minor anionic constituents (F, Cl, P) were not determined.

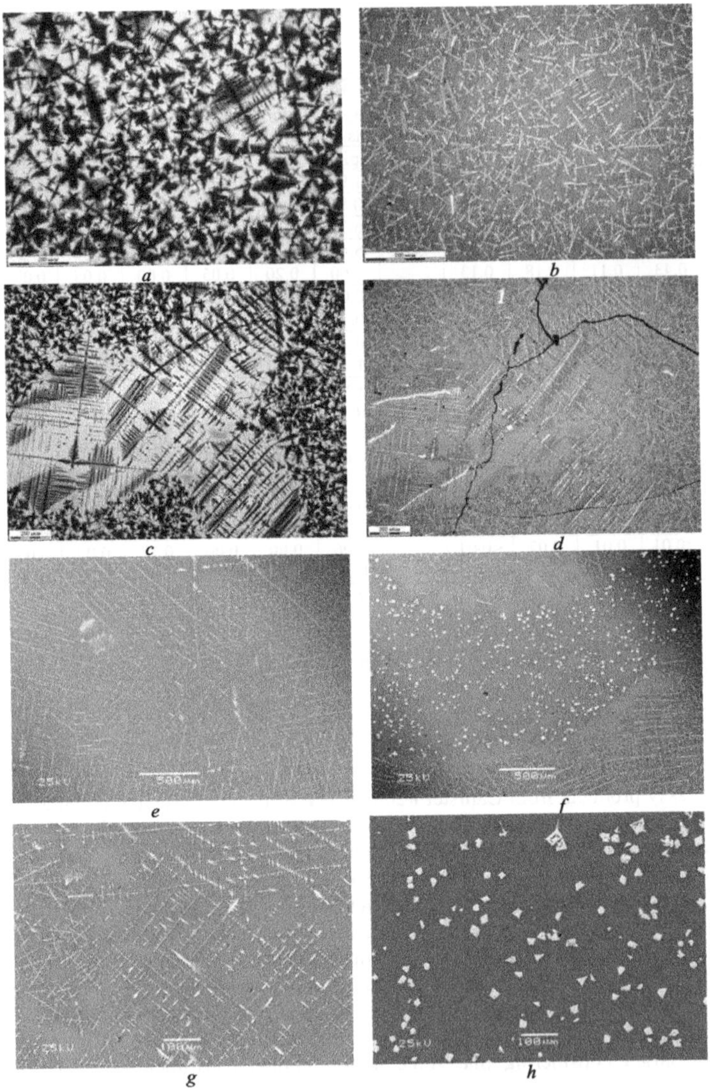

Figure 3. Photomicrographs from optical microscopy of the samples from canister #6 (a, b) and DV of the cold crucible (c, d) in transmitted (a, c) and reflected (b, d) light at parallel Nicols (scale bars are 200 μm) and SEM Images of the Glassy Product from Canister #6 (e – general view, f – border zone, g – zone of skeleton-type crystals, h – zone of individual crystals).

Formulae of the spinel phases in the sample from Canister #20 recalculated from their chemical compositions (Table III) were as follows:

- in the zone of the skeleton-type crystals:
  $(Mg0.04Mn0.06Fe0.13Ni0.77Zn0.10)^{2+}(Fe1.71Al0.15Cr0.04)^{3+}O_4$ (Spinel-1),
  $(Mg0.03Mn0.07Fe0.09Ni0.74 Zn0.07)^{2+}(Fe1.68Al0.13Cr0.19)^{3+}O_4$ (Spinel-2);
- in the zone with individual crystals:
  $(Mg0.04Mn0.07Fe0.08Ni0.73Zn0.08)^{2+}(Fe1.81Al0.18Cr0.01)^{3+}O_4$ (Spinel-1),
  $(Mg0.05Mn0.07Fe0.09Ni0.71Zn0.08)^{2+}(Fe1.64Al0.14Cr0.22)^{3+}O_4$ (Spinel-2),
  $(Mg0.01Mn0.06Fe0.14Ni0.73Zn0.06)^{2+}(Fe1.79Al0.07Cr0.14)^{3+}O_4$ (Spinel-3).

All the spinel types were cubic with a space group *Fd3m* (Z=8). Their lattice parameters determined from XRD data (Figure 1) are given in Table IV. As seen from Figure 1, the major reflection on the XRD pattern due to the spinel structure phase in the product sampled from Canister #6 was split into three peaks corresponding to the three different spinel varieties. Significant asymmetry of the peak and occurrence of "shoulders" from both lower and higher angles of 2-theta demonstrated incomplete phase formation under non-equilibrium spinel crystallization conditions from the melt yielding spinel grains with variable chemical composition. Spinel grains in the product from Canister #20 are more uniform in chemical composition and only two well-resolved peaks were observed on the XRD pattern. The sample from the DV cooled for the longest time in the de-energized cold crucible under much more equilibrium conditions contained spinel with rather uniform chemical composition. As a result, the XRD peak was only slightly asymmetric with respect to its position on the angular range and the lattice parameter tended to some averaged values.

Table IV. Averaged Lattice Parameters of the Spinel Varieties in the Glassy Products

Product	Spinel	d_{max}, Å	*a*, Å
Canister #6	Spinel-2	2.520	8.358
	Spinel-3	2.518	8.351
	Spinel-1	2.513	8.335
Canister #20	Spinel-2	2.526	8.378
	Spinel-1,3	2.515	8.341
"Dead volume"	Spinel	2.518	8.351

Determination of the Degree of Crystallinity

The degree of crystallinity of the vitrified SB4 waste surrogate sampled from Canisters #6, #20 and DV of the large-scale cold crucible was measured using their SEM images (three replicate SEM microphotographs) processed by a Corel Photo-Paint software.

The degree of crystallinity of the product sampled from Canister #6 was slightly lower than that of the product from Canister #20 and DV of the cold crucible. As expected, the highest degree of crystallinity was observed in the product sampled from the DV (13.3 vol % in the zone with the skeleton-type crystals and 10.7 vol % in the zone with individual crystals). Average degree of crystallinity may be estimated as ~12 vol %. This value is a little bit higher than that observed in the product obtained in the bench-scale cold crucible but at higher waste loading (60 wt %). The degree of crystallinity of the product with ~50 wt % waste loading obtained in the smaller cold crucible was estimated to be ~5-7 vol %. The slower cooling rate in the larger sized cold crucible favored an increase in the degree of crystallinity and larger crystallite size.

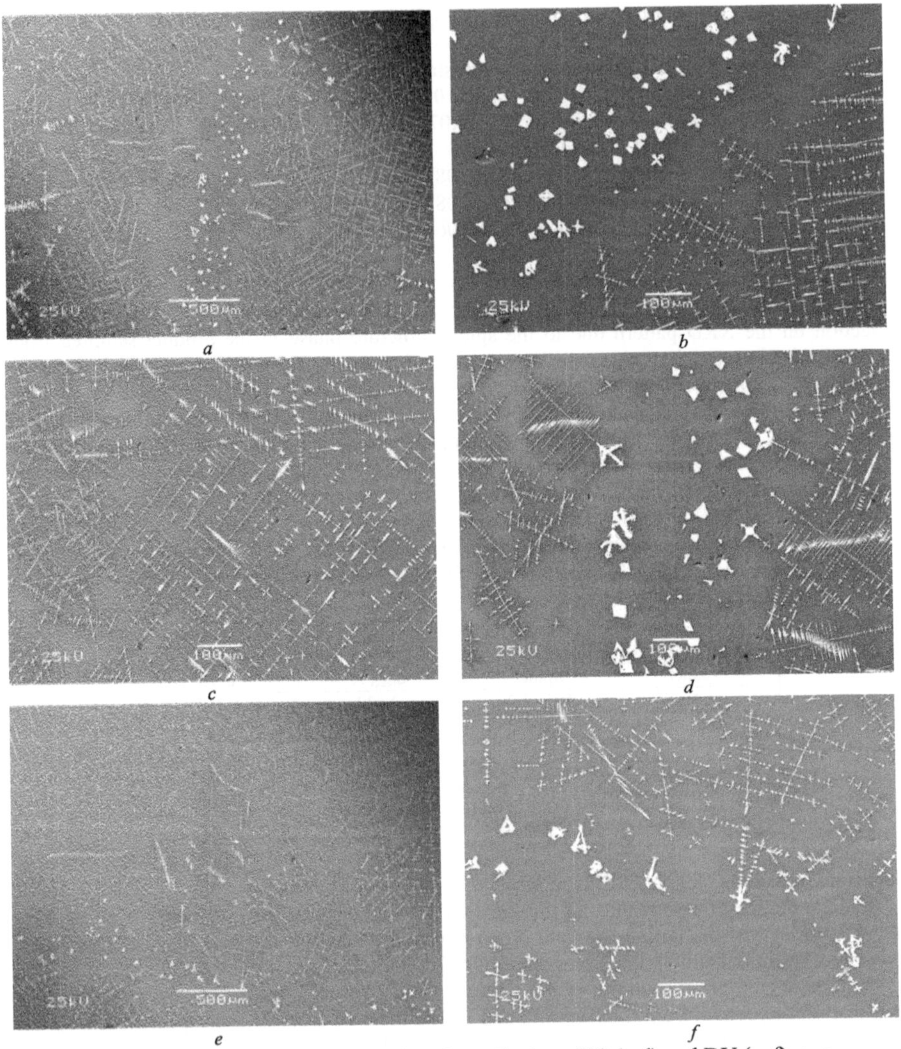

Figure 4. SEM Images of the Glassy Product from Canister #20 (a-d) and DV (e, f). a, e – general view; b, f – border zone between zones with the skeleton-type and individual crystals; c – zone of the skeleton-type crystals; d – zone of individual crystals.

Infrared Spectroscopy Results

Infra-red spectra of the vitrified SB4 waste surrogate within the range of 1600-400 cm^{-1} consisted of a very strong absorption band at 850-1150 cm^{-1}, a strong band at 400-550 cm^{-1}, weaker bands at 1350-1450 cm^{-1} and 700-750 cm^{-1}, and a weak band (shoulder) at 1250-1300 cm^{-1} (Figure 5). The bands with maxima at 1395-1410 cm^{-1} and ~1270 cm^{-1} were components of twice degenerated asymmetric valence ν_3 O—B—O vibration in the boron-oxygen triangular BO_3 units. The bands at ~720 cm^{-1} and ~655 cm^{-1} were components of twice degenerated asymmetric deformation δ_4 O—B—O vibrations [9]. The strongest band with a maximum at ~1000 cm^{-1} was due to asymmetric O—Si—O vibrations in the tetrahedral SiO_4 units. This band is normally a superposition of the bands due to O—Si—O vibrations in the SiO_4 units with various numbers of bridging oxygen ions connecting SiO_4 tetrahedra in the network: four (Q^4) – at ~1100-1150 cm^{-1}, three (Q^3) – at ~1050-1100 cm^{-1}, two (Q^2) – at ~1000-1050 cm^{-1}, one (Q^1) – at ~950-1000 cm^{-1}, and zero (Q^0) – at ~900-950 cm^{-1}. The wavenumber range of 400-550 cm^{-1} corresponded to deformation vibrations in the SiO_4 units [10,11].

The occurrence of the bands with maxima at ~1400 cm^{-1} and ~1270 cm^{-1} indicated significant fraction of trigonally coordinated boron in all the glasses studied. Position of the maximum of the strongest band due to O—Si—O vibrations at ~990-1000 cm^{-1} pointed to Q^2 and Q^1 units prevailing in the glass network. The band with a maximum at ~720 cm^{-1} was a superposition of the bands due to asymmetric deformation δ_4 O—B—O and symmetric ν_1 O—Si—O vibrations. The latter may be present in IR spectra only if symmetry of the SiO_4 units is lower than T_d (C_{3v} or C_{2v}) [10]. The Q^4 and Q^0 units had T_d symmetry whereas the Q^1 and Q^3 units had C_{3v} and the Q^2 units had C_{2v} symmetry. Taking into account the relatively low silica content in the glassy products (≤44 wt % in the vitreous phase) only a minor fraction of the Q^3 units may be present; therefore, the glass network was built from rather short metasilicate chains.

Minor BO_4 tetrahedral units may be integrated in these chains. The bands due to O—B—O vibrations in the BO_4 units and Si—O—B^{IV} vibrations bonding SiO_4 and BO_4 tetrahedra were positioned within the same range (950-1050 cm^{-1}) as O—Si—O bonds in the SiO_4 units and cannot be clearly distinguished. However, in high Al_2O_3 content glasses, boron is predominantly three-coordinated with respect to oxygen. For all the glasses studied, the ψ_B factor allowing evaluation of relative fractions of three- and four-coordinated boron and determined as:

$$\psi_B = \{([Na_2O]+[K_2O]+[BaO])+0,7([CaO]+[SrO]+[CdO]+[PbO])+ 0,3([Li_2O]+[MgO]+[ZnO])-[Al_2O_3]\}/[B_2O_3]$$

was lower than zero (behavior of iron in glasses was similar to that of aluminum). At this value all the boron should be three-coordinated whereas major aluminum – four-coordinated and minor aluminum – six-coordinated [12].

As followed from the IR spectra, vitreous phases had similar structure in all the glassy materials at 40 to 60 wt % SB4 waste loading. The position of the maximum in the IR spectra of the band due to O—Si—O bonds in the SiO_4 units of the glass with ~60 wt % waste loading was slightly shifted to lower wavenumbers (~990 cm^{-1}) indicating some lower degree of connectedness of the glass network due to higher fraction of the Q^1 units with one bridging oxygen ion.

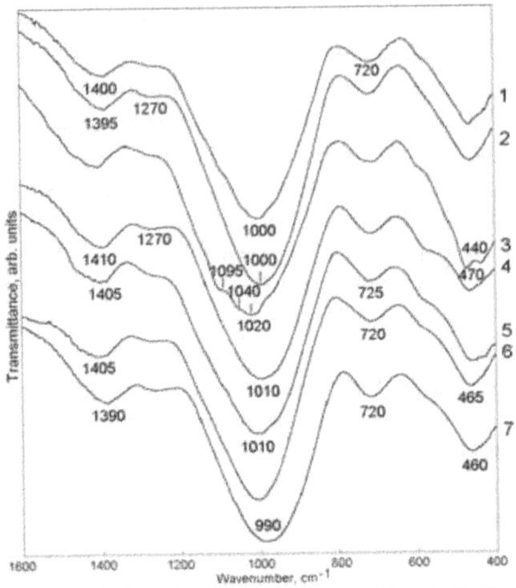

Figure 5. Infrared spectra of the glassy materials with 40 (1,2), 50 (3-6) and 60 wt % (7) waste loading produced in the lab- (1,3), bench- (2,4,7), and large-scale (5,6) cold crucibles.

The bands due to vibrations of Fe^{3+}—O bonds were positioned within the ranges of 550-650 cm^{-1} (Fe^{IV}—O) and 300-400 cm^{-1} (Fe^{VI}—O) [13]. They were overlapped with strong bands due to deformation vibrations in the SiO_4 groups. Just a weak shoulder within the range 550-600 cm^{-1}, which can be attributed to vibrations of the Fe^{IV}—O in the glass network, was observed in IR spectra. Absence of narrow bands in the above-mentioned ranges indicated low content of the spinel structure phases that was consistent with the XRD and SEM data.

The bands due to vibration of Al^{IV}—O bonds in the glass network were within the range of 700-850 cm^{-1} [13] and overlapped with the band due to symmetric valence vibrations in the SiO_4 units.

Leach Resistance of Vitrified Waste

Leaching data obtained by PCT-A procedure are given in Table II and can be compared with the data for the samples prepared in small crucibles in a laboratory furnace. Normalized release values for all the elements were approximately 10 to 30 times lower than those from the Environmental Assessment (EA) Standard glass and comparable to the Approved Reference Material (ARM) Standard.

It should be noted, that at the same waste loading (50 wt %) normalized release values for Li and B from the glassy material produced in the large-scale cold crucible was found to be higher by a factors of 1.5 to 2 than that from the material produced in the bench-scale cold crucible. Na and Si normalized release values from both the materials were similar. The materials from the bench-scale cold crucible were produced from chemicals simulating Frit 503-R4 whereas the materials from the large-scale cold crucible were produced from actual Frit 503-R4. It can be suggested that the lithium-borate groups maintain their structure in the final glass yielding nearly congruent release of both constituents as lithium borates.

CONCLUSIONS

Vitrification of SB4 waste surrogate at relatively low waste loading (~40 wt %) yielded predominantly amorphous product containing trace amounts of unreacted quartz and spinel-type crystallites. An increase in waste loading to 50 wt % and then 60 wt % resulted in formation of spinel structure crystals (trevorite, zincochromite and their solid solutions). No nepheline was found, even at 60 wt % waste loading in glass. The amount of crystalline phase in the glassy materials increased with the increase of waste loading and cold crucible dimensions. The maximum content of crystalline phase in the material produced in the large-scale cold crucible was estimated to be ~12 vol %.

The anionic motif of the glass network was composed of rather short metasilicate chains with end-units having three non-bridging oxygen ions and high borate constituent with predominantly three-coordinated boron. Durability measurements of all the glasses demonstrated low release of major elements (B, Li, Na, Si) and the release of the elements were 10 to 30 times lower than those from the Environmental Assessment (EA) Standard glass.

ACKNOWLEDGEMENTS

The work was performed under Contract between US DOE/SRNL and Daymos Ltd./SIA Radon. The authors thank Radon personnel who participated in the vitrification tests of the SB4 waste surrogate. Special thanks are given to Mr. K.D. Gerdes for financial support from US DOE, Ms. N.P. Penionzkiewic (SIA Radon) for recording of XRD patterns and Mr. B.S. Nikonov (IGEM RAS) for his help in SEM/EDS study of the samples.

REFERENCES

1. S.L. Marra, R.J. O'Driscoll, T.L. Fellinger, J.W. Ray, P.M. Patel, and J.E. Occhipinti. DWPF Vitrification – Transition to the Second Batch of HLW Radioactive Sludge, *Waste Management '99. Conf.* February Tucson, AZ, ID 48-5. CD-ROM.
2. A.P. Kobelev, S.V. Stefanovsky, O.A. Knyazev, T.N. Lashchenova, A.G. Ptashkin, M.A. Polkanov, E.W. Holtzscheiter, and J.C. Marra, Results of a 50% Waste Loading Vitrification Test Using the Cold Crucible Melter for Savannah River Site, *Waste Management '06 Conf.* February 26 – March 2, 2006. Tucson, AZ, ID 6127. CD-ROM.
3. American Society for Testing and Materials (ASTM). *Standard Test Methods for Determining Chemical Durability of Nuclear, Hazardous, and Mixed Waste Glasses: The Product Consistency Test (PCT)*, ASTM C1285-97, West Conshohoken, PA, 1997.
4. S.V. Stefanovsky, A.P. Kobelev, V.V. Lebedev, M.A. Polkanov, A.G. Ptashkin, O.A. Knyazev, and J.C. Marra, Cold Crucible Vitrification of Uranium-Bearing High Level Waste Surrogate, *Mat. Res. Soc. Symp. Proc. 2008* (*Scientific Basis for Nuclear Waste Management XXXI*, Sheffield, UK, 16th – 21st September 2007), Paper 5P11 (in press).
5. S.V. Stefanovsky, A.P. Kobelev, V.V. Lebedev, M.A. Polkanov, A.G. Ptashkin, O.A. Knyazev, and J.C. Marra. Vitrification of Savannah River Site SB4 Waste Surrogate in the Radon Bench-Scale Cold Crucible Unit, *Waste Management 2008 Conf.* February 24-28, 2008. Phoenix, AZ, ID 8035. CD-ROM.
6. A.P. Kobelev, S.V. Stefanovsky, V.V. Lebedev, M.A. Polkanov, V.V. Gorbunov, A.G. Ptashkin, O.A. Knyazev, J.C. Marra, and K.D. Gerdes. Full-Scale Cold Crucible Test on Vitrification of Savannah River Site SB4 HLW Surrogate, *The MS&T/ACerS 110th Annual Meeting*, October 5-9, 2008. Pittsburgh, PA (this volume).

7. J.C. Marra, D.K. Peeler, T.B. Edwards, K.M. Fox, J.D. Vienna, A. Fluegel, S.V. Stefanovsky, and A.S. Aloy, Glass Formulation to Support Melter Testing to Demonstrate Enhanced High Level Waste Throughput, *Mat. Res. Soc. Symp. Proc. 2007* (*Scientific Basis for Nuclear Waste Management XXXI.* Sheffield, UK, 16th-21st September 2007), in press.
8. S.L. Marra and C.M. Jantzen, *Characterization of Projected DWPF Glasses Heat Treated to Simulate Canister Centerline Cooling (U)*. WSRC-TR-92-142, Rev. 1, 1993. 40 p.
9. V.A. Kolesova, Vibrational Spectra and the Structure of Alkali Borate Glasses, *Glass Physics and Chemistry* (Russ.) Vol. 12, No. 1, 1986, p. 4-13.
10. K. Nakamoto, *Infrared Spectra of Inorganic and Coordination Compounds*, John Wiley & Sons, Inc. New York – London, 1963.
11. J. Wong, C.A. Angell, *Glass Structure by Spectroscopy*, Marcel Dekker, Inc. New York – Basel, 1976.
12. A.A. Appen, *Chemistry of Glass* (Russ.), Khimiya, Leningrad, 1974.
13. I.I. Plysnina, *Infrared Spectra of Minerals*, MGU, Moscow, 1977.

ACCELERATED WEATHERING OF WASTE GLASS AT 90°C WITH THE PRESSURIZED UNSATURATED FLOW (PUF) APPARATUS: IMPLICATIONS FOR PREDICTING GLASS CORROSION WITH A REACTIVE TRANSPORT MODEL

E. M. Pierce* and D. H. Bacon
Energy & Environment Directorate, Pacific Northwest National Laboratory, P.O. Box 999, MS: K3-62, Richland, WA 99354
*corresponding author: Eric.Pierce@pnl.gov, Phone: (509) 372-4555, FAX: (509) 372-4600

ABSTRACT

The interest in the long-term durability of waste glass stems from the need to predict radionuclide release rates from the corroding waste form over geologic time-scales. Several long-term test methods have been developed to accelerate the glass-water reaction [drip test, vapor hydration test, product consistency test-B, and pressurized unsaturated flow (PUF)]. Currently, the PUF test is the only method that can mimic the unsaturated hydraulic properties expected in a subsurface disposal facility and simultaneously monitor the glass-water reaction. PUF tests are being conducted to accelerate the weathering of glass and validate the model parameters being used to predict long-term glass behavior. The model parameters are obtained by conducting a series of single-pass flow-through experiments and are used to parameterize the chemical affinity rate equation to predict the effect of rate influencing variables on glass dissolution. A one-dimensional reactive chemical transport simulations of glass dissolution and secondary phase formation during a 1.5-year long PUF experiment was conducted with the subsurface transport over reactive multi-phases (STORM) code. Results show that parameterization of the computer model by combining direct laboratory measurements and thermodynamic data provides an integrated approach to predicting glass behavior over geologic-time scales. The results from these bench-scale experiments will be extended to a set of field lysimeter experiments (i.e., experiments on buried glass at the Hanford Site) that have been in progress for 6 years on several glass specimens.

INTRODUCTION

The immobilization of radioactive waste into a vitrified matrix (i.e., glass) has been selected by the U. S. Department of Energy (DOE) as the best available treatment technology for a portion of the high-level waste (HLW) and low-activity waste (LAW) being stored at the Hanford Site, southeastern Washington. Currently, DOE plans to dispose of the glasses made from waste at Hanford at two U. S. locations: (1) the LAW glass will be stored on-site at the Integrated Disposal Facility (IDF) and (2) the HLW glass will be disposed at the proposed mined geologic repository in Yucca Mountain, Nevada. Vitrifying the LAW is expected to generate over 1.6×10^5 m^3 of glass [1]. The volume of immobilized low-activity waste (ILAW) at Hanford is the largest in the DOE complex and is one of the largest inventories (approximately 0.89×10^{18} Bq total activity) of long-lived radionuclides planned for disposal in a low-level waste (LLW) facility globally. Before the ILAW can be disposed, DOE must conduct a credible safety analysis for the IDF by providing estimates of radionuclide release rates from the engineered portion of the disposal facility (source term). These estimates are expected to be based on chemical reactions that occur in the near-field and are controlled by the dissolution of the vitrified matrix. Therefore to provide credible estimates, the effect of potential rate influencing variables on the dissolution of glass must be understood and incorporated into models in order to effectively predict the glass-water reaction over the period of regulatory concern (approximately 10 000 years). This technical strategy [3-5], currently being used in LAW performance assessment (PA) for the IDF [3,6], requires the use of a reactive chemical transport code [e.g., Subsurface Transport Over Reactive Multi-Phases (STORM)] that integrates the results obtained from bench-scale laboratory test methods and results from long-term accelerated weathering tests to predict the long-term weathering of glass. Equation (1), developed by Åagaard and Helgeson [7] and applied to glass by Grambow [8], is the kinetic rate equation that is used in the STORM code to predict the rate of glass dissolution and subsequent radionuclide release as a function of time and space. This equation is based on Transition State Theory (TST) [9] and relates the effects of: pH, temperature, saturation state of the system, and the activities of rate enhancing or inhibiting species on glass dissolution.

$$r = \bar{k}\, v_i\, a_{H^+}^{\pm\eta} g(I) \exp\left(\frac{-E_a}{RT}\right)\left[1-\left(\frac{Q}{K_g}\right)^{\sigma}\right]\prod_j a_j^{\eta_i},\ i=1,2,...N \qquad (1)$$

where r is the dissolution rate [g/(m^2·d)], $\bar{k}$ is the forward rate constant [g/(m^2·d)], v_i is the mass fraction of element i, a_{H^+} is the hydrogen ion activity, a_j is the activity of the jth aqueous species that acts as an inhibitor or catalyst, η is the power law coefficient, E_a is the activation energy (kJ/mol), R is the gas constant [kJ/(mol·K)], T is the temperature (K), Q is the ion activity product (unitless), K_g is the pseudo-equilibrium constant for the rate controlling phase or phases (unitless), and σ is the Temkin coefficient ($\sigma = 1$).

For the IDF-PA, data collection has been focused on measuring and quantifying the effects of environmentally relevant and sensitive parameters (e.g., effect of pH, temperature, and solution composition) that are needed to predict, with a high level of confidence, the long-term behavior of glass. It is important to note that this methodology is similar to the strategy being used to analyze the Drigg LLW site in the UK [10,11]. Each laboratory test method; namely, single-pass flow-through (SPFT) [3,12] and PUF [2,13], is used to evaluate different aspects of the glass-water reaction. In the work reported here, we provide an example of how results from the afore-mentioned laboratory test methods are evaluated and used to predict the release of elements from a 1.5-year long accelerated weathering experiment conducted at 90°C on a prototypic ILAW glass with the STORM code. Although PUF experiments can accommodate other components of the engineered system (sediment/soil, steel container, etc.), additional components were not included in the experiments discussed here. Several of these components are apart of the field lysimeter experiments and their potential impact on ILAW glass performance will be as part of these field experiments.

EXPERIMENTAL MATERIALS AND METHODS

Glass composition, characterization, preparation, and surface area

The target composition of a prototypic immobilized low-activity waste (ILAW) glass, LAWA44, is given in Table 1. This glass was prepared by mixing measured amounts of dried reagent-grade chemicals in an agate mill. The mixtures were melted in a platinum (90%) rhodium (10%) crucible, and the molten glass was poured onto a cool stainless steel plate. After being quenched, the ILAW glasses were heat treated by placing the glass in a preheated oven at 930°C and then cooling at 21°C/hr. This cooling rate is consistent with the computed thermal cooling profile for the centerline of the proposed ILAW container (1.2 m × 1.2 m × 1 m).

The glass specimens used in the dissolution experiments were prepared by crushing glass samples in a ceramic ball mill and sieving with ASTM sieves. The <149 to >75 μm diameter size fraction was used for SPFT tests and <250 to >170 μm diameter for the PUF test. Prior to use, the powders were washed in 18 MΩ-cm ultra pure de-ionized water (DIW), cleaned ultrasonically in DIW, rinsed in ethanol, and dried at 90°C (±2°C) in a oven. Each glass sample, before and after testing, was stored at room temperature (~23°C) in a desiccator that contained $CaSO_4$.

Table 1. Composition (Mass%), particle density, surface area, and free energy of hydration for the selected ILAW Glasses.

Oxide	LAWA44
Al_2O_3	6.20
B_2O_3	8.90
CaO	1.99
Fe_2O_3	6.98
K_2O	0.50
Li_2O	NA
MgO	1.99
Na_2O	20.0
Nd_2O_3	NA
NiO	NA
SiO_2	44.55
TiO_2	1.99
ZnO	2.96
ZrO_2	2.99
[a]Others	0.95
Total	100.0
Formula Weight, g/mol	66.96
Density, g/m^3	$(2.698 \pm 0.008) \times 10^6$
ΔG_{hyd}, kJ/mol	-58.5

[a]Others include minor amounts of Cl, F, MoO_3, P_2O_5, Re_2O_7, and SO_3.

The specific surface area of each sample was calculated with a geometric formula, Eq. (2) from McGrail et al. [14],

$$S = \frac{3m}{\rho r} \tag{2}$$

where S is the surface area in m^2, m is the mass of glass in g, ρ is the glass density in g/m^3 measured with a Micromeritics He-gas pycnometer], and r is average radius in m. This formula assumes that the particles are spherical, size of the grains are normally distributed, and surface pits, cracks, and other forms of surface roughness do not affect the surface area. Although all three of these assumptions may not be valid, results from experiments with LAW glass monoliths [15] and natural glass samples [16] suggest the geometric surface area best represents the overall glass surface area. Therefore, all the reported rates in this study are normalized to the geometric surface area.

Single-pass flow-through test method and solution analysis

Dissolution experiments were conducted with the SPFT apparatus [12]. For additional details on this experimental system see Pierce et al. [12] and the references contained therein.

All SPFT experiments were conducted at pH values from 7.0 to 12.0 or at various $a[SiO_2\ (aq)]$ at 90°C (±2°C) and at the forward rate of dissolution. For an explanation of the importance of conducting these experiments at the forward or maximum dissolution rate see Pierce et al [12] and the references contained therein. The solution temperature inside of a blank reactor was measured with an *in situ* thermocouple. The flow rate was determined by weighing the volume of solution collected at each sampling time (i.e., gravimetrically) and variations were typically less than 5%. Once constant conditions (i.e., buffers, flow rates, temperature, etc.) were established, the pre-weighed crushed glass specimens (ranging from 0.25 g to 1.00 g per reactor) were placed into test reactors. Experiments were run until steady-state effluent solution conditions prevailed, which typically occurred after 3 to 8 reactor volumes. All influent and effluent solutions were analyzed for Al, B, Na, and Si, via inductively-coupled plasma optical emission spectroscopy (ICP-OES), and pH at select sampling points. The accuracy of the pH measurement was ±0.02 pH units.

The solutions used to control the pH in the SPFT experiments were prepared by adding measured amounts of the organic buffer *tris* hydroxyl-aminomethane (THAM) to DIW and adjusting the solution to the desired pH value with 15.8 M HNO_3 or 1 M LiOH. The THAM buffer range is between pH 7 to 10; therefore the alkaline solutions, pH range 11 and 12, were prepared by adding LiOH and LiCl to DIW and adjusting the solution to the desired pH value with 15.8 M HNO_3 or 1 M LiOH. Table 2 contains a summary of the *in situ* pH values computed at each test temperature with the EQ3NR code [version 7.0, [17]]. It is important to take into account the change in pH that occurs at different temperatures when computing dissolution rates from SPFT data as the *in situ* pH can vary by as much as 1.5 pH units over the temperature range from 23° to 90°C.

Influent silicon concentration was varied from dilute to saturated with respect to amorphous silica [SiO_2(am)], based on the results from calculations using EQ3NR [17] for select experiments. It is important to note that the solubility behavior of SiO_2(am) changes with temperature; therefore, the target amount of Si added to each solution was adjusted to correspond to the experimental temperature being interrogated, resulting in Si solution concentrations that ranged from 15 to 140 ppm. These solutions were prepared by dissolving analytical grade silicic acid powder ($SiO_2 \cdot H_2O$) in a solution of 0.05 M THAM buffer and heating the mixture in a 90°C oven for no less than three days to facilitate complete dissolution of the $SiO_2 \cdot H_2O$ powder. Upon complete dissolution, each solution was removed from the oven, allowed to cool, and pH adjusted (target pH = 9.0) using aliquots of 15.8 M HNO_3 or 1 M LiOH.

Table 2. Composition of solutions used in SPFT experiments. Solution pH values above 23°C were calculated with EQ3NR Code V7.2b database.

Solution	Composition	pH@	
		23°C	90°C
1	0.05 M [a]THAM + 0.047 M HNO_3	7.01	5.55
2	0.05 M THAM + 0.02 M HNO_3	8.32	6.89
3	0.05 M THAM + 0.0041 M HNO_3	8.99	7.72
4	0.05 M THAM + 0.003 M LiOH	9.99	8.52
5	0.0107 M LiOH + 0.010 M LiCl	11.00	10.06
6	0.0207 M LiOH + 0.010 M LiCl	12.02	10.70
7	0.05 M THAM + [b]Si	9.00	8.27

[a]THAM = *tris* hydroxymethyl aminomethane buffer.
[b]Correspond to the pH at Si saturation with respect to SiO_2(am).

Quantification of SPFT dissolution rate

Dissolution rates, based on steady-state concentrations of elements in the effluent, are normalized to the amount of the element present in the sample by the following formula:

$$r_i = \frac{(C_i - C_{i,b})q}{f_i S_{geo}} \tag{3}$$

where r_i is the normalized glass dissolution rate based on the release of element i [g/(m^2·d)], C_i is the concentration of element i in the effluent (g/m^3), $C_{i,b}$ is the average background concentration of element i in the effluent (g/m^3), q is the flow rate (m^3/s), f_i is the mass fraction of element i in the glass (dimensionless), and S_{geo} is the geometric surface area of the sample (m^2). The average background concentration was typically below the estimated limit of quantification (i.e., 10 to 30 µg/L for Al, 25 to 50 µg/L for B, 500 µg/L for Na, and 500 µg/L for Si). The 2σ experimental uncertainty for SPFT-measured dissolution rates were determined by error propagation and resulted in uncertainties of approximately ±30% (or ±0.2 log units when reported as $\log_{10}$ rates).

For all experiments (e.g., SPFT and PUF) discussed in this paper, boron was used as the index for glass dissolution. Boron is considered the most reliable index for glass dissolution because it serves as a glass network former, like Al and Si, but is not sequestered by alteration phase(s) after being released from the glass. Unlike B, other elements may be retained in alteration phases during the glass-water reaction. Therefore, an element with a lower normalized release rate, in comparison to B indicates the element is being retained by an alteration phase or phases, whereas a higher normalized release rate suggests preferential release, such as alkali-hydrogen ion exchange.

Pressurized unsaturated flow apparatus test method and solution analysis

The PUF apparatus allows for accelerated weathering experiments to be conducted under hydraulically unsaturated conditions, thereby mimicking the open-flow and transport properties expected to occur in the IDF environment while allowing the corroding glass to achieve a final reaction state. The PUF apparatus provides the capability to vary the volumetric water content from saturation to 20% or less, minimize the flow rate to increase liquid residence time, and operate at a maximum temperature of 99°C. The PUF column operates under a hydraulically unsaturated condition by creating a steady-state vertical water flow, while maintaining uniform water content throughout the column; by using gravity to assist in drainage; and by maintaining a constant pressure throughout the column. Constant pressure is maintained with a porous Ti plate and gas pressure. For additional details on the PUF system see Pierce et al. [2,13] and the corresponding references contained therein.

The PUF column was packed with 33.35 g of crushed and cleaned LAWA44 glass, which resulted in a fill volume of 12.36 cm^3 and void volume of 9.36 cm^3. The mass difference between the packed and empty column was used to calculate the initial porosity of approximately 0.43 ±0.03 (unitless). After packing, the column was vacuum saturated with DIW at ambient temperature. A temperature controller was then programmed to heat the column to 90°C in approximately 1 hr (1°C min^{-1}). The column initially was allowed to desaturate by gravity drainage during heating and was also vented periodically to maintain an internal pressure less than the bubble pressure of the porous plate. After reaching 90°C, the influent valve was opened, and influent was set to a flow rate of 2 mL d^{-1}. The influent reservoir containing DIW was maintained at ambient temperature and periodically refilled during the 500 day test period. Column venting was set to occur once an hour, so the partial pressure of O_2 and CO_2 could remain relatively constant. Effluent samples were collected into tared vials from which samples were extracted and acidified for elemental analysis with inductively coupled plasma-optical emission spectroscopy (ICP-OES) and -mass spectrometry (ICP-MS).

Quantification of PUF dissolution rate

The results from chemical analyses on collected effluent samples are used to calculate a normalized release rate according to McGrail et al. [18]

$$r_i = \frac{4\varepsilon Q(c_{iL} - c_{ib})}{\theta s(1-\varepsilon)\rho\pi d^2 L f_i} \qquad (4)$$

where r_i is the normalized rate of ith element [g/(m^2·d)], c_{iL} is the effluent concentration of ith element (g/m^3), c_{ib} is the background concentration of ith element (g/m^3), d is the column diameter (m), f_i is the mass fraction of ith element in the glass (unitless), L is the column length (m), Q is the volumetric flow rate (m^3/d), s is the specific surface area of the glass sample (m^2/g), ε is the porosity (unitless), ρ is the glass density (g/m^3), and θ is the volumetric water content (unitless). Similar to SPFT rates, the 2σ experimental uncertainty for PUF-measured dissolution rates were determined by propagation of error and resulted in uncertainties of approximately ±30% (or ±0.2 log units when reported as $\log_{10}$ rates).

STORM simulations

The STORM simulator [19] couples unsaturated flow, chemical reaction, and contaminant transport to simulate corrosion of the LAWA44 glass in the PUF column [20]. This simulator was developed by coupling Sub-surface Transport Over Multi-Phases (STOMP), a non-isothermal multiphase flow simulator [21], with AREST-CT Version 1.1, a reactive transport and porous medium alteration simulator [22,23]. The underlying mathematics in STORM is contained in a set of coupled, nonlinear, partial differential equations that represent conservation equations for energy, mass, and momentum. These equations describe the rate of change of the solute concentrations of pore water in a variably saturated, non-isothermal porous medium. The capabilities of STORM include: (1) kinetic dissolution of glass, (2) kinetic precipitation and dissolution of secondary phases, (3) aqueous equilibrium speciation, (4) gas-aqueous equilibria, (5) two-phase flow (water and air), and (6) dynamic hydraulic properties. Assuming the parameters in Equation (1) are all known, the kinetic dissolution of glass is calculated by determining the pH and ion activity product, Q, at each grid node for each time step to build-up a time and spatially dependent picture of glass dissolution during the PUF experiment. This approach implies that under these experimental conditions the activity of rate influencing variables is minimal. STORM runs efficiently in parallel on multi-core workstations and supercomputers, shortening execution times. The verification studies for STORM are documented in the STORM user's guide [19].

The PUF column was modeled as a vertical, one-dimensional domain, 7.62 cm long, divided into 100 grid cells. The cross sectional area of the column was assumed to be 2.856 cm^2. The upper boundary condition assumed was a constant water influx of 0.70175 cm/day, which, multiplied by the cross-sectional area gives a flow rate of 2 mL/day. The column temperature was assumed to be a constant 90°C. The inflowing and initial water in the column is assumed to be DIW with all total species concentrations equal to 1×10^{-10} mol/kg and pH equal to 7.

Secondary phases are solids that precipitate from a supersaturated aqueous solution. A list of potential secondary phases that form from long-term weathering experiments with the various waste glass formulations and from modeling the solution chemistry observed in experiments with the EQ3/6 code is provided in [24,25]. The secondary phase and corresponding log K at 90°C added to geochemical database for the simulations were Anatase (TiO_2, log K = -8.0), Baddeleyite (ZrO_2, log K = -10.0), Clinochlore-14Å [$Mg_5Al_2Si_3O_{10}(OH)_8$, log K = -66.4], Diaspore ($AlHO_2$, log K = -4.7), and Zinc hydroxide (γ-$Zn(OH)_2$, log K = -14.5). The log K values were corrected for temperature and taken from the EQ3/6 data0.com.R8 database [26]. Clinochlore-14Å, of unknown composition, was the only phase positively identified by XRD analysis of the reacted glass. The other phases used in the simulations were predicted to form by reaction progress modeling of PCT experiments on LAWA44 glass at 90°C with the EQ3/6 computer code [27].

The aqueous geochemical reactions, not listed, were identified by simulating the dissolution of waste glass in DIW at 90°C with the EQ3/6 code package [27]. All data were obtained from the EQ3/6 data0.com.R8 database [26]. The EQ3/6 software was used to extract a subset of aqueous (and solid) species from the large thermodynamic database that were relevant for the reactive transport simulations.

RESULTS

SPFT results

Although other factors can impact the glass-water reaction, the two factors expected to have the greatest influence under conditions expected in the IDF are solution pH and $a[SiO_2\ (aq)]$. Figure 1 shows the steady-dissolution rate, indexed by B release, at 90°C as a function of pH(*T*). These results illustrate that as pH increases the overall glass dissolution rate also increases for LAWA44. This direct relationship between the dissolution rate and pH, going from the neutral to alkaline pH range, is typical of natural glasses [16] as well as other LAW glass formulations [28]. Conducting a linear regression on the data gave a slope $\eta = 0.41 \pm 0.06$.

Also shown in Figure 1, are results from experiments conducted with input solutions doped with Si (ranging from 15 to 140 ppm) at pH(23°C) = 9.0 and 90°C. Before discussing these results, it is important to determine the aqueous Si speciation at 90°C to correctly account for the change in Si speciation with temperature in the analysis. Therefore, the solution speciation of dissolved Si species was computed with the aid of the geochemical code EQ3NR [17]. These modeling calculations predicted that greater than 87% of the total dissolved Si is SiO_2 (aq) with the remaining 13% as $HSiO_3^-$.

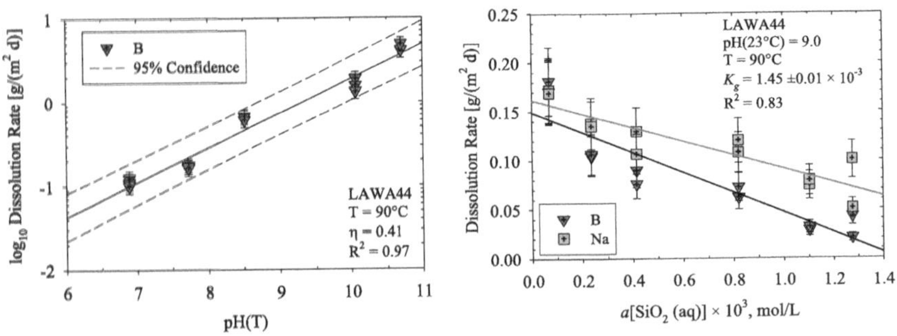

Figure 1. Effect of solution pH (left) and $a[SiO_2\ (aq)]$ (right) on the normalized release of B at 90°C for LAWA44 Glass. Effect of $a[SiO_2\ (aq)]$ on the normalized Na release is also shown.

The normalized release rates, based on B and Na release, decrease with increasing $a[SiO_2(aq)]$ (Figure 1). This decrease occurs because the chemical potential difference between the glass and contacting solution decreases. In the case of a mineral or solid phase, this effect can be expressed mathematically as the chemical affinity of reaction,

$$A = RT \ln\left(\frac{K}{Q}\right) = -\Delta G_r \quad (5)$$

where A is the chemical affinity (kJ/mol), ΔG_r is the free energy of reaction (kJ/mol), R is the gas constant [J/(mol·K)], T is the temperature (K), K is the equilibrium constant (unitless), and Q is the ion activity product (unitless) [19].

Chemical affinity is a measure of the departure from equilibrium. However in this system, chemical affinity is difficult to define because glass cannot be in equilibrium with aqueous solution. Nonetheless the release of elements from the glass matrix decreases as the chemical potential between the glass and contacting solution decreases. This effect is referred to as a condition of pseudo-equilibrium (K_g). The effect of chemical affinity on the glass dissolution rate under alkaline conditions has been attributed to an increase in the dissolved Si concentration in the contacting solution [20,21]. A previous approach that has been used to describe the effect of chemical affinity is to use the value for K of a simple SiO_2 phase, such as chalcedony or SiO_2 (am) [22], and allow this value to represent K_g in the model. In other words, one way to approximate the K_g for silicate-based glass is to use K of a mineral that is structurally and chemically similar. Here we estimate K_g [$(1.45 \pm 0.01) \times 10^{-3}$ mol/L] by conducting a linear regression on the dissolution rate data, based on B release. The value for K_g represents the $a[SiO_2 (aq)]$ when the glass dissolution rate decreases to a residual rate. This value is greater than the equilibrium constant for SiO_2 (am) (5.31×10^{-3} mol/L) but lower than quartz [SiO_2 (c)] (6.64×10^{-4} mol/L).

PUF results

Over the 500-day duration, the glass corrosion rate (not shown) has slowed from an early-time rate of approximately 0.2 to 0.01 g/(m^2·d). This rate is similar to the rate observed in SPFT experiments conducted in solutions saturated with respect to SiO_2 (am). In comparison to previous PUF tests, the B release rate for LAWA44 glass is anywhere from 8 to 40 times lower than the Hanford LAW Product (HLP) glass series; HLP-9 glass [0.1 to 0.08 g/(m^2·d)], HLP-10 glass [0.1 to 0.08 g/(m^2·d)], and HLP-31 [0.4 [g/(m^2·d)], and 10× lower than LAWABP1 glass [0.1 g/(m^2·d)] [28]. Therefore, these PUF test results indicate that LAWA44 glass performs well and is one of the more durable ILAW glasses tested to date.

At approximately day 135, the DIW influent in the PUF test with LAWA44 glass was changed out for the 50 mg/L humic acid solution. No change in electrical conductivity or solution pH occurred immediately after the change to the humic acid solution. Additionally, there was no change in the dissolution rate, although the higher Na content of the humic acid solution is clearly detected over the injection period (from day 135 to day 165). Essentially identical results were obtained with the fulvic acid solution, i.e. no statistically significant change in dissolution rate was observed. These results are in accord with what was observed in SPFT experiments with the same organic acids [28].

After termination of the PUF test with LAWA44 glass, the reacted solids were sub-sampled as found (loose and moist particles) at 1 to 5 mm intervals. These samples were analyzed for moisture content; by drying in glass vials at room temperature in a vacuum desiccators containing $CaSO_4$ desiccant. Once dried the samples were analyzed for alteration phases with X-ray diffraction (XRD). The background subtracted XRD patterns (not shown) taken from the upper region of the column (highest moisture content) contain low intensity broad reflections typical of clay minerals. The diffraction peaks appear to come from a first order basal reflection of 6.2° 2θ (14.1Å). Chlorite, a phyllosilicate mineral, has a basal series of diffractions peaks based on a first order reflection of 14.2 Å, with the odd ordered peaks having weak intensities as the iron content increases [29]. Chlorite minerals commonly have a considerable range of substitutions for both trivalent and divalent ions, hence the phase ID of clinochlore should be considered qualitative in terms of the chemical makeup of the clay mineral. In contrast to the sample removed from the upper region of the column, a sample taken further downstream (lowest moisture content) shows only a very thin alteration layer. High magnification scanning electron microscope images show sparse precipitates located on portions of the layer but these were not identifiable crystalline phases in the XRD analysis.

DISCUSSION

A large volume of SPFT results on LAWA44 at 90°C have been collected to derive the input parameters necessary to conduct long-term calculations with the STORM computer code. By incorporating these parameters and using Equation (6), where $\vec{k}$ = 1.3 x 10^4 g/(m^2 d), the elemental concentration in the effluent solutions exiting the PUF experiment was predicted with the STORM code. The estimated $\vec{k}$ is determined with SPFT data collected over a range of temperatures and pH values, see Pierce et al. [24] for more details. The value for the log K_g was adjusted to match the observed effluent B concentrations (Figure 2) during the simulations from a value of log K_g = -2.84 to a value of -2.1, which is 5.5 times higher than the value predicted in SPFT experiments (Figure 2).

$$r = \vec{k}\,10^{(0.45)(\mathrm{pH})}\left(1-\left(\frac{Q}{0.0079}\right)^{\sigma}\right) \quad (6)$$

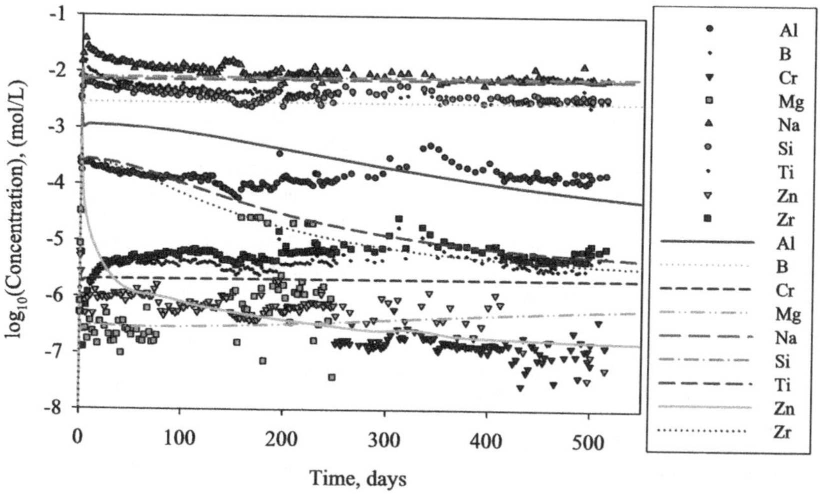

Figure 2. Log_{10} Concentration (mol/L) for Selected Elements Released from PUF Tests with LAWA44 Glass. Data points represent measured results and lines represent predicted results from the STORM simulation.

Results from the analyses of effluent solution samples along with the 1-D STORM simulation is illustrated in Figure 2 for the major and a few minor elements contained in the LAWA44 glass. In general Figure 2 shows that measured and predicted values for B, Na, and Si and to some extent Mg are within the experimental error of one another for the majority of the experiment, especially at the later times. Unlike the afore-mentioned elements, Al, Cr, Ti, Zn, and Zr show a significant difference between the measured and predicted values. This discrepancy between measured and predicted for the majority of these elements is because of the kinetic constraints that exist as aqueous species are converted to amorphous precursors which then transform to the more stable crystalline parent phase. The log K values taken from EQ3/6 and used in the simulator are for the crystalline parent mineral, which under these dynamic conditions takes time to form. Given the limitation of these calculations and analytical uncertainty of the measured concentrations, the overall agreement between measured and predicted values is reasonable, especially at the later times.

SUMMARY

Parameterization of a kinetic rate equation was performed through a combination of laboratory experimentation, extraction of relevant data from the literature (principally thermodynamic data), and parameter estimation. This kinetic equation was added to a reactive chemical transport code and used to predict element release from a 1.5-year long accelerated weathering experiment conducted with the PUF apparatus on a prototypic ILAW glass, LAWA44. The reactive chemical transport simulation takes into account the coupled effects of fluid flow and glass-water reactions on the chemistry of fluids percolating through the PUF experiment. The fluid chemistry is coupled with a kinetic rate equation that describes the response of the glass dissolution rate to changes in fluid composition during the PUF experiment, all computed as a function of time and space. This kinetic rate equation assumes (1) the dependence of dissolution and precipitation rates on departure from equilibrium are based on arguments and assumptions of TST and (2) the driving force for the transformation of unstable to stable minerals is governed principally by the magnitude of displacement from thermodynamic equilibrium. Although additional model refinements are needed, these results are an example of the type of data needed to provide a sound scientific basis for determining the long-term release rates of radionuclides from LAW glasses over a period of regulatory concern. The results from these bench-scale experiments will be extended to a set of field lysimeter experiments (i.e., experiments on buried glass at the Hanford Site) that have been in progress for 6 years on several glass specimens.

REFERENCES

[1] R. J. Puigh Flour Federal Services, HNF-4950, Rev. 1, (**1999**).
[2] E. M. Pierce, B. P. McGrail et al. *Applied Geochemistry* **2007**, *22*, 1841-1859.
[3] B. P. McGrail, D. H. Bacon et al. *Journal of Nuclear Materials* **2001**, *298*, 95-111.
[4] B. P. McGrail, D. H. Bacon et al. Pacific Northwest National Laboratory, PNNL-11834 Rev.1, (**2000**).
[5] B. P. McGrail, D. H. Bacon et al. Pacific Northwest National Laboratory, PNNL-14362, (**2003**).
[6] F. M. Mann, R. J. Puigh et al. U. S. Department of Energy, DOE/ORP-2000-24, Rev.0, (**2001**).
[7] P. Aagaard and H. C. Helgeson. *American Journal of Science* **1982**, *282*, 237-285.
[8] B. Grambow. Material Research Symposium Proceedings **1985**, 44, 15-27.
[9] H. Eyring. *J. Chem. Phys.* **1935**, *3*, 107-114.
[10] J. S. Small, P. N. Humphrey et al. In *Scientific Basis for Nuclear Waste Management XXIII*; Smith, R. W. and Shoesmith, D. W., Eds.; Materials Research Society: Boston, Massachusetts, (**2000**); Vol. 608, pp 129-134.
[11] P. K. Abraitis, F. R. Livens et al. *Applied Geochemistry* **2000**, *15*, 1399-1416.
[12] E. M. Pierce, E. L. Richards et al. *Environmental Chemistry* **2008**, *5*, 1-13.
[13] E. M. Pierce, B. P. McGrail et al. *Nuclear Technology* **2006**, *155*, 133-148.
[14] B. P. McGrail, W. L. Ebert et al. *Journal of Nuclear Materials* **1997**, *249*, 175-189.
[15] B. P. McGrail, J. P. Icenhower et al. Pacific Northwest National Laboratory, PNNL-13381, (**2000**).
[16] D. Wolff-Boenisch, S. R. Gislason et al. *Geochimica Cosmochimica Acta* **2004**, *68*, 4843-4858.

[17] T. J. Wolery Lawrence Livermore National Laboratory, UCRL-MA-110662 PT III, (**1992**).
[18] B. P. McGrail, J. P. Icenhower et al. Pacific Northwest National Laboratory, PNNL-14145, (**2002**).
[19] D. H. Bacon, M. D. White et al. Pacific Northwest National Laboratory, PNNL-14783, (**2004**).
[20] D. H. Bacon and B. P. McGrail Pacific Northwest National Laboratory, PNNL-15198, (**2005**).
[21] M. D. White and M. Oostrom Pacific Northwest National Laboratory, PNNL-11217, (**1996**).
[22] Y. Chen, B. P. McGrail et al. In *Scientific Basis for Nuclear Waste Management XX*; Gray, W. J. and Triay, I., Eds.; Materials Research Society: Pittsburgh, (**1997**).
[23] Y. Chen, D. W. Engel et al. Pacific Northwest Laboratory, PNL-10692, (**1995**).
[24] E. M. Pierce, B. P. McGrail et al. Pacific Northwest National Laboratory, PNNL-14805, (**2004**).
[25] E. M. Pierce, B. P. McGrail et al. Pacific Northwest National Laboratory, PNNL-15126, (**2005**).
[26] S. A. Daveler and T. J. Wolery LLNL, UCRL-MA-110662 PT II, (**1992**).
[27] T. J. Wolery Lawrence Livermore National Laboratory, UCRL-MA-110662 PT I, (**1992**).
[28] B. P. McGrail, J. P. Icenhower et al. Pacific Northwest National Laboratory, PNNL-13043, Rev.2, (**2001**).
[29] D. M. Moore and R. C. Reynolds; Oxford University Press: New York, (**1989**).

THE PRODUCT CONSISTENCY TEST (PCT): HOW AND WHY IT WAS DEVELOPED

Carol M. Jantzen and Ned E. Bibler
Savannah River National Laboratory
Aiken, South Carolina 29808

ABSTRACT

The Product Consistency Test (PCT), American Society for Testing Materials (ASTM) Standard C1285, is currently used world wide for testing glass and glass-ceramic waste forms for high level waste (HLW), low level waste (LLW), and hazardous wastes. Development of the PCT was initiated in 1986 because HLW glass waste forms required extensive characterization before actual production began and required continued characterization during production (≥25 years). Non-radioactive startup was in 1994 and radioactive startup was in 1996. The PCT underwent extensive development from 1986-1994 and became an ASTM consensus standard in 1994. During the extensive laboratory testing and inter- and intra-laboratory round robins using non-radioactive and radioactive glasses, the PCT was shown to be very reproducible, to yield reliable results rapidly, to distinguish between glasses of different durability and homogeneity, and to easily be performed in shielded cell facilities with radioactive samples. In 1997, the scope was broadened to include hazardous and mixed (radioactive and hazardous) waste glasses. In 2002, the scope was broadened to include glass-ceramic waste forms which are currently being recommended for second generation nuclear wastes yet to be generated in the nuclear renaissance. Since the PCT has proven useful for glass-ceramics with up to 75% ceramic component and has been used to evaluate Pu ceramic waste forms, the use of this test for other ceramic/mineral waste forms such as geopolymers, hydroceramics, and fluidized bed steam reformer mineralized product is under investigation.

INTRODUCTION

The first U.S. facility and the world's largest facility to immobilize high-level nuclear waste in durable borosilicate glass began radioactive operation in 1996 at the Savannah River Site (SRS). The product specifications on the glass waste form produced in the SRS vitrification facility, known as the Defense Waste Processing Facility (DWPF), required extensive characterization of non-radioactive waste glass simulants made in pilot scale facilities before actual production began in the engineering scale DWPF. In addition, glass quality needed to be continuously characterized during production to ensure that the glass made in 1996 was as durable as the glass made in 2008 or as durable as glasses to be made in 2020.

To aid in the HLW glass characterization, a glass durability (leach) test was needed that was easily reproducible, could be performed remotely on highly radioactive samples, and could yield results rapidly. A test known as the Product Consistency Test (PCT) was developed between 1986 and 1994. The initial basis for the PCT was a crushed glass durability test known as the Materials Characterization Center test #3 (MCC-3) that was performed with constant agitation. Some of the attributes of the MCC-3 test, other ASTM tests, and a Corning Glass Works (CGW) crushed glass test were all incorporated into the PCT. The features of each test that had the potential to optimize the following criteria were considered:

- sensitivity of the test method to glass composition and homogeneity
- minimum time necessary to demonstrate product quality
- ease of sample preparation for radioactive glass
- precision of the test results
- acceptance by waste form developers and repository projects

The PCT is a crushed glass test performed at 90°C performed under static conditions (during initial testing the constant agitation used in MCC-3 was determined not to be necessary). The PCT became an ASTM consensus standard in 1994[1] after thorough review by ASTM Committee C26 which consisted of waste form developers, repository representatives, and Nuclear Regulatory Commission (NRC) representatives.

In 1997, the scope was broadened to include hazardous as well as mixed (radioactive and hazardous) glasses. In 2002, the scope of the PCT was broadened to encompass glass-ceramic waste forms. Glass-ceramics are currently being recommended for second generation nuclear wastes yet to be generated in the nuclear renaissance, e.g. the Global Nuclear Energy Partnership (GNEP): glass-ceramics for GNEP are being referred to as Glass Composite Materials (GCM).[2] When dealing with glass ceramics, devitrified glasses, or GCM's of any type, one must deal with the release of certain radionuclides from the ceramic or crystalline phases versus the amorphous glass. The strategy for doing this is to choose a unique element from each phase (glass and crystal) in the final waste form. For example, in glass bonded sodalite a unique element and/or radionuclide is monitored from the glassy phase, the halide phases, and the sodalite phase. For example:

- Si, Al, Na, Li (from sodalite and glass phases)
- B (glass phase)
- Cl (sodalite and halite phases)

The only manner in which a crystalline phase does not have to be separately monitored is if the phase is shown, by extensive study, to be inert and not contain radionuclides, e.g. iron bearing spinels in DWPF glasses. The rationale for determining which unique element or radionuclides should be monitored from each phase, is detailed in ASTM C1285 for glass bonded ceramics and devitrified HLW glasses.

The PCT was designed to relate waste form processing to waste form performance. It is a short-term test method yielding a response that is sensitive to both the effects of processing variables on the waste form and its chemical durability. Which attributes or constituents should be tracked will depend on the waste form and its degradation mechanism. The phase composition of multi-phase waste forms may be important to performance, or only the concentration of a key radionuclide or hazardous species.

In the case of homogeneous borosilicate HLW glasses, acceptable performance is defined as an acceptably low dissolution rate, which is controlled by maintaining the glass composition within an acceptable range. The approach can be represented in terms of the following relationships:

$$\text{process control} \leftrightarrow \text{composition control} \leftrightarrow \text{dissolution rate control} \leftrightarrow \text{performance control} \leftrightarrow \text{acceptable performance} \quad (1)$$

SCOPE OF THE PCT

The current scope of the PCT procedure states:

"These product consistency test methods A and B evaluate the chemical durability of homogeneous glasses, phase separated glasses, devitrified glasses, glass ceramics, and/or multiphase glass ceramic waste forms hereafter collectively referred to as "glass waste forms" by measuring the concentrations of the chemical species released to a test solution."

Based on the HLW waste acceptance process described below, the original scope of the PCT was to determine the chemical durability of homogeneous and devitrified HLW glasses by measuring the concentrations of the chemical species released from a crushed glass to a test solution. The test scope covered both radioactive and simulated waste glasses and their devitrified products formed during canister cooling. Between 1994 and 1996 the Mixed Waste Focus Area (MWFA) was actively pursuing vitrification of mixed wastes,[3,4,5,6] e.g. those wastes that are both Resource Conservation Recovery Act (RCRA) hazardous and radioactive. In addition, vitrification of hazardous mining wastes[7,8] and mill tailings[9] was being investigated. Members of the C26.13 and C26.07 ASTM subcommittees requested that the scope of the PCT be broadened to include mixed waste glasses and hazardous waste glasses. The revised scope was incorporated in 1997.

Since the PCT procedure was already applicable to devitrified glasses, Argonne National Laboratory – West (ANL-W) suggested that the scope be broadened to accept their Ceramic Waste Form (CWF) which is a glass bonded sodalite containing about 25 wt% borosilicate glass and 75% ceramic phases (primarily sodalite and NaCl).[10,11,12]

SIGNIFICANCE AND USE OF THE PCT

- The Product Consistency Test methods (A and B) provide data useful for evaluating the chemical durability of glass waste forms as measured by elemental release. Accordingly, it may be applicable throughout manufacturing, research, and development.
- Test Method A can specifically be used to obtain data to evaluate whether the chemical durability of glass waste forms have been consistently controlled during production (see Table I).
- Test Method B can specifically be used to measure the chemical durability of glass waste forms under various leaching conditions, for example, varying test durations, test temperatures, ratio of sample surface area (*S*) to leachant volume (*V*), and leachant types (see Table I). Data from this test may form part of the larger body of data that are necessary in the logical approach to long-term prediction of waste form behavior (see ASTM C 1174).[13]

THE PCT AND THE HIGH LEVEL WASTE (HLW) ACCEPTANCE PROCESS

Production of canistered borosilicate waste forms by waste form producers at SRS and West Valley began ~12 years before the submission of the first license application for the first federal geologic repository was being written.[14] Thus, to ensure that the DWPF and West Valley waste forms would be acceptable at the federal repository and to allow production to begin, the Department of Energy, defined a Waste Acceptance Process for DWPF and West Valley canistered waste forms.

Table I. Applications and Test Conditions for PCT-A and PCT-B Tests

	PCT Test Method A	PCT Test Method B
Type of Waste Form	Radioactive, Mixed, Simulated, Hazardous	Radioactive, Mixed, Simulated, Hazardous
Usage	During production for rapid analysis; for waste compliance	Scoping tests; Crystallization studies; Comparative waste form evaluation
Test Vessel	Unsensitized Type 304L stainless steel; vessels rated to >0.5 MPa	Unsensitized Type 304L stainless steel or Teflon vessels rated to >0.5 MPa
Test Duration	7 days ±2%	7 days ±2% or varying times
Leachant	ASTM Type I water	ASTM Type I water or other solutions
Condition	Static	Static
Minimum Mass	≥1 g	≥1 g
Particle Size	U.S. Standard ASTM - 100 to + 200 mesh	U.S. Standard ASTM - 100 to + 200 mesh or other sizes which are <40 mesh
Leachant Volume	10 ±0.5 cm^3/gram of sample mass	10 ±0.5 cc/gram of sample mass or other volume/sample masses
Temperature	90 ± 2°C	90 ± 2°C or other temperatures (note any changes in reaction mechanism)

As part of the Waste Acceptance Process, the DOE-repository program developed Waste Acceptance Product Specifications (WAPS) for the borosilicate waste forms. The WAPS identifies the characteristics that the waste forms must have in order to be compatible with the repository. DWPF must comply with the specifications to ensure that the canistered waste forms will be acceptable for disposal.

The WAPS specifications most relevant to public health and safety are those relating to release of radionuclides. WAPS Specification 1.3 relates to the ability of the vitrification process to consistently control the final waste form durability, i.e., the stability of the glass against attack by water:

1.3 Specification for Product Consistency

"The producer shall demonstrate control of waste form production by comparing, either directly or indirectly, production samples to the Environmental Assessment (EA) benchmark glass."[15]

1.3.1 Acceptance Criterion

"The consistency of the waste form shall be demonstrated using the Product Consistency Test (PCT). For acceptance, the mean concentrations of lithium, sodium and boron in the leachate, after normalizing for the concentrations in the glass, shall each be less than those of the benchmark glass described in the Environmental Assessment for selection of the DWPF waste form[16]...One acceptable method of demonstrating that the acceptance criterion is met, would be to ensure that the mean PCT results for each waste type are at least two standard deviations below the mean PCT results of the EA glass."

Lithium, sodium, and boron releases were compared to the rate at which various radionuclides were released. For example, in high level borosilicate waste glass, Tc^{99}, present at ~4.1 x 10^{-4} weight % in the waste form, was shown[17,18,19,20,21,22,23,24] to be released at the same maximum normalized glass release as boron, lithium, and sodium. Since Tc^{99} is released more rapidly than any other radionuclide from HLW glass, monitoring the boron, lithium, and/or sodium release provides a measure of the the maximum radionuclide releases expected for HLW glass. Therefore, for borosilicate glass waste forms, the leachates are routinely analyzed for boron, lithium, and sodium if these elements are present at > 1 mass % in the glass. Additional mechanistic information about high level borosilicate waste glass durability is gained by analyzing for other elements present at > 1 weight % in the glass.

In addition, data was needed on how crystallization of borosilicate waste glasses impacted the overall durability and radionuclide release. These specifications required extensive characterization of the glass product both before and after production. Before production, the DWPF had to extensively characterize the chemical durability of simulated and actual waste glasses to demonstrate that the DWPF could produce an acceptable product. After production began, the DWPF had to confirm[25,26,27] that the glass produced did, in fact, satisfy the specifications for product consistency and radionuclide release.

It should be noted that the determination of the linkage between Tc^{99} release and the boron, lithium, and/or sodium release required years of intense study.[17,18,19,20,21,22,23,24] This linkage will not be the same for every waste form tested and new marker elements and/or the determination of the radionuclide release itself may be required for new waste forms. This strategy and required testing was detailed in the ASTM 2002 revision of the PCT.

DEVELOPMENT OF THE PCT

In 1986 the Savannah River Laboratory (SRL) now the Savannah River National Laboratory (SRNL), undertook the development of a leach test specifically designed to establish conformance with the WAPS. The primary objective of the test would be to confirm that the DWPF glass product was consistently acceptable from a durability perspective. Such a leach test would be easily reproducible, capable of being performed remotely on highly radioactive samples of glass and able to yield reliable results rapidly. Several standard leach tests were examined, with a wide variety of test configurations. Tests examined included those used widely in the nuclear industry for a variety of waste forms, such as the Materials Characterization Center (MCC-1 and MCC-3)[28] test protocols, as well as ASTM tests for commercial glass (ASTM C-225)[29] and container glass classified as municipal waste (ASTM D3987).[30] The MCC tests were of lengthy test duration (~28 days) while the ASTM and Corning Glass Works (CGW) tests were of short test duration (1 hour, 1 day, or 2 day). Some tests were room temperature, some 80°C, some 90°C, and one was run at 121°C in an autoclave. The tests were screened based on the following criteria:

- Sensitivity of the test to glass quality parameters, such as composition and homogeneity. The response of the test must be dominated by the waste form. Tests which are not glass dominated are not adequate measures of waste form product quality. For example, tests have been designed to measure glass performance under repository conditions or conditions of rapid flow. These tests are dominated by repository parameters such as groundwater chemistry or flow rate or both.

- Minimum test duration necessary to demonstrate product quality with a high degree of precision. Use of a short test during production provides rapid confirmation of waste form product quality. The extensive characterization of the product before the startup of the DWPF required thousands of samples. Long-term tests would have made it difficult to get the information needed in a timely manner and more susceptible to the effects of inadvertent errors, such as power losses.
- Feasibility of remote performance of the test. Once radioactive operations began the test was performed on highly radioactive samples in shielded cell facilities with manipulators. Thus the sample preparation and test procedure were kept mechanically simple.
- Precision of the test. The test must be precise enough so that during production the possibility of obtaining incorrect indications of glass quality is minimized.
- Acceptance of test results by the federal repository.

Based on the preliminary screening of the MCC, ASTM, and CGW tests no one test completely fulfilled these objectives. However, the MCC-3 test protocol came closest to satisfying most of the criteria. Thus, a limited text matrix variant of the MCC-3 test method using the shorter test durations and static conditions characteristic of the ASTM and CGW tests was used as a starting point for development of the PCT (Version 1.0).

Based on the results of the SRNL internal round robin (described below) and technical reviews of the test protocol by experts in other laboratories, the PCT protocol was modified (Version 2.0). The parameters necessary for determining glass quality with a high degree of precision were optimized.

Based on the results of the SRNL hosted external round robin (described below), the PCT protocol was again modified (Version 3.0). Versions 3.0 to 7.0 were submitted to ASTM subcommittee C26.13 (Repository Waste Package Interactions) for consensus review between 1990 and 1994. Version 5.0 was balloted at the C26.13 subcommittee level with no negative ballots. Several affirmatives with comments were addressed in Version 6.0. The revised Version 6.0 of the PCT was balloted at the C26 (Nuclear Fuel Cycle) full committee and subcommittee level with no negative ballots. Several affirmatives with comments were addressed in Version 7.0 which was balloted at the full Society level with no negative ballots. The PCT became an ASTM standard in late 1994.

HLW Glass: SRNL Internal Round Robin[31]

An internal SRNL round robin was held in 1987 using the initial (Version 1.0) test protocol. Three researchers participated in the round robin. Three glasses were used. One was partially crystallized, one was phase separated, and one was homogeneous). The round robin had three primary objectives:

- To determine the effects of various test parameters (duration, agitation, radiation, vessel material, and filtration) on test results.
- To select a set of test responses which were both reliable and sensitive to glass quality parameters (composition and homogeneity).
- To provide initial estimates of the achievable within-laboratory precision of the test.

The PCT was found to be sensitive to both glass composition and homogeneity. Consistent relative glass durability could be achieved in only one day. However, optimum precision was achieved at seven days. The 7 day precision (for triplicate tests) for any one investigator was 2-3%. Variations between investigators were 5-7%. The PCT could be performed remotely, with the same precision. The results showed no significant effect of radiation of glass durability and demonstrated that similar results were obtained whether Teflon® or stainless steel vessels were used for non-radioactive glasses. The most sensitive and precise indicators of glass quality were found to be B, Li, Na and Si. If K was present at greater than 2 wt% in the glass, it was also a good indicator of glass quality. Filtration of leachate samples improved the precision and was found to be necessary. Agitation was found to be unnecessary for the 7 day test duration.

HLW Glass: Multi-Laboratory External Round Robin[32]

From 1988 to 1989 a seven laboratory round robin was performed using Version 2.0 of the PCT procedure. The purpose of the round robin was to better determine the inter- and intra-laboratory precision and accuracy of the PCT protocol, for use in establishing product compliance. The participants were selected based on their experience in glass testing, and included the Materials Characterization Center (MCC), Argonne National Laboratory (ANL), Catholic University of America (CUA), Corning Engineering Laboratory Services (CELS), Pacific Northwest National Laboratory (PNNL), SRNL, and the University of Florida (UF).

Four glasses were used in the multi-laboratory round robin. These included a National Institute of Science and Technology (NIST) reference glass (SRM-623), an MCC standard reference glass (ARM-1), and two glasses which were based on possible compositions to be produced in the DWPF. The latter two glasses were fabricated and analyzed by CELS.

The MCC prepared test kits for each of the participants. The MCC crushed and sieved samples of glass, cleaned leach vessels, provided filters, sample handling equipment, and the test protocol (Version 2.0). In addition to the test materials, 100 mL of a multi-element standard solution for chemical analysis was also supplied. This allowed estimation of the effect of lab-to-lab variability in chemical analysis, an independent measurement of the variability in actually performing the PCT protocol.

Each of the laboratories tested all four glasses. Each laboratory tested at least one sample of each glass per week, for three consecutive weeks. Each laboratory tested one of the simulated waste glasses in triplicate during the first week. Each laboratory analyzed its own leachate as well as the multi-element solution standard. Leachate samples and the multi-element solution standard were analyzed for Na, Li, K, Al, Si, Fe, B, F, Cl, NO_3, and SO_4. The pH of all solutions was also measured.

The results of this external round robin confirmed the results of the SRNL round robin. The PCT was determined to be a precise indicator of glass quality. After statistical analysis of the results, the MCC concluded that a laboratory experienced in performing the PCT (i.e., one able to control the test precision) would be able to discriminate between glasses whose leach response differed by $\leq$10% (based on B, Na, Si, and Li), at the 95% confidence level. This was approximately four times superior to the results from around robin of the MCC-1 test also conducted by the MCC. The single researcher precision varied between 1.8-2.3%. The pooled total within and between laboratory precision varied from 7.5-10%.

The glasses tested in the multi-laboratory round robin had not been "washed" of electrostatically adhering fines (Figure 1). During the analyses of the results of the multi-laboratory round robin it was determined that the fines that eletrostatically cling to the larger (100-200 mesh particles) gave a non-reproducible contribution to the PCT response and so a "washing protocol" to remove the electrostatic fines needed to be added to the PCT protocol. The washing protocol was incorporated into Version 2.1 of the PCT procedure.

HLW Glass: Multi-Laboratory Confirmatory Testing on Radioactive Glasses[33]

Version 2.1 of the PCT protocol was used for confirmatory testing with radioactive glasses. Two laboratories (ANL and SRNL) participated in the radioactive round robin. All of the PCT operations which involved the glass were performed remotely with master slave manipulators in shielded cell facilities. This included grinding, sieving, and washing the glasses, assembling the test apparatus (see Figure 2), and leaching in a 90±2°C oven. Stainless steel vessels were used for the radioactive testing.

The test solutions only contained a small amount of radioactivity at the end of the testing. This allowed the leachate to be removed from the shielded cells, sampled in a radioactive hood and analyzed. The results measured by ANL and SRNL for the B, Na, Li, and Si agreed within 10%, providing confirmation of the results of the multi-laboratory non-radioactive round robin.

HLW Glass-Ceramic Waste Forms (CWF): Multi-Laboratory External Round Robin[10,11,12]

ANL-W sponsored a four laboratory round robin that generated within-laboratory and between laboratory RSD data for glass-bonded ceramics which was incorporated into the PCT in 2002. The within-laboratory RSD for boron and lithium from the borosilicate glass was 8.1% and 5.3%, respectively. Sodium from the borosilicate glass and the sodalite and other minor sodium containing phases had a %RSD of 13.4%. The between-laboratory relative standard deviation for boron and lithium from the borosilicate glass was 19% and 12%, respectively. Sodium from the borosilicate glass and the other sodium containing phases had a %RSD of 18%.

PCT TEST DISCRIMINATION

An independent study[34] was performed at PNNL to evaluate the discrimination of the PCT test by varying the test duration (3 vs. 7 days), varying the SA/V ratio (5, 10 and 100 mL/g), washing of fines (with and without), and filtering (with 0.45μm and without). All four factors were found to influence the solution pH and the elemental releases from the glasses. The SA/V ratio and leachate filtration were found to have the largest effect on the precision of the elemental releases. Larger SA/V ratios (more glass to less solution) and leachate filtration improved the discrimination of the test. Washing of the fines suggested that this procedure might dampen the discrimination ability of the test but washing of the fines was found to improve the precision of the replicates.

The results of the PCT can be expressed in different units, e.g. NC_i (g/L), NL_i (g/m^2), or NR_i (g/m^2•day) where "i" is the element or radionuclide of interest in the waste form. Each unit is representative of either the grams of glass dissolved or leached in the PCT based on the element "i". The calculation of NC_i (g/L) does not require the surface area to be measured or calculated but assumes that similar waste forms are being compared, e.g. that the waste form has the same density and surface roughness as the waste form that it is being compared to. However, when reporting the PCT results in units of NL_i (g/m^2) and/or NR_i (g/m^2•day) a surface area term is used in the calculation. For these units of reporting, the manner in which the surface area is determined (calculated or measured) is extremely important.

Figure 1. Adherent fines on – 100 to + 200 mesh glass particles before washing (Top). Glass particles after washing in ASTM Type I Water and Ethanol (Bottom). Ethanol only is specified when water soluble phases are present.

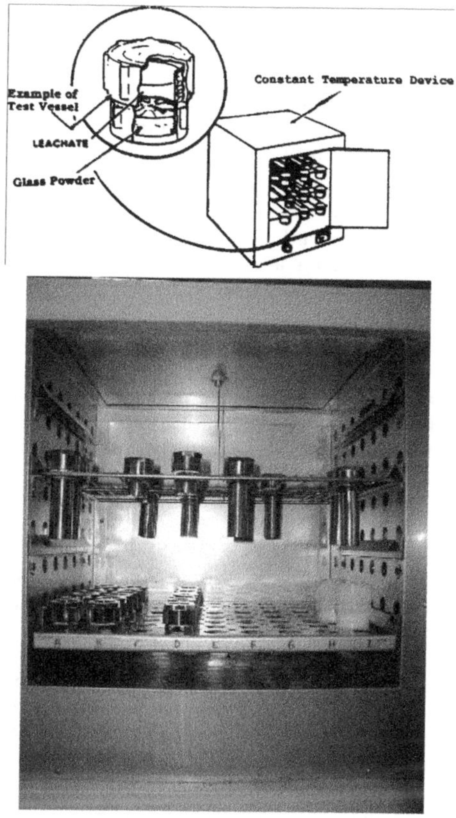

Figure 2. Test apparatus showing both stainless steel and Teflon® vessels.

RELATION OF PCT RESULTS TO OTHER DURABILITY TEST RESPONSES

The surface area (S) to volume (V) of leachant ratio times the test duration in days, a parameter known as (S/V)•time, is used as a test acceleration parameter and has widely been used to relate the response of a variety of laboratory glass durability tests to each other. For example:

- short-term crushed glass tests (PCT and MCC-3) have been related to long-term crushed glass tests (PCT and MCC-3)[35,36,37,38] by increasing S, decreasing V, or increasing time

- short and long-term monolith tests (PCT and MCC-1) have been related to short and long-term crushed glass tests (PCT and MCC-3) [37,38,39]
- short-term crushed glass tests (PCT) have been related to long-term burial tests[40]
- long-term crushed glass tests (PCT) have been related to shorter term, higher temperature, Vapor Hydration Test (VHT) responses, e.g. the HLW Environmental Assessment (EA) glass reaches the same stage of durability within 56 days at 20,000 m^{-1} or >313 days at 2000 m^{-1} when tested by PCT at 90°C[39] or within 6 days when tested by VHT at 200°C[39]
- the forward rate of the short term crushed glass test (PCT) has been shown to be an upper bound for accelerated durability behavior (the return to the forward rate or Stage III leaching behavior) [14]

The relation of the different test responses demonstrates that the test responses are related mechanistically by the acceleration factors being used in the leaching protocols. When performing comparative test results care must be taken[41] during data interpretation since different pH values are achieved during static testing at different S/V ratios than in dynamic testing in buffer solutions. This affects the reaction rates and must be accounted for when comparing the results of the various durability tests.

COMPARISON OF PCT GLASS RESPONSE TO OTHER WASTE FORMS

When it is necessary to compare the response of ceramic/mineral or GCM waste forms to glass, the PCT is often the initial test used because of its simplicity, response, and reproducibility. If the ceramic/mineral or GCM waste form has significant surface roughness this presents variability in the PCT response as well as for other crushed waste form tests such as the Single Pass Flow Through Test (SPFT), and the Pressure Unsaturated Flow-through (PUF) test. McGrail[42] has recommended that the geometric methodology of the determination of surface area given in the PCT test protocol (ASTM C1285) and SPFT (ASTM C1662) protocol is the correct way to assess the surface area of vitreous waste forms. He has also recommended that the BET surface area is the correct way to assess the surface area of mineral products due to their additional surface roughness in SPFT and PUF tests.[43] McGrail has also used the pore size measurement and calculation of surface area for the estimation of the surface area for SPFT testing of the foamy glass created by bulk vitrification,[44] i.e. there is no uniform surface area measurement methodology.

When the PCT was used for the ANL CWF (glass bonded sodalite) the geometric methodology of determination of surface area was used[45,46,47] except when the particle size was too small to determine via sieving, then the Gaussian particle size distribution was measured[48] by particle size analysis. For other applications of the PCT to Pu ceramics,[49,50,51] geopolymers,[52] grout,[53], hydroceramics,[54] and fly ash,[55] the geometric surface area has been used.

Pareizs, et. al.[56] and Jantzen, et. al.[57] initiated a comparison of geometric versus measured (Microtrac and BET) surface area for various mineral waste forms with high surface roughness and these comparisons are in progress. The appropriate surface area determination, as a function of surface roughness, is an area that needs further investigation for all the crushed waste form durability tests such as the PCT, the SPFT, and the PUF test.

CONCLUSIONS

Over the last 21 years the PCT has been shown to be a robust and short duration test. The approach can be represented in terms of the following relationships:

process control ↔composition control ↔dissolution rate control ↔ performance control ↔acceptable performance

While the current scope of the PCT does not currently include multiphase ceramic waste forms, progress has been made in establishing some of the relationships above, e.g. the process control ↔composition control ↔dissolution rate control, so that such applications can now begin to be considered. This will involve determining the best way to measure the S/V parameter for mineral waste forms versus glass for the PCT, the SPFT, and the PUF test. The 2008 revision of the PCT is currently used world wide for testing of glass and glass-ceramics and has been used successfully on ceramic waste forms although this application is not specifically covered in the current scope, significance and use statements in the standard.

REFERENCES

1. Standard Test Methods for Determining Chemical Durability of Nuclear Waste Glasses: The Product Consistency Test (PCT) Standard C1285, Annual Book ofASTM Standards Vol. 12.01, Amer. Society for Testing and Materials, West Conshohocken, PA (2008).
2. M.T. Peters, R.C. Ewing, and C.I. Steefel, "GNEP Waste Form Campaign Science & Technology and Modeling & Simulation Program: Roadmap with Rationale & Recommendations," U.S. DOE Report GNEP-M50-3040-303 and GNEP-M50-3030-101 (March, 2008).
3. Moghissi, A.A., Blauvelt, R.K., Benda, G.A., and Rothermich, N.E. (Eds.), Proceed. of the Second Intl. Mixed Waste Symposium, Baltimore, MD, Univ. of Maryland (1993).
4. Frank, C.W. and Steele, J.L., Spectrum 94 Nucl. and Haz. Waste Mgt., Am. Nuclear Soc. (1994).
5. Moghissi, A.A., Blauvelt, R.K., Benda, G.A., and Rothermich, N.E (Eds.), Proceed. of the Third Intl. Mixed Waste Symposium, Baltimore, MD, Univ. of Maryland (1995).
6. Post, R.G. and Wacks, M.E. (Eds.), Proceed. of Waste Mgt. '95, '96, '97, University of Arizona, Tucson, AZ (1995-1997).
7. Jantzen, C. M., Pickett, J.B. and Schumacher, R.F. "Mining Industry Waste Remediated for Recycle by Vitrification: A Case Study," WSRC-MS-99-00469, Canadian Env. Protection Mag. (July 1999).
8. Jantzen, C.M., Schumacher, R.F. and Pickett, J.B. "Mining Industry Waste Remediated for Recycle by Vitrification," Environmental Issues and Waste Mgt. Technologies, D.R. Spearing, et.al. (Eds.), Am. Ceram. Soc., Westerville, OH, VI, 65-74 (2001).
9. Jantzen, C.M. Pickett, J.B., and Richards, R.S., "Vitrification of Simulated Fernald K-65 Silo Waste at Low Temperature," Ceram. Trans., v. 93, 107-116 (1999).
10. Morss, L.R., Stanley, M.L., Tatko, C.D., Ebert, W.L. "Corrosion of Glass Bonded Sodalite as a Function of pH and Temperature," Sci. Basis for Nucl. Waste Mgt., XXIII, Mat. Res. Soc., Pittsburgh, PA, 733-738 (2000).
11. Sinkler, W., O'Holleran, T.P., Frank, S.M., Richmann, M.K., Johnson, S.G. "Characterization of A Glass-Bonded Ceramic Waste Form Loaded with U and Pu," Sci. Basis for Nucl. Waste Mgt., XXIII, R.W. Smith and D.W. Shoesmith (Eds.), Mat. Res. Soc., Pittsburgh, PA, 423-429 (2000).
12. Moschetti, T., Sinkler, W., Disanto, T., Hois, M.H., Warren, A.R., Cummings, D., Johnson, S.G., Goff, K.M., Bateman, K.J. , Frank, S.M. "Characterization of a Ceramic Waste Form Encapsulating Radioactive Electrorefiner Salt," Sci. Basis for Nucl. Waste Mgt., XXIII, Mat. Res. Soc., Pittsburgh, PA, 577-582 (2000).
13. ASTM C-1174, Annual Book of ASTM Standards, Vol. 12.01: Standard Practice for Prediction of the Long-Term Behavior of Materials, Including Waste Forms, Used in Engineered Barrier Systems (EBS) for Geologic Disposal of High–Level Radioactive Waste.

14. Ebert, W. "Defense High Level Waste Glass Degradation," Office of Civilian Radioactive Waste Management Analysis/Model, ANL-EBS-MD-000016, Rev.00 ICN01 (Dec. 2000).
15. Jantzen,C.M., Bibler, N.E., Beam, D.C. and Pickett, M.A., "Characterization of the Defense Waste Processing Facility (DWPF) Environmental Assessment (EA) Glass Standard Reference Material," U.S. DOE Report WSRC-TR-92-346, Rev. 1, Westinghouse Savannah River Company, Aiken, SC (February 1993).
16. U.S. Department of Energy, "Environmental Assessment-Waste Form Selection for SRP High-Level Waste," U.S. DOE Report DOE/EA-0179, Washington, DC (1982).
17. Bates, J. K., Lam, D. J., and Steindler, M. J., "Extended Leach Studies of Actinide-Doped SRL 131 Glass," Sci. Basis for Nucl. Waste Mgt., VI, North-Holland, NY, 183–190 (1983)
18. Bibler, N. E. and Jurgensen, A. R., "Leaching Tc-99 from SRP Glass in Simulated Tuff and Salt Groundwaters," Sci. Basis Nucl. Waste Mgt., XI, Mat. Res. Soc., Pittsburgh, PA, 585–593 (1988).
19. Bradley, D. J., Harvey, C. O., and Turcotte, R. P., "Leaching of Actinides and Technetium from Simulated High-Level Waste Glass," U.S. DOE Report PNL-3152, Pacific Northwest Laboratory, Richland, WA (1979).
20. Fillet, S., Nogues, J., Vernaz, E., and Jacquet-Francillon, N., "Leaching of Actinides from the French LWR Reference Glass," Sci. Basis for Nucl. Waste Mgt., IX, Mat. Res. Soc., Pittsburgh, PA, 211–218 (1985).
21. Bazan, F., Rego, J., and Aines, R. D., "Leaching of Actinide-doped Nuclear Waste Glass in a Tuff-Dominated System," Sci. Basis for Nucl. Waste Mgt., X, Mat. Res. Soc., Pittsburgh, PA, 447–458 (1987).
22. Vernaz E. Y. and Godon, N., "Leaching of Actinides from Nuclear Waste Glass: French Experience," Sci. Basis for Nucl. Waste Mgt., XV, Mat. Res. Soc., Pittsburgh, PA, 37–48 (1992).
23. Ebert, W. L., Wolf, S. F., and Bates, J. K., "The Release of Technetium from Defense Waste Processing Facility Glasses," Sci. Basis for Nucl. Waste Mgt., XIX, Mat. Res. Soc., Pittsburgh, PA, 221–227 (1996).
24. McGrail, B. P., "Waste Package Component Interactions with Savannah River Defense Waste Glass in a Low-Magnesium Salt Brine," Nuclear Technology, 1986, pp. 168–186.
25. Bibler, N.E., Ray, J.W., Fellinger, T.L., Hodoh, O.B. Beck, R.S. and Lien, O.G., "Characterization of the Radioactive Glass Currently Being Produced by the DWPF at the Savannah River Site," U.S. DOE Report WSRC-MS-97-00617, Waste Management 98.
26. Fellinger, T.L. and Bibler, N.E., "Results of the Chemical Composition and the Product Consistency Test for the DWPF Macro Batch 2 Glass Pour Stream Sample Taken During the Pouring of Canister S01142," U.S. DOE Report WSRC-RP-2000-00281, Westinghouse Savannah River Co., Aiken, SC (2000).
27. Bibler, N.E., Fellinger, T.L., and Hodoh, O.B. "DWPF Glass Results for the Analysis of a Pour Stream Sample Taken During Pouring of the 61st Canister (Canister S00482)," U.S. DOE Report WSRC-RP-98-00054, Westinghouse Savannah River Co., Aiken, SC (1998).
28. Nuclear Waste Materials Handbook-Waste Form Test Methods, J.E. Mendel (Compiler), U.S. DOE Report DOE/TIC-11400, Materials Characterization Center, Battelle Pacific Northwest Laboratories, Richland, WA (1983).
29. ASTM C-225, Annual Book of ASTM Standards, Vol. 15.02: Standard Test Methods for Resistance of Glass Containers to Chemical Attack.
30. ASTM C3987, Annual Book of ASTM Standards, Vol. 11.04: Standard Test Method for Shake Extraction of Solid Waste with Water.
31. Jantzen, C.M. and Bibler, N.E. "Product Consistency Test (PCT) and Test Protocol," U.S. DOE Report DPST-87-575, Savannah River Laboratory, Aiken, SC (1987).
32. Piepel, G.F., Jones, T.E., Eggett, D.L., Mellinger, G.B., "Product Consistency Test Round Robin Conducted by the Materials Characterization Center – Summary Report," U.S. DOE Report PNL-6967, MCC, Battelle Pacific Northwest Laboratories, Richland, WA (1989).
33. Bibler, N.E. and Bates, J.K. "Product Consistency Leach Tests of Savannah River Site Readiaoctie Waste Glasses," Sci. Basis for Nucl. Waste Mgt., XIII V.M. Oversby and P.W. Brown (Eds.), Mat. Res. Soc., Pittsburgh, PA, 327-338 (1990).
34. Shade, J.W. and Piepel, G.F. ""Evaluation of Experimental Factors that Influence the Application and Discrimination Capability of the Product Consistency Test," U.S. DOE Report PNL-7530 (June 1991); also released as an attachment to Jantzen, C.M., "Evaluation of Experimental Factors that Influence the Application and Discrimination Capability of the Product Consistency Test," U.S. DOE Report, WSRC-TR-90-526, Rev. 1 (October 1990).

35. Ribet, S., Muller,I.S., Pegg, I.L., Gin, S. and Frugier, P. "Compositional Effects on the Long-Term Durability of Nuclear Waste Glasses: A Statistical Approach," Sci. Basis for Nuclear Waste Mgt. XXVIII, J.M. Hanchar, et. al. (Eds.), Mat. Res. Soc., Pittsburgh, PA, 309-314 (2004).
36. Muller,I.S.,Ribet,S. Pegg, I.L., Gin,S. and Frugier, P. "Characterization of Alteration Phases on HLW Glasses After 15 Years of PCT Leaching," Ceram. Trans., 176, Am. Ceram. Society, Westerville, OH, 191-199 (2006).
37. Jantzen, C.M., "Prediction of Glass Durability as a Function of Environmental Conditions," Proceedings of the Symposium on Materials Stability and Environmental Degradation, A. Barkatt et al. (Eds.), Mat. Res. Soc., Pittsburgh, PA, 143-159 (1988).
38. Jantzen, C.M. "Prediction of Glass Durability as a Function of Glass Composition and Test Conditions: Thermodynamics and Kinetics," Proceedings of the First Intl. Conf. on Advances in the Fusion of Glass, Am. Ceram. Society, Westerville, OH, p.24.1-24.17 (1988).
39. Bates, J.K., Buck,E.C., Dietz, N.L., DiSanto, T., Ebert, W.L., Emery, J.W., Fortner, J.A., Hafenrichter, L.D., Hoh, J.C., Luo, J.S., Nunez, L., Surchik, M.T., Wolf, S.F. and Wronkiewicz, D.J. "ANL Technical Support Program for DOE Office of Env. Mgt.," U.S. DOE Report ANL-96/11, Argonne National Laboratory, Argonne, IL (July, 1996).
40. C.M. Jantzen, D.I. Kaplan, N.E. Bibler, D.K. Peeler, and M.J. Plodinec, "Performance of a Radioactive High Level Waste Glass After 24 Years Burial," J. Nucl. Mat., 378, 244-256 (2008).
41. W.L. Ebert, J.K. Bates, C.R. Bradley, E.C. Buck, N.L. Dietz, and N.R. Brown, "The Long-Term Alteration of Borosilicate Waste Glasses," Ceram. Trans., V. 39, 333-340 (1994).
42. McGrail, B.P.; Ebert, W.L.; Bakel, A.J.; Peeler, D.K. "Measurement of Kinetic Rate Law Parameters on a Na-Ca-Al Borosilicate Glass for Low-activity Waste." J. Nucl. Mat. 249:175-189 (1997).
43. McGrail, B.P., Schaef,H.T., Martin,P.F., Bacon, D.H., Rodriguez, E.A., McCready, D.E., Primak, A.N., and Orr, R.D., "Initial Evaluation of Steam-Reformed Low Activity Waste for Direct Land Disposal," U.S. DOE Report PNWD-3288 (2003).
44. B.P. McGrail, E.M. Pierce, H.T. Schaef, E.A. Rodriguez, J.L. Steele, A.T. Owen, and D.M. Wellman, "Laboratory Testing of Bulk Vitrified and Steam-Reformed Low-Activity Forms to Support a Preliminary Assessment for an Integrated Disposal Facility," U.S. DOE Report PNNL-14414 (2003).
45. Lewis, M.A., Ebert, W.L., and Simpson, L.J., "Comparison of the Corrosion Behaviors of the Glass-Bonded Sodalite Ceramic Waste Form and Reference HLW Glasses," Ceram. Trans. V. 107, 281-288 (2000).
46. Ebert, W.L., Lewis, M.A. and Johnson, S.G., "Monitoring the Consistency of Multiphase Waste Forms," Am. Nucl. Soc. Winter Meeting (2001).
47. Ebert, W.L., Lewis, M.A. and Johnson, S.G., "Application of PCT to the EBR II Ceramic Waste Form," DOE Spent Nuclear Fuel Meeting, Charleston, SC (2002).
48. Lewis, M.A., Fischer, D.F., and Smith, L.J., "Salt-Occluded Zeolites as an Immobilization Matrix for Chloride Waste Salt," J. Am. Ceram. Soc. 76[11], 2826-32 (1993).
49. Bakel, A.J., Buck, E.C., Wolf, S.F., Ebbinghaus, B, Bates, J.K. and Chamberlain, D.B., "Corrosion of a Pu-Doped Zirconolite-Rich Ceramic," Ceram. Trans. V.87, 551-560 (1998).
50. Buck, E.C., Ebbinghaus, B., Bakel, A.J., and Bates, J.K., "Characterization of a Plutonium-Bearring Zirconolite-Rich SYNROC," Sci. Basis Nucl. Waste Mgt., XX, Mat. Res. Soc., Pittsburgh, PA,1259-66 (1997).
51. Bakel, A.J., Mertz, C.J., Hash, M.C. and Chamberlain, D.B., "The Long-term Corrosion Behavior of Titanate Ceramics for Pu Disposition: Rate-Controlling Processes," Sci. Basis Nucl. Waste Mgt. XXIII, Mat. Res. Soc., Pittsburgh, PA,387-391 (2000).
52. Ly. L, Vance, E.R., Perera, D.S., Aly, Z. and Olufson, K., "Leaching of Geopolymers in Deionised Water," AZojomo (Journal of Materials Online), V.3, article 3826 (May 2007).
53. Herbst, A.K., :Experimental Test Plan for Grouting H-3 Calcine," ICP/EXT-06-01166 (January 2006).
54. Bao, Y., Grutzeck, M.W. and Jantzen C.M. "Preparation and Properties of Hydroceramic Waste Forms Made with Simulated Hanford Low Activity Waste (LAW)," J. Am. Ceramic Soc., 88 [12], 3287-3302 (2005).
55. Bumrongjaroen, W., Muller, I., Schweitzer, and Livingston, R., "Application of Glass Corrosion Tests to the Reactivity of Fly Ash," World of Coal Ash Conference (May 2007).
56. Pareizs, J.M., Jantzen, C.M., Lorier, T.H., "Durability Testing of Fluidized Bed Steam Reformer (FBSR) Waste Forms for High Sodium Wastes at Hanford and Idaho," U.S. DOE Report WSRC-TR-2005-00102 (July 2005).
57. Jantzen, C.M., Pareizs, J.M., Lorier, T.H. and Marra, J.C. "Durability Testing of Fluidized Bed Steam Reforming (FBSR) Products," Ceramic Transactions V. 176, 121-137 (2006).

Green Technologies for Materials Manufacturing and Processing

GREEN PROCESS FOR RECOVERY OF COPPER

Kevin N. Barber and Allen W. Apblett
Department of Chemistry
Oklahoma State University
Stillwater, OK, 74078.

ABSTRACT

The production of metals or metal oxides from sulfide ores is problematic since it entails "roasting" of the ores to convert them to oxides that can subsequently be reduced to metals. Roasting produces large quantities of sulfur dioxide and other pollutants that must be trapped. There are several alternative approaches in which the sulfide ions are oxidized to sulfur or sulfur oxyanions leading to solubilization of the metal ions. It is typical that such reactions produce copper sulfate, the starting point for the "molybdate process" for rapid and environmentally-friendly separation of copper as copper oxide or hydroxide from solution. The process entails reaction of copper-containing solutions with a soluble molybdate salt (preferably sodium or potassium) to precipitate an amorphous copper molybdate hydroxide with a copper content of 36.6% by weight. Subsequent treatment with sodium hydroxide leads to quantitative recovery of copper as copper hydroxide and regeneration of the sodium molybdate precipitating reagent.

INTRODUCTION

The direct conversion of metal sulfides to metals via carbothermal reduction or other processes is problematic. Therefore, conventional pyrometallurgical approaches to refining of sulfide ores usually include a roasting step to convert them to oxides that are more readily reduced[1]. Unfortunately, roasting produces sulfur dioxide as a by-product a gas that contributes to acid rain. While modern smelters capture this gas and turn it into sulfuric acid (at an additional cost) that can be used in the extraction process, there is a significant interest in alternative greener processes that have lower energy needs and reduced environmental impact[2].

The focus of this investigation is copper ore refining but the processes are applicable to other transition metal sulfide ores. The most common copper ore, accounting for 50% of copper production is chalcopyrite, $CuFeS_2$. There are also a number of other sulfide ores (such as covellite, CuS, and chalcocite, Cu_2S) as well as some oxide ores such as malachite $(Cu_2(CO_3)(OH)_2$ and cuprite, Cu_2O[3]. The oxide ores are the most economic to refine since they are readily leached by sulfuric acid, usually using a heap or dump leach process with subsequent isolation of copper by the SX-EW process, a combination of solvent extraction and electrowinning[4]. Notably, the process described below could be used to isolate copper hydroxide from the initially formed sulfuric acid leachate. In the case of sulfide ores, hydrometallurgical processes have also been developed as alternatives to pyrometallurgy[5]. An excellent example is using a bacterial oxidation process to oxidize the sulfides to sulfuric acid, which also leads to simultaneous leaching with sulfuric acid[6]. As with the oxide ores, the SX-EW process can be used to recover the copper from the pregnant leach solution. The same result can be realized abiotically using sulfuric acid accompanied by aerial oxidation.

EXPERIMENTAL

All reagents were commercial products (ACS Reagent grade or higher) and were used without further purification. Water was purified by reverse osmosis followed by deionization.

Bulk pyrolyses at various temperatures were performed in ambient air in a digitally-controlled muffle furnace using approximately 1 g samples, a ramp of 10 °C/min and a hold time of 4 hr. X-ray powder diffraction (XRD) patterns were recorded on a Bruker AXS D-8 Advance X-ray powder diffractometer using copper K_α radiation (λ=1.5418 Å). Crystalline phases were identified using a search/match program and the PDF-2 database of the International Centre for Diffraction Data [7]. X-ray fluorescence spectroscopy (XRF) was performed on a Thermo Electron QuanX XRD spectrometer.

Reaction of Sodium Molybdate with Copper Sulfate

A solution of $Na_2MoO_4•2H_2O$ (5.0 mmol) dissolved in 5 ml of water and a solution of $CuSO_4•5H_2O$ (2.5 moles) in 5 ml of water were prepared in separate vials. The contents of each vial were poured into a nylon centrifuge tube causing the immediate formation of a green precipitate. Each vial was rinsed twice with 1 g of water and the rinses were added to the centrifuge tube. The reaction mixture was mixed on a roller mill for one hour and then the solid was separated by centrifugation. The supernatant liquid was poured off and the solid was resuspended in 10 ml of water and than centrifuged again. The washing process was repeated twice more with water and the three times with acetone to expedite the drying of the product. Finally, the product was dried in a 60 °C oven to constant mass to yield 0.431 g of the pale green copper molybdate hydroxide product.

Several variations on this experiment were performed in which the ratios of the reagents were changed and the mixture was subjected to heating.

Isolation of Copper Oxide by the Molybdate Process

A reaction was performed between 1.2106 g (mmol) of sodium molybdate dihydrate and 0.6242 g (2.500 mmol) of copper sulfate pentahudrate as described above. The resulting pale green copper molybdate was then treated with 5.03 g of 0.997 N sodium hydroxide. The resulting mixture was stirred with a glass rod leading to an immediate reaction to produce an aqua blue solid. After 5 minutes the mixture was centrifuged and the sodium hydroxide treatment was repeated. The solid was then washed twice with 10 ml of water followed by centrifugation. The $Cu(OH)_2$ product was then quantitatively transferred to a pre-weighed 150 ml beaker by rinsing with water. The water was evaporated at 115°C and then the residue was converted to copper(II) oxide by heating to 250°C for 6.5 hour. The yield of black CuO was 0.1986 g (2.497 mmol) corresponding to a copper recovery of 99.87%.

RESULTS AND DISCUSSION

Initial testing of the new process for refining of copper was performed with aqueous solutions of copper sulfate and sodium molybdate. When solutions of these two compounds are mixed, a pale mint green precipitate immediately forms. Experiments were performed in which the reaction mixture was either filtered without heating or was refluxed for a period of time up to one hour. It was found that the recovery of copper was quantitative after heating for 15 minutes but was only 92.7% without the heating step when the reagents were combined in a 1:1 molar ratio. Alternatively, refluxing was unnecessary when a larger excess of sodium molybdate was used such as 2 moles sodium molybdate per mole of copper sulfate. XRF spectroscopy (Figure 1) showed that the only metals present in the precipitate are copper and molybdenum. Most importantly, no sodium was detectable. Gravimetric analysis for copper (see below) showed a content of 36.6%. Thermal gravimetric analysis (Figure 2) showed that the product was a hydrate that loses water in two steps: a gradual loss of 1.48% from room temperature to 245 °C and a sharper water loss of 3.03% between 245 and 356 °C. Based on the yield of the precipitated product and the analytical information, a formula of $Cu(MoO_4)_{0.6}(OH)_{0.8}$ was calculated. X-ray powder diffraction indicated that the product was predominantly amorphous with only very weak reflections for lindgrenite, a rare copper molybdate mineral with the formula $Cu_3(MoO_4)_2(OH)_2$. Notably, the product produced in this investigation is richer in copper and water content than lindgrenite. Interestingly, a previous investigation[8] found chemically-pure crystalline lindgrenite to be the product from the reaction of aqueous molybdate with copper sulfate, nitrate, or chloride under varying concentrations and proportion of the reagents but that is clearly not the case in this investigation. The copper sulfate product in the previous investigation was found to have a copper content of 34.76% and a water content of 0.97%, significantly different from the 36.6% and 4.51%, respectively, found for the product in this investigation. It is suspected that the difference arises from variation in the contact time of the solution with the precipitate. That is, the initially formed precipitate undergoes crystallization by a dissolution/reprecipitation mechanism that results in crystalline lindgrenite with the correct formula. Since in ore processing, the precipitate is likely to be separated rapidly, the increase of copper content of the precipitate and reduced molybdate demand of the amorphous product is attractive.

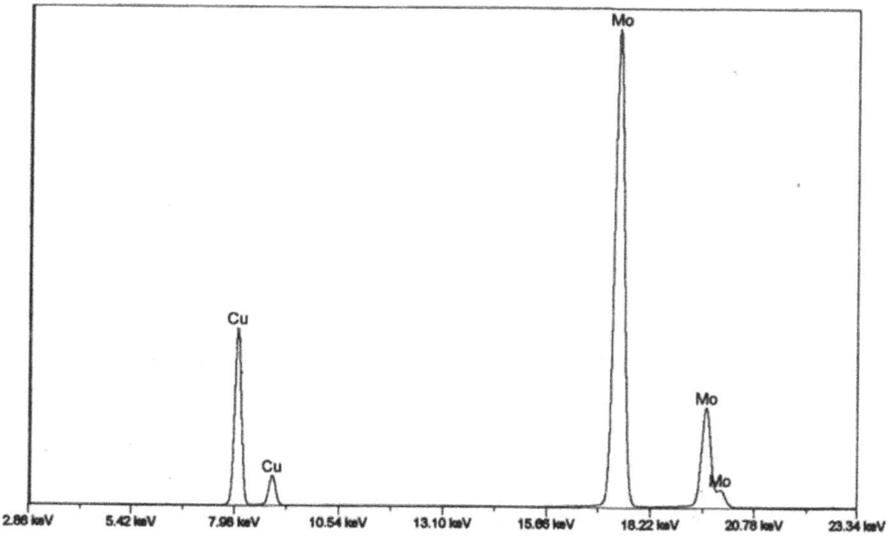

Figure 1. XRF Spectrum for $Cu(MoO_4)_{0.6}(OH)_{0.8}$

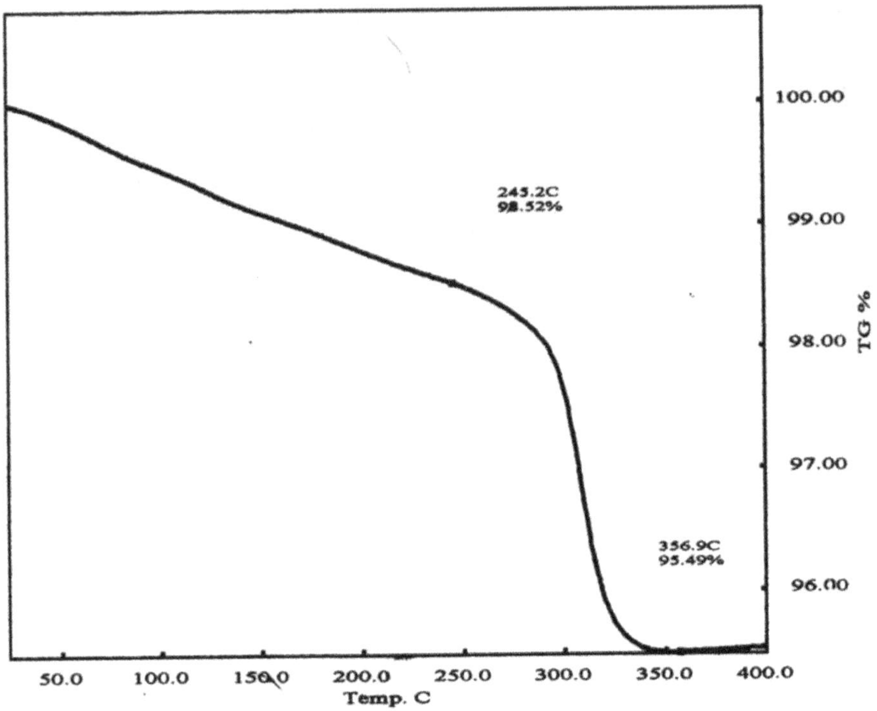

Figure 2. Thermal Gravimetric Trace for $Cu(MoO_4)_{0.6}(OH)_{0.8}$

The second step in the proposed process is the isolation of copper as copper hydroxide. We have previously demonstrated a process for uranium separation from water a synthetic uranium molybdate is separated into solid ammonium uranate and aqueous ammonium molybdate by treatment with ammonium hydroxide[9]. This reagent would not be suitable as a reagent for separating copper and molybdenum oxide since the strong complexation of copper ions by ammonia will solubilize copper hydroxide. Therefore, sodium hydroxide was used in this investigation since it both removes the complexation problem and also lends itself to the cyclic process because sodium molybdate is regenerated. Exposure of aqueous sodium hydroxide to the copper molybdate precipitate to aqueous sodium hydroxide leads to immediate reaction to produce bright blue copper hydroxide. Filtration and washing of the copper hydroxide with water and heating at 250 °C yielded copper oxide in 99.87% yield based on the initial amount of copper sulfate used. Thus, the overall process depicted in Figure 3 is extremely efficient. The only caveat is that sodium sulfate is produced as a by-product but the use of potassium salts instead of sodium is possible and the eventually isolated potassium sulfate could be used in fertilizer applications.

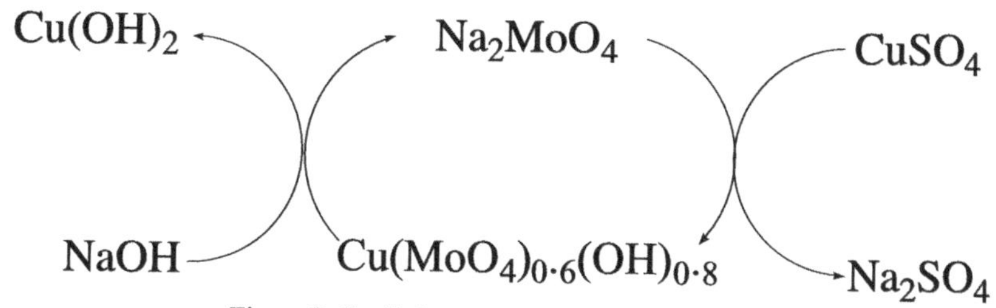

Figure 3. Cyclic Process for Isolation of Copper

CONCLUSIONS

The experiments with copper sulfate are a good model for copper solution derived from treatment of oxide or basic carbonate salts with sulfuric acid and for simple copper sulfide salts after oxidative leaching with sulfuric acid. It is possible to get nearly quantitative (99.87%) recovery of copper from the cyclic process with the copper being isolated as copper hydroxide or, after dehydration, copper oxide. The copper hydroxide is very useful for preparing copper salts by reaction with acids. Both copper hydroxide and copper oxide can be readily converted to copper metal or to other copper-containing ceramics.

ACKNOWLEDGMENTS

We are grateful for initial support of this research provided by the OSU Environmental Institute's Center for Water Research. The National Science Foundation is acknowledged for providing the funds for the X-ray powder diffractometer used in this investigation.

REFERENCES

[1] R. R. Moskalyk, A. M. Alfantazi, Review of Copper Pyrometallurgical Practice: Today and Tomorrow" *Minerals Engineering* **16**, 893-919 (2003).
[2] W. Dicinoski, A Revolution in Copper Recovery *Mining Magazine* **182**, 258, 260, 262 (2000).
[3] C.H. Samans, *Engineering Metals and their Alloys* MacMillan 1949
[4] M. S. Prasad, V. P. Kenyen and D. N. Assar. Development of SX-EW Process for Copper Recovery-An Overview. *Mineral Processing and Extractive Metallurgy Review*, **8**, 95-118, (1992)
[5] J. Peacey, X.-J. Guo, E. Robles, "Copper Hydrometallurgy-Current Status, Preliminary Economics, Future Direction and Positioning Versus Smelting" *Transactions of Nonferrous Metals Society of China* **14**, 560-568 (2004).
[6] G. Rossi, Biohydrometallurgy of Copper Ores: State of the Art and Prospects *Mines & Carrieres: Les Techniques*, 10-21 (1992)
[6] Powder Diffraction File (PDF-2) (International Centre for Diffraction Data, Newtown Square, PA).
[8] R.P. Dowdy, G.A. Kunz, and H.E. Sauberlich, Effect of a Copper-Molybdenum Compound Upon Copper Metabolism in the Rat *Journal of Nutrition* **99**, 491-496 (1969)
[9] B.P Kiran and A. W. Apblett "Process for Selective Removal and Concentration of Actinides and Heavy Metals from Water" *Ceramic Transactions*, **155**, 371-380 (2004).

EXPLOITING HALL-PETCH STRENGTHENING FOR SUSTAINABILITY

R. Heard
Carnegie Mellon University
Pittsburgh, PA, USA

U. Erb
University of Toronto
Toronto, Ontario, Canada

G. Palumbo
Integran Technologies Inc.
Toronto, Ontario, Canada

ABSTRACT

Sustainable engineering embraces conservation of materials and energy. Nanomaterials, and in particular nano-metal materials, can play a significant role in reducing the negative impact that we have on our planet through the exploitation of properties derived from the Hall-Petch relationship. Designers can reduce section mass in moving parts thus enhancing energy efficiency and reducing greenhouse gas emissions (e.g., MetaFuse™ nanometal-polymer hybrids for lightweight automotive components). They can also design for longer product life-cycles, thus reducing scrap rates and landfill requirements. Moreover, Hall-Petch strengthening of environmentally benign alloy systems can lead to the replacement of incumbent toxic materials and processes (e.g., nano CoP coatings as a replacement for hexavalent chromium plating; nano-copper coaxial composite wire for the replacement of toxic Beryllium alloys). In this paper, an overview is provided of these "sustainable" nanomaterials which are rapidly finding market acceptance.

INTRODUCTION

Sustainable engineering practice as it applies to materials must encompass the conservation of resources and energy. In the past, engineering over-design has been a large contributor in the over-use of materials. Progress made in modeling of required and actual performance issues for engineered products has resulted in improved design and material savings through physical reduction in component volume or weight due to section optimization. The contribution the reduction of material makes to sustainability goals is further augmented through the reduction in energy directly associated with the production of the material from extraction through final disposal including the material and energy (real and hidden) of products and material used to acquire, maintain, and dispose of the material.

To reduce materials and energy and move toward sustainable practices, designers and engineers must take advantage of gains made in material properties to further reduce section size and weight. Nanotechnology and specifically nano-metals offer superior properties in simple metal systems that can be exploited to reduce section size, extend life, and replace often more complicated metallic alloys. Each of these offers a step toward sustainability.

DESIGN CONSIDERATIONS

Although the design process can be complicated, it must result in materials being selected and dimensions established so that the part meets the objective of the design based on satisfaction of the applied constraints. The traditional design constraints that apply to an objective are often mechanical but can also be economic, aesthetic etc. To design within the framework of sustainability, factors that include environmental impact and resource management must also be considered as constraints. To guide designers, there are models available for estimation of environmental and sustainability impacts, many based on life cycle analysis (LCA) methodology.[(1) (2) (3)]

Life cycle analysis for a product is a measure of the impact the product has on the environment through its' useful life. In the best case, a complete life cycle analysis for any product would start with all the raw materials and energy sources in the earth and end with all materials back in the earth or at least somewhere in the environment but not part of the original product.[(4)] This inventory should include all energy, materials and equipment necessary for acquiring the materials. [(5)]

Materials applications find themselves belonging to one of two broad categories, materials used in moving systems, or materials used in static systems. For moving systems applications, the primary environmental impact is often dominated by energy consumption during the use phase (example: moving vehicle parts), and therefore, material density (weight) is a primary design concern. If there is a mechanical constraint to the application, then other material properties such as strength, stiffness (modulus), endurance, etc. also need to be considered along with the density (for example strength/density ratio). Selection criteria for materials based on this approach have been put forth in detail by Ashby. [(6)]

For static applications of materials, production energies often dominate. Materials in this classification would be those used in infrastructure or in applications where the material use is not significantly influencing the rate or quantity of energy the product consumes over it's life (example: bridge materials, structural materials used for common goods, bicycles etc.). Satisfaction of the mechanical constraint by a material mechanical property usually dictates the design and the section size required. Improvements in material mechanical properties then lead to a reduction in section and thus material savings. (Note: an exception must be acknowledged for certain static applications where efficiency of the materials influences the energy consumption of the static product (example: housing materials). In these special cases, production energies do not dominate and optimization of the material property, not a reduction in section is required)

To comply with sustainability objectives in either static or moving applications, we find that we must consider resource availability, optimize material yields, be able to create parts with minimal manufacturing, ensure the material contributes to more efficient operation during service (reduction of energy and extending life), then finally, be recyclable. In practice, these constraints should be used to screen the material candidates that satisfy the mechanical constraints.

ADVANTAGES OFFERED BY NANOCRYSTALLINE METALLIC MATERIALS

In many applications, improved mechanical properties can help designers and engineers move toward sustainability objectives. The selection and use of nano-materials can contribute through exploitation of the property improvements attributed to the Hall-Petch relationship, relating strength improvement to grain refinement in convention materials.(7)(8) It has been established that the Hall-Petch relationship extends into the nano regime for most metals and is useful until a notable deviation from the general relationship occurs at very small grain sizes.(9) The grain size associated with the onset of this deviation from the classic Hall-Petch has been observed when grain sizes is reduced near 10 nm.(10)(11)(12) It is not the intent of this paper to examine the mechanisms that result in this deviation, but rather, to investigate design opportunities using the Hall-Petch relationship as it approaches this point of deviation.

There are properties that are not significantly influenced by grain size reduction in the nano scale regime down to the grain size where the deviation occurs. These are saturation magnetization, heat capacity, thermal expansion, and Youngs modulus. (13)

As predicted by the Hall-Petch relationship, material properties that are traditionally related to strength are strongly influenced by grain size, therefore, improvements in hardness, yield strength, tensile elongation, wear rate, Taber wear index, coefficient of friction and electrical resistivity have been reported. (14-20) Consistent with these would be the expected decrease in mechanical properties such as ductility or elongation. This has been observed in tensile testing of nanocrystalline nickel, however, greater tensile ductility has been reported for other electrodeposited nano-crystals including copper, cobalt and nickel-iron alloys. (21,22,23,24)

Nano-grained materials used in this work have been produced using a pulse plating technique developed and refined for many metal systems. Pulse plating electrodeposition can be used for pure metals, alloys, metal matrix composites, ceramics, semiconductors and conductive polymers.(25,26) Applications can include platings for decorative wear, corrosion, erosion and oxidation resistance, but also for build up of large thick-walled (> 1 mm) electro-formed products, or for specific functional coatings for magnetic and electronic applications. Material development has been driven by specific industry needs, and as a result, experience has been built on the nickel, iron, cobalt and copper metallic systems. These base materials have been produced as pure metals, as alloys, or as composites either augmented with dispersed second phase materials (Type 1 Composites) or as layer structures (Type 2 Composites).

Three examples for design solutions based on the use of nano-grain metallic properties improvements in are presented below. They have been classified into Metal-Metal Composites, Metal -Polymer Composites and Metal-Ceramic Composites to emphasize the ability to use the nano-structured materials to create new design solutions.

METAL – METAL COMPOSITES

In moving systems, weight reduction is a primary goal for energy conservation. Direct metal replacement is possible as a design goal for many parts (replacement of high density metal/alloy with one of lower density) especially if the constraints impose few restrictions. If parts have more complicated constraints, or no adequate material for direct substitution exists, new solutions concepts are required. One such problem is presented in moving systems by the electrical conductors used for either primary energy or control/monitoring signals. These are typically copper bearing materials with high density and thus a reduction in weight of these conductor components becomes a major design goal.

Often in these moving systems, electrical wire is designed on the basis of both conductivity and strength, where strength is usually the primary design parameter. It is not uncommon in critical applications, to increase the gauge of a critical section of wire in order to guarantee its structural integrity. In a case where a two-fold (2x) diameter increase is made on the basis of breaking strength reinforcement (example 28 AWG (American Wire Gauge standard) to 22 AWG wire), there is an a 400% increase in overall weight. Specifically for signal wire, the cross section of the wire is usually determined by the strength requirements because the copper (conductivity) required to carry the signal can be quite small. Typically the strength criteria is reached by either the addition of alloy elements which degrade the conductivity to a lower but still acceptable level or by increasing the copper cross section beyond that which is actually necessary for conductivity needs. In some applications, the application of aluminum wire offers a design solution because of its lower density. Aluminum however, suffers from a ductility decrease as it ages in service. (27)

It is paramount that conductors offer little in resistance to current flow, thus conductivity must be maintained when strengthening a conductor material, regardless of the strengthening strategy employed. As noted above, the current practice to achieve higher strengths in copper is through alloying, or work hardening, both which result in reduced conductivity due to the introduction of electron scattering sites throughout the material. Despite this drawback, many conductors in high performance applications are alloys such as copper beryllium CS95. (28)

Direct replacement of a copper beryllium alloy with nanocrystalline copper offers a possible design solution however, this solution requires a trade off of the mechanical enhancement of the copper through Hall-Petch strengthening and the degradation of the conductivity by the electron scattering phenomena that occurs as the population of grain boundaries is increased. (28) Calculations show that nanocrystalline copper with a grain size of 200 nm offers similar material properties when compared to the CS95 alloy, however processing a fully nanocrystalline wire by electroplating is not possible because the process requires a substrate. The use of a thin copper wire as the core substrate provides a solution, however the use of this wire immediately reduces the properties of a 100% nanocrystalline product by the area fraction.

The design solution which seeks to provide high strength and reasonable conductivity lies where no single material satisfies both the conductivity and the strength constraints, therefore a composite structure can be envisioned where the strength requirement can be provided by one material and the conductivity by the second material (copper). Arranged as a composite (Type 2), the nanocrystalline metal can be electroplated on a high conductivity pure copper substrate core as shown in Figure 1. The calculated properties for this composite conductor along with benchmark data for pure copper, beryllium copper alloy CS95, and aluminum are summarized in Table I. The composite nano-nickel / copper wire offers improved tensile strength and strength/weight ratio so that a 19-strand 26 AWG nano-metal coated conductor could effectively replace a larger diameter 19-strand 24 AWG CS95 copper alloy conductor, with respective breaking loads of 165 N and 151 N, and provide a conductor weight reduction of approximately 35%. The nano-cobalt and copper composite offers even higher strength but with a larger economic penalty, yet sufficient design flexibility exists to boost the conductivity at the expense of conductor strength by simply reducing the proportion of nano-metal coating to copper and vice versa. For practical purposes, the final construction of a conductor could likely consist of multiple strands of the co-axially coated material (e.g. 19 concentric strands for a size 26 AWG conductor). Additionally, nanocrystalline Ni or Co based coatings provide good corrosion protection, eliminating the need to apply a secondary coating (usually a thin layer of tin, silver, or nickel in order to impart corrosion resistance to the copper). The use of nanocrystalline metals to provide the strengthening also eliminates beryllium from the environment and material cycle.

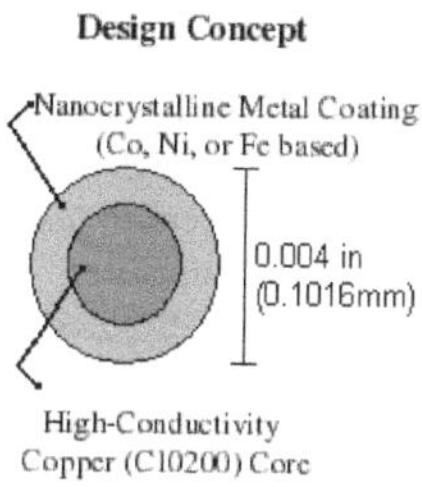

Figure 1. Design of Type 2 Metal – Metal composite conductor.

Table I – Calculated Properties of Conductor Materials

Conductor Material	Tensile Break Strength, MPa	Conductivity, %IACS	Ductility, Min. %EL	Strength-to-Weight, MPa/(g/cm^3)
Copper	200	100	25	22
Beryllium Copper Alloy CS95**	655	63	6	74
Aluminum***	125	57	6	47
Nano-Cu (~200nm)	665	80	21	75
44 vol.% Nano-Ni on Copper	766	63	7	89.1
44 vol.% Nano-CoP on Copper	1072	59	6	120

Calculated based on known properties of constituent metals; **(27) ; ***(28)

METAL – POLYMER COMPOSITES

Just as metal systems can be combined into Type 2 composite structures, metal and polymer systems offer a similar opportunity. The properties of these composites will be dependent on the individual materials used to construct the composite, on any synergistic properties, and on the physical design of the macro scale structure. In these composites, the nano-metallic component contributes the strength and properties of the metallic system (including the Hall-Petch relationship to be exploited) while the polymer provides a low density substrate which has easy formability and properties expected from polymeric systems.

Electroplating technology can be used for deposition of nano-metallic surfaces on polymeric materials resulting in a Type 2 or sandwich composite, constructed with the outer layer (face) of the composite being the harder metallic material and the core the polymer substrate, although an opposite arrangement is possible.

The properties of a Type 2 composite are often complicated due to the anisotropic properties and the large difference in properties between a strong, stiff, metallic face and a softer more pliant core. These structures, when loaded in the direction parallel to the layers have the loads taken up by the stronger material and failure is closely tied to the failure of the strong material. Composite properties across the section are much different (perpendicular to the layered structure) and failure is tied more to the mechanism of failure, being bending, compression , shearing, etc.

The nano-metal coating on polymeric materials results in significant improvement in strength, impact and flexural properties. Day has reported the increase in flexural modulus, impact energy and tensile strength properties of a Zytel PA66 polymer coated with 100 micron thick layer of nano Fe-Ni. [(29)] In addition to the bulk material properties, surface and special properties such as hardness (scratch resistance), electrical and thermal conductivity, magnetic shielding, creep strength, paint-ability and environmental resistance are also improved, possibly extending the application limits for polymeric materials.

One significant advantage of Type 2 composite construction using the electroplating technique is the flexibility to construct composites that are porous structures. Figure 2 is a carbon foam substrate that has been plated with a nano crystalline Ni-Fe alloy. The structure is extremely light and is designed as an energy absorption layer for moving vehicles.

Exploitation of performance improvements offered by nano-metal-polymer composite structures has been quickly adapted by the sports equipment market (tennis rackets, golf club shafts). Application of these composites as a substitute for more dense metallic parts in transportation systems is also highly attractive from a design view point. Not only can designers take advantage of the strength provided by the nano-metal layer, but the metallic layer also permits polymers to be applied in applications where heat loading might have previously limited their use. [(29)]

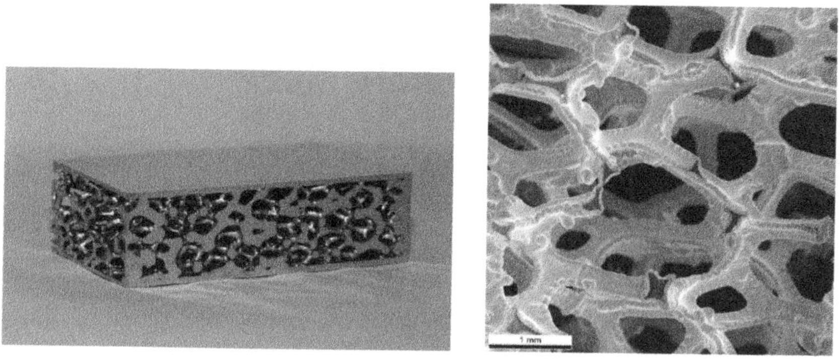

Figure 2 NanoFoam™: Thickness: 12.5mm; >90% open porosity; Ni-10%Fe (15nm avg. grain size) on carbon pre-form, top and bottom solid face plates are NanoPlate®: 2 ultra- high strength nanocrystalline Ni-10%Fe (15nm avg. grain size) sheets 1 mm thick . Weight per unit area: ~14.68 kg/m^2

METAL - CERAMIC COMPOSITES

Metal-Ceramic composites of Type 2 and Type 1 can be fabricated by electrodeposition of metallic systems on ceramic or by entrapment of ceramic particles in the electroplated metal during a co-deposition process. This co-depositing of a Type 1 composite offers materials that have less anisotropy than Type 2 composites engineered by material selection.

In Type 1 composites, the matrix consists of metal with nano-size grains that provide the improved general properties. The second phase particles within the matrix can be selected to provide various functionality, for example, harder particles may be used for wear resistance, materials with high thermal conductivity can be used for thermal property enhancement, or softer, lubricious materials can be used for providing dry lubrication. These composites can be fabricated by co-depositing either conventional large particles [30,31,32,33] or newly developed nano-powders, nano-tubes or nano-fibers. [34, 35]

The use of second phase particles to create a composite coating allows designers to further exploit the properties of nano grain size metals. Instead of being restricted by the properties of the metal alone, composite layers can be engineered to provide unique solutions. As an example, hard, wear resistant layers can be deposited that surpass the performance of many traditional materials, keeping materials in service longer thereby extending material life and meeting green design goals. Figure 3 illustrates the improved wear resistance for electroplated nickel phosphorus-boron carbide composites over tool steel or hard chrome plating.

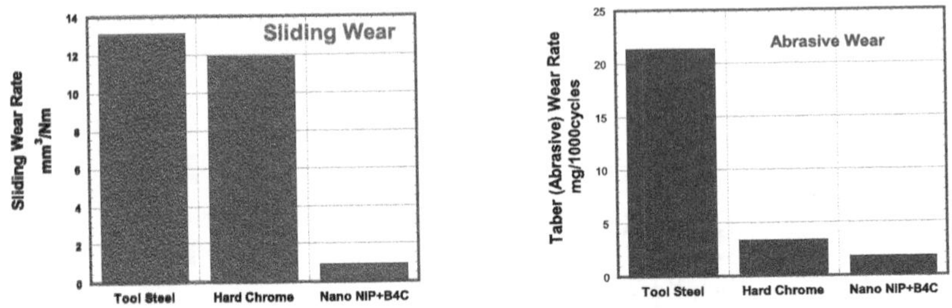

Figure 3: Wear Rates for Tool Steel, Hard Chromium and Nano NiP-B_4C composite coatings

It has been reported that incorporating up to ~30 vol.% of ceramic phase had relatively little effect on the hardness of nickel phosphorus nano-composite, regardless of whether the particles themselves are relatively soft (e.g. h-BN and TiO_2) or relatively hard (e.g. SiC and B_4C), figure 4 [(36)] The overall microhardness of the nano-composites was therefore, largely determined by the microhardness of the nanocrystalline metal matrix. Under abrasive wear conditions however, clear differences were noted between the materials (figure 5a) and when plotting the Taber wear index (TWI) against the particle hardness, the dominant factor in determining the abrasive wear resistance was the hardness of the ceramic particle itself, figure 5b.[(36)] An interesting consequence of these results is that it may be possible to use the ceramic reinforcing phase to provide the necessary abrasive wear resistance, while the nano-metal matrix could be selected on other factors such as corrosion resistance, soft-magnetic properties, etc. in order to produce multi-functional nano-materials. [(36)]

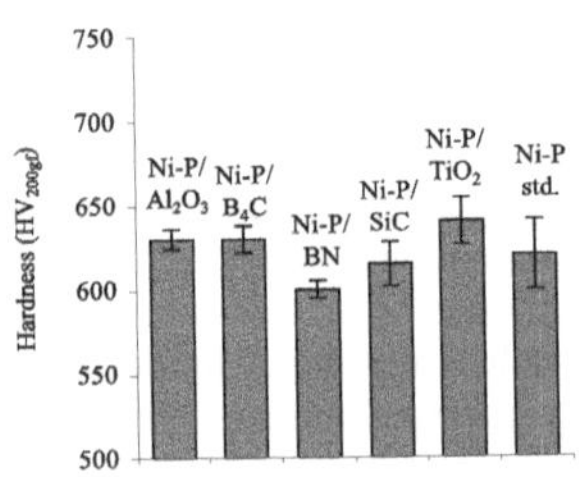

Figure 4 Nanometallic – Ceramic hardness values [(36)]

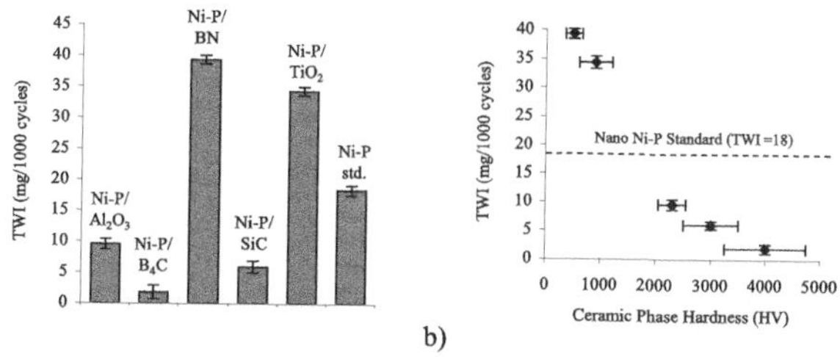

Figure 5 Nanometallic – Ceramic composite Taber wear performance [36]

Work has also advanced Co-W nanocrystalline alloys that incorporate industrial diamonds as a second phase. This material shows improved hardness and improved thermal conductivity through the deposited layer. Coatings of a cobalt - tungsten alloy with high hardness for wear resistance coupled with good thermal properties are being researched as possible replacements for electrolytic chromium under certain conditions. The properties of the CoW coatings are compared to hard chrome in Table II.

Table II Properties of CoW and CoW-diamond coatings compared to hard chrome.

Property	CoW	CoW-diamond	Hard Chrome
Composition	1.7wt%W	1.9wt%W	-
Grain Size	< 20 nm	< 20 nm	-
Microhardness (VHN_{100})	480-580	400-500	700-900
Ductility	-	< 6%	<0.1%
Coefficient of Friction	0.41 (steel ball)	0.43 (steel ball)	0.85 (steel ball)
Cu Ball Scar	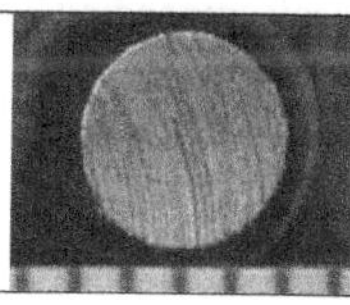	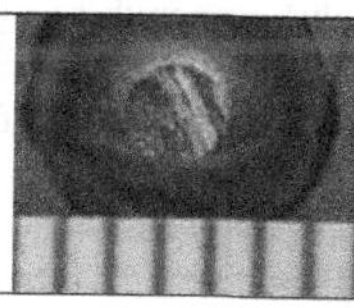	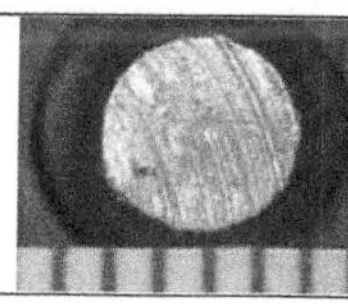
Taber Wear Index (mg/1000 cycles)	30	5	3-4
Adhesion	Pass	Pass	-

CONCLUSIONS

New composite materials, fabricated from simple electroplating of nano-metals, offer enhanced properties derived from the Hall-Petch relationship. These nano-metals are produced from pure metals or simple alloys and can combined with other metals, polymers or ceramics as Type 1 or Type 2 composites. The resulting composites offer improved properties to designers wishing to take a more sustainable approach, minimizing section weight, section thickness and other environmental issues during the material selection process.

ACKNOWLEDGMENTS

The authors would like to recognized the efforts of Jon McCrea, Iain Brooks, Ian Winfield, and funding from The Natural Sciences and Engineering Research Council of Canada, and the Pennsylvania Nanotechnology Commercialization Center.

REFERENCES

[1] C. Hendrickson, A. Horvath, S. Joshi and L.B. Lave, Economic Input-Output Models for Environmental Life Cycle Analysis, *Environmental Science & Technology*, **32**, 184A-191A, (1998)

[2] M. Goedkoop, S. Effting, M. Collignon, The Eco-indicator 99, A Damage Oriented Method for Life Cycle Impact Accessment, Manual for Designers (April 2000), *http://www.pre.nl*

[3] ISO 14001:2004, *Environmental management systems - Requirements with guidance for use*, 2nd Ed., Geneva, Switzerland.

[4] M. Curran, Life cycle Assessment: Principles and Practices, Scientific Applications International Corporation (SAIC), contract under U.S. Environmental Protection Agency , EPA/600/R-06/060 , May 2006 , http://www.epa.gov/NRMRL/lcaccess/pdfs/600r06060.pdf

[5] S. Svoboda, Note on Life Cycle Analysis, National Pollution Prevention Center for Higher Education , Ann Arbor, MI , March 1995, http://www.umich.edu/~nppcpub/resources/compendia/CORPpdfs/CORPlca.pdf

[6] M. Ashby, *Materials Selection in Mechanical Design,* 3rd Ed. Butterworth-Heinemann, Oxford UK, 2005

[7] E. Hall, The deformation and ageing of mild steel: III Discussion of results. *Proc. Phys. Soc. London.* **B64**, 747–753, (1951)

[8] N. Petch, The cleavage of polycrystals., *J. Iron Steel Inst.*, 25–28, (1953)

[9] C. Koch, J. Narayan, The Inverse Hall-Petch Effect - Fact or Artifact? *Mat. Res. Soc. Symp.* , Materials Research Society, **634**, B5.1 – B5.11, (2001)

[10] T. Nieh, J. Wadsworth, Hall–Petch relation in nanocrystalline solids. *Scripta Met. Mater.*, **25**, 955–958 , (1991)

[11] H. Kim, M. Bush, Effects of grain size and porosity on the elastic modulus of nanocrystalline materials, *Nanostructured Materials*, **11**, 361-367, (1999)

[12] M. Dao, L. Lu, R. Asaro, J. De Hosson, E. Ma, Toward a quantitative understanding of mechanical behavior of nanocrystalline metals, *Acta Metallurgica*, **55**, 4041 – 4065, (2007)

[13] U. Erb, K. T. Aust and G. Palumbo, Electrodeposited Nanocrystalline Metals Alloys and Composites, in *Nanostructured Materials*, 2nd ed., C.C. Koch (ed.), William Andrew Publ., Norwich, NY, 235-292, (2007)

[14] Gonzalez, F., Brennenstuhl, A. M., Palumbo, G., Erb, U., and Lichtenberger, P. C., Electrodeposited Nanostructured Nickel for In-situ Nuclear Stream Generator Repair, *Mat. Sci. Forum*, 831-836, (1996)

[15] G. Palumbo, F. Gonzalez, A. Brennenstuhl, U. Erb, W. Shmayda, P. Lichtenberger, In-situ nuclear steam generator repair using electrodeposited nanocrystalline nickel, *Nanostr. Mat.*, **9**, 737-746, (1997)

[16] A. El-Sherik, U. Erb, G. Palumbo, K. Aust, Deviations from Hall-Petch Behavior in As-Prepared Nanocrystalline Nickel, *Scripta Metall. Mater.*, **27**, 1185 – 1188, (1992)

[17] A. El-Sherik, U. Erb, *Nickel-Cobalt 97, Applications and Materials Performance* (F. Smith, et al. eds.), The Metallurgical Society of CIM, Montreal, (1997)

[18] N. Wang, Z. Wang, K. Aust, U. Erb, Room temperature creep behavior of nanocrystalline nickel produced by an electrodeposition technique, *Mat. Sci. Eng. A*, **237**, 150-158, (1997)

[19] D. Jeong, *Abrasive Wear Behaviour of Electrodeposited Nanocrystalline Materials,* Ph.D. Thesis, University of Toronto, Ontario, Canada, 2003

[20] M. Aus, B. Szpunar, U. Erb, A. El-Sherik, G. Palumbo, K. Aust, Electrical resistivity of bulk nanocrystalline nickel, *J. Appl. Phys.*, **75**, 3632-3634, (1994)

[21] Y. Shen, L Lu, Q. Lu, Z. Jin, K. Lu, Tensile properties of copper with nano-scale twins, *Scripta Mater.,* **52**, 989-994, (2005)

[22] A. Karimpoor, U. Erb, K. Aust, Z. Wang, G. Palumbo, *Mat. Sci. Forum,* **386**, 415-420, (2002)

[23] A. Karimpoor, U. Erb, K. Aust, G. Palumbo, High strength nanocrystalline cobalt with high tensile ductility, *Scripta Mater.,* **49**, 651-656, (2003)

[24] J. McCrea, G.Palumbo, G. Hibbard, U. Erb, Properties and Applications for Electrodeposited Nanocrystalline Fe-Ni Alloys, *Rev. Adv. Mater. Sci.,* **5**, 252-259, (2003)

[25] M. Schlesinger, M. Paunovic, (eds.), *Modern Electroplating*, 4th ed., John Wiley & Sons, NY, (2000)

[26] A. Brenner, *Electrodeposition of Alloys-Principles and Practice*, Academic Press, NY, (1963)

[27] J. Ballenghien, A. Synodinos, *Fly By Aluminum Cables and Associated Contact Technology* AEISS, (2003)

[28] E. Fisk, Developments in Alloy Conductors, *Wire and Cable Technology International*, September, 24-28, (2003)

[29] M. Day, Nanometal-Polymer Hybrid, *Advanced Materials and Processes*, April, 25 – 28, (2008)

[30] U. Erb, A. El-Sherik, G. Palumbo, K. Aust, Synthesis, structure and properties of electroplated nanocrystalline materials, *Nanostr. Mat.*, **2**, 383-390, (1993)

[31] G. Palumbo, U. Erb, J. McCrea, D. Hibbard, I. Brooks, F. Gonzalez, K. Panagiotopoulos, Electrodeposited Nanocrystalline Coatings for Hard-Facing Applications, *AESF SUR/FIN Proc.*, 204-212, (2002)

[32] A. Zimmerman, D. Clark, K. Aust, U. Erb, Pulse electrodeposition of Ni-SiC nanocomposites, *Mater. Letti.*, **52**, 85-90, (2002)

[33] A Zimmerman, G. Palumbo, K. Aust, U. Erb, Mechanical Properties of nickel silicon carbide nanocomposites, *Mat. Sci. Eng.* A, **328**,137-146, (2002)

[34] R. Oberle, M. Scanlon, R. Cammarata, P. Searson, Processing and hardness of electrodeposited Ni/Al_2O_3 nanocomposites, *Appl. Phys. Lett.*, **66**,19-21, (1995)

[35] X. Chen, F. Cheng, S. Li, L. Zhou, D. Li, Electrodeposited nickle composites containing carbon nanotubes, *Surf. Coat. Tech.*, **55**, 274-278, (2002)

[36] G.D. Hibbard, J.L. McCrea, and G. Palumbo, Microstructural Design of Abrasive-Wear Resistant Nanocomposites, *Processing and Fabrication of Advanced Materials XVI*, M. Gupta, T.S. Srivatsan, S.K. Thakur (eds.) Research Publishing Services, Singapore, 147-156, (2007)

NOVEL PROCESS DEVELOPMENT WITH CONTINUOUS CASTING AND PRECISE FORGING FOR AL-SI ALLOYS TO PRODUCE AN ENGINE PISTON

O. Umezawa[1], H. Takagi[2], T. Sekiguchi[3], T. Yamashita[2], and N. Miyamoto[3]
[1] Dept. of Mater. Sci. & Eng., Yokohama National University, Yokohama, 240-8501, Japan
[2] Toyama Alloy Co., Imizu, Toyama, 934-8515, Japan
[3] Miyamoto Industry Co., Shiotani, Tochigi, 329-2442, Japan

ABSTRACT

A new vertical semi-continuous casting process using an adiabatic and rapid cooling mold has been developed to produce a billet with smooth surface for eutectic or hyper-eutectic Al-Si-Cu-Mg-(Ni, Fe, Mn) alloys. Without peeling and hot-extrusion processes the cast billet with 83 mm in diameter has been successfully applied to warm forging. Although primary Si crystals, intermetallic compounds and α dendrite structure were mostly refined under high cooling condition in the casting, coalesced Si crystals or compounds were still remained in the center part. A working and annealing step has been installed due to increase formability of the cast with a modification of its microstructure. An isothermal forging process with greener lubricant and precise mold technologies has been developed and a prototype of motorbike engine piston was produced. The forged alloys showed higher fatigue strength and creep resistance than conventional materials at high temperatures. This process is promised to provide an up-grade solution to the recycle of secondary aluminum alloy.

INTRODUCTION

The products generally consist of many kinds of materials. On the other hand, it has been required to save resources and energy, and to recycle materials. Engineering subjects for the ideal recycling system in metals are to reduce impurity content from scrap melt, to immunize metals against impurities or make metals innocuous from impurities, and finally to replace metals with inherently recyclable metals. The concept of *ecomaterials* (environmental conscious materials) has been proposed in Japan since 1992.[1] The consideration of environmental issues for all materials was deemed new and significant. The *ecomaterials* are associated with; minimal health hazards, minimal harmful emissions and wastes, minimal energy requirement, maximal recyclability and minimal material resource depletion, optimal physical properties and best technical performance. Traditional metals are pressed to meet new performance targets, with higher quality, at lower cost, with environmentally benign process and with globally source. Such paradigm shift in the metals industry has proceeded.[2] Through discussion in the disassembly, recovery and waste treatment stages, furthermore, it is concluded that the developments of product design for easy disassembly and of easier recyclable material design are needed. That means that novel process technologies building high performance and/or service with low environmental burden are necessary. Those challenges facing *ecomaterials* may make an important role to develop the global eco-society. In particular, the secondary materials accepted for recycling should be considered to immunize their impurities by a microstructural control, since no disassemble operation fully renews them.

In the case of aluminum, the cascade of material flow is suitable for recycling, because producing aluminum from sources such as bauxite consumes large amount of energy. The application of secondary (scrap) metals, however, has been almost limited to cast materials, and the dilution by raw material has been inevitable. The reason why Si and Fe elements are problem in aluminum alloy recycle stage is to lead to very poor workability of aluminum production, and difficult to remove them, although Si, Fe and Cu are major detrimental elements in recycling aluminum products. Al-Si-Cu-Mg alloys are one of the major cast materials, and include almost no dislike elements for the *ecomaterial*. In addition, removal of Si from scrapped aluminum products is difficult and higher costly as well as that of Fe. Al-Si system having plural phases with low mutual solid solubility are effectively the *in-situ* metal-metal composites which mean one of simplified recycle models and give an adequate balance with properties such as strength, elongation, lightweight, good wear resistance and low thermal expansion. To control balanced properties without detrimental elements for the *ecomaterials*, fine microstructure with plural phases, *mesocomplex structure*, is one of candidates for the alloy design.[3-4]

Therefore, novel processes to refine microstructure of Al-Si-Cu-Mg-(Ni, Fe, Mn) alloys for high performance forging products have been developed and mechanical properties of the treated materials were characterized.

PROCESSES TO DEVELOP FINE MICROSTRUCTURE

To improve mechanical properties of Al-Si-X cast materials, microstructural modifications have been commonly achieved by the addition of elements such as Sr and P into the melt, or by the hot-forging and long-time solution heat-treatment. Nevertheless, the ductility of the treated materials is not enough, and they can hardly be applied to wrought materials. Coarse Si crystals and/or coarse intermetallic compounds cause poor ductility and often give an origin of fatigue crack initiation site as well as inclusion. To improve both workability and mechanical properties, thus, the Si crystals and compounds must be refined to avoid the sample fracture due to their cracking. The most promising method of refining microstructures is rapid solidification. Spray-forming and thixoforming have been used for refining the Si in Al-Si alloys. However, those methods are not amenable to mass production, so the present work deals with a novel semi-continuous casting and thermomechanical treatment for Al-Si-X cast alloys.

Direct chill (DC) casting process has been widely adopted to produce aluminum cast billets. However, the billets need to peel their surface followed by hot-extrusion. Yamashita *et al.*[5] has developed an adiabatic graphite mold casting (HIM) process to produce A6000 series alloys billets with a smooth surface. The cast billets directly provide forging materials and eliminate peeling and hot-extrusion process. Furthermore direct cooling of billets results in avoidance of segregation and improvement of mechanical properties. Therefore, it reduces the cost of the forging aluminum materials approximately 20% in the case of automotive suspension arms, and lightens them. Thus the direct cooling process with adiabatic mold was chosen to apply the Al-Si billet casting.

However, the heavy cold-work cannot be applied to the Al-Si alloy casts, because it causes severe cracking in the primary Si crystals and coarse intermetallic compounds. Generally, hot-working refers to deformation carried out under conditions of temperature and strain rate. Since recovery processes occur substantially during the deformation process, large strains can be achieved with essentially no strain hardening. It results in a decrease in the energy required to deform the metal and an increased ability to flow without cracking. Thus the higher working temperature is, the less cracked Si crystals are detected. Severe cracking in Si crystals occurs under cold-working operations. The cracks may lead to sample fracture, but be useful in refining the Si crystals. In order to avoid fracture, the cold-working operations must be carried out steps with intermediate annealing. Umezawa *et al.*[3-4] have proposed a repeated thermomechanical treatment (RTMT) to produce a heavily deformable hyper-eutectic Al-Si-X alloys.

Fragmentation of Si crystals and compounds and their dispersion in the matrix were achieved through repeated cold-working and annealing steps. This concept of microstructual modification was placed in the prior treatment step of isothermal forging instead of hot-extrusion.

SEMI-CONTINUOUS CAST BILLET WITH SMOOTH SURFACE

Development of Casting Technology

A new vertical semi-continuous casting process equipped with an adiabatic and rapid cooling mold (HI-RCM) has been developed to produce billets of 83 mm in diameter.[6] Figure 1 shows the schematic illustration of new casting equipment. Direct water-cooling at lower edge of graphite mold is adopted to increase cooling rate in solidification, although no cooling of the mold was done in the HIM process.

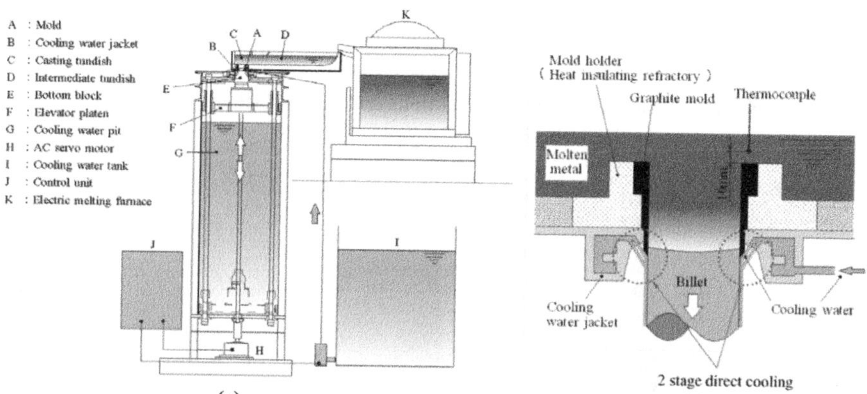

(a) (b)

Figure 1. Schematic illustration of vertical semi-continuous casting equipment (a) and magnified image of adiabatic and rapid cooling mold (b).

Influence of casting conditions on surface roughness of the billet was examined. The higher casting rate to pull down the billet was, the lower roughness was. But there was an upper-limit of casting rate to keep molten metal in the mold as shown in Figure 2. At higher flow rate of cooling water, smooth surface of the billet was available in lower casting rate. Figure 3 represents the outlook of A4032 (Al-12.7Si-0.9Cu-1.0Mg-0.5Fe-0.9Ni in mass%) alloy billets produced by semi-continuous casting methods. Surface roughness of DC billet (204 mm in diameter) was about 1 mm. In the HI-RCM process it was decreased less than 1.6 μm in average at the casting rate of 180 mm/min.[6]

Figure 4 shows the longitudinal section of A4032 HI-RCM billets with different casting rate. Chill layer with periodical triangle structure appears in the slow cooling condition and results in lapping surface shown in Figure 3(c). The triangle structure was neither a cold-shut in the DC casting nor a ripple in the hot-top casting.[7-8] It may be related with a segregation in partial slow cooling zone or re-melted zone in the mold. The increase of casting rate also makes liquid-solid interface lowered in the center part as shown in Figure 5.

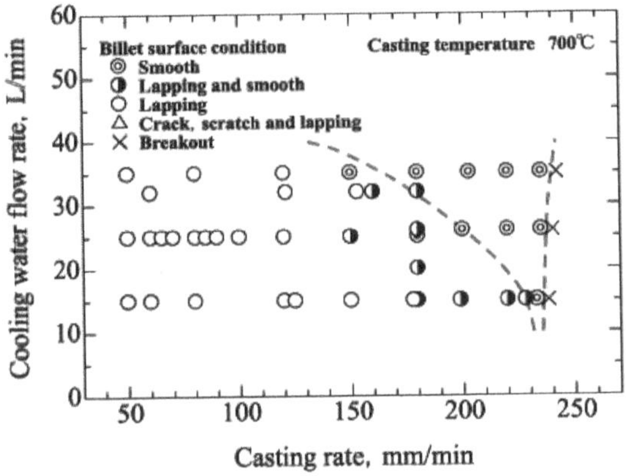

Figure 2. Influence of casting conditions on billet surface for A4032 under the casting temperature of 973 K.

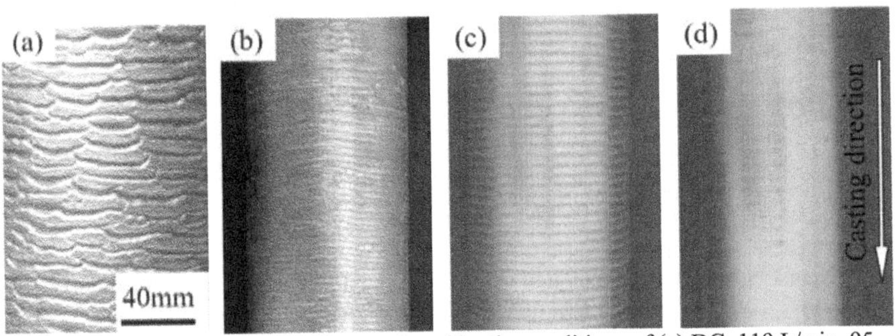

Figure 3. Outlook of A4032 billets produced by the conditions of (a) DC: 119 L/min, 95 mm/min,
(b) HIM: 25 L/min, 100 mm/min, (c) HI-RCM: 25 L/min, 100 mm/min and
(d) HI-RCM: 25 L/min, 180 mm/min.

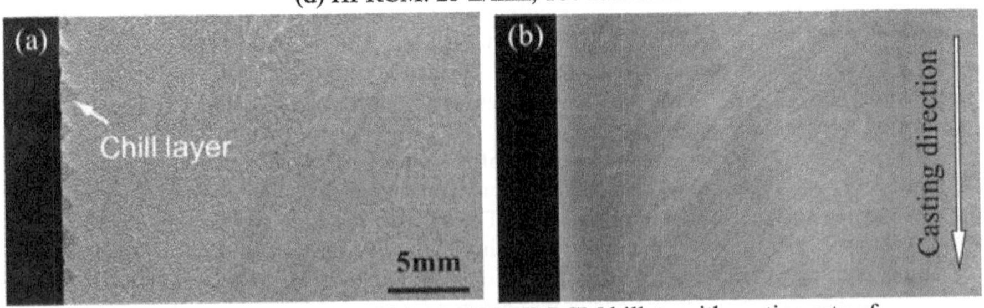

Figure 4. Macrostructure of the A4032 HI-RCM billets with casting rate of
(a) 100 mm/min and (b) 180 mm/min.

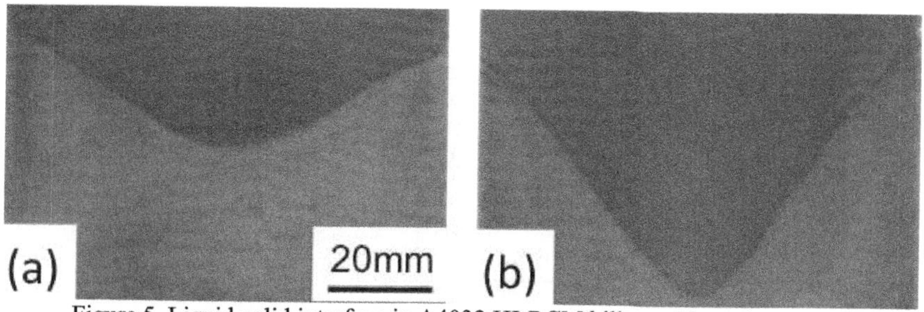

Figure 5. Liquid-solid interface in A4032 HI-RCM billets with casting rate of (a) 100 mm/min and (b) 180 mm/min.

Test Alloys and Their Microstructure

Seven kinds of eutectic or hyper-eutectic Al-Si-Cu-Mg-(Ni, Fe, Mn) alloys listed in Table I were chosen to provide test billets to produce a prototype of motorbike engine piston by forging. In order to increase high temperature strength the alloys contain several percent of transition metals such as Cu, Ni, Fe and Mn.

Figure 6 shows microstructure near surface of the billets. Fine primary Si (about 10 μm in diameter) and eutectic phases are dispersed. Secondary dendrite arm spacing (DAS) was several μm, which was much lower than that of DC cast. On the other hand, coalesced Si crystals and/or compounds were observed in the center part of the billets for hyper-eutectic alloys as shown in Figure 7. In particular coarse AlSiFeMn compounds appear in UTM404 and 405 alloys (Figs. 7(c) and (d)). In order to avoid them total amount of Fe and Mn should be less than 1 mass%. DAS was about 15 μm in the center part. Fine rod-like AlSiCuNiFe compounds as well as eutectic Si were dispersed in the matrix. No remarkable segregation was detected in all billets. For higher Si alloy e.g. Al-17Si-3Cu-1Mg-1Ni, furthermore, fine structure was obtained with casting temperature above 1023 K.[9] Therefore, new HI-RCM vertical semi-continuous casting process is suitable to provide the billets for forging.

Table I. The chemical compositions of test alloys (mass%)

Alloys	Si	Fe	Cu	Mn	Mg	Cr	Ni	Zr	Ti	P	Ca	Al
UTM401	14.0	0.15	3.0	-	1.0	-	1.0	0.1	0.01	0.012	-	Bal.
UTM402	12.0	0.15	3.0	-	1.0	-	1.0	0.1	0.01	0.012	-	Bal.
UTM403	12.0	0.15	3.0	-	1.0	0.1	1.0	0.1	0.01	0.012	-	Bal.
UTM404	14.0	1.0	3.0	1.0	1.0	-	-	0.1	0.01	0.012	-	Bal.
UTM405	12.0	1.0	3.0	1.0	1.0	-	-	0.1	0.01	0.012	-	Bal.
UTM406	12.0	0.6	3.0	0.4	1.0	0.1	1.0	0.1	0.01	0.012	-	Bal.
UTM407	12.0	0.6	3.0	0.4	1.0	0.1	1.0	0.1	0.01	-	0.012	Bal.

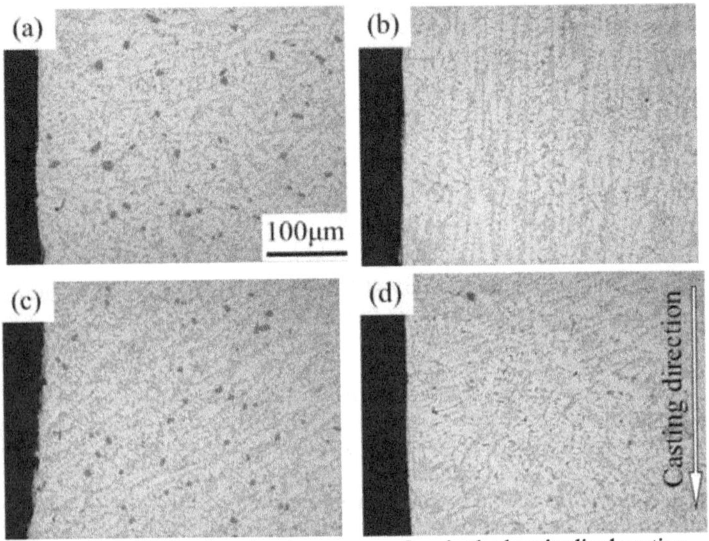

Figure 5. Solidification structure near surface in the longitudinal section for the billets of (a)UTM401, (b)UTM402, (c)UTM404 and (d)UTM405.

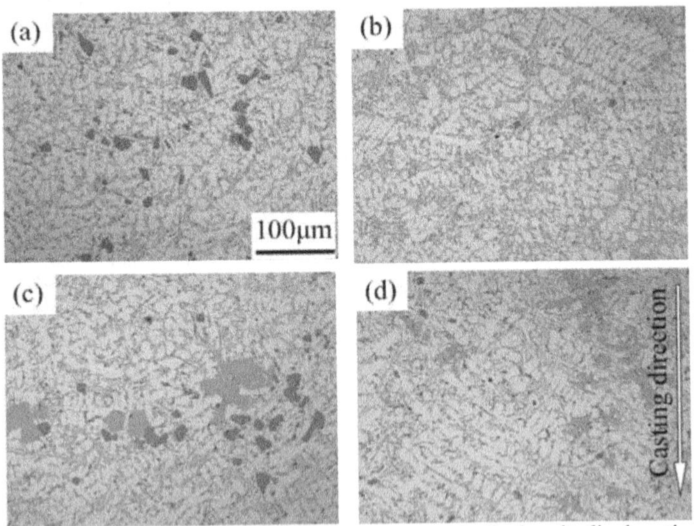

Figure 6. Solidification structure at the center part in the longitudinal section for the billets of (a)UTM401, (b)UTM402, (c)UTM404 and (d)UTM405.

Fatigue Strength and Crack Generation

Load-controlling fatigue tests for the billets of UTM401-405 alloys were carried out with stress ratio, R ($\sigma_{min}/\sigma_{max}$) = 0.01, in a sine wave at 296 K and 523 K. Influence of primary Si crystals and/or compounds on fatigue strength were examined. There was no big difference of tensile strength by the position in the billets and the alloys. For the hyper-eutectic alloys, however, distribution of primary Si crystals and/or compounds strongly affected to the fatigue strength as shown in Figure 7.

Fatigue crack initiation sites and fracture surfaces were analyzed by scanning electron microscopy. The crack length, crack width and distance from specimen surface to center of ellipse were determined, where the direction of crack length was parallel to the initial crack propagating direction. Maximum fatigue crack size in Stage II was taken from the ripple mark on fracture surface. Thus both initial and maximum crack sizes were experimentally given. A linear fracture mechanics program, SCAN,[10] was adopted to evaluate the ΔK of a subsurface crack. The calculated stress intensity range resulted in the modeling of fatigue crack growth by integral of the Paris equation as follows:

$$\frac{da}{dN} = C(\Delta K)^m \tag{1}$$

where C and m are constants. The constants in the Eq.(1) were chosen as $C=8.08\times10^{-13}$ and m=5.03. Those were resulted from crack propagating tests of AC8A cast alloy in the reference.[11] Then the crack propagating life, Np, was estimated by the crack growth modeling. The number of cycles to failure, Nf, was obtained in the experimental. Based on the estimation coalesced compounds gave the fatigue crack initiation site as indicated by arrow in Figure 8. In order to improve fatigue strength of the alloys, further dispersion of Si crystals and compounds by forging process is required.

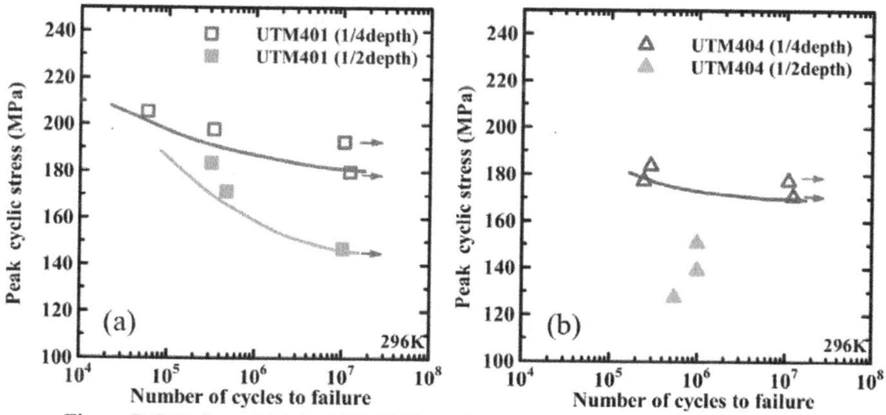

Figure 7. S-N data of (a) UTM401 billet and (b) UTM404 one at 296 K (R=0.01).

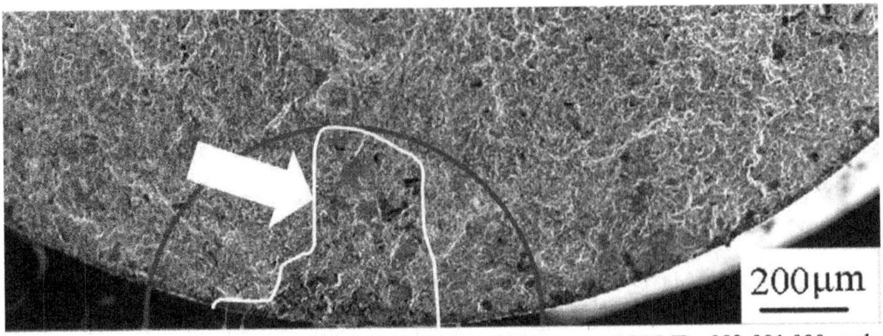

Figure 8. Fatigue crack initiation sites in UTM404: at 296 K, σ_{max}=139 MPa, Nf=981,000 cycles. Arrow shows coalesced second phases giving the crack initiation site. A half ellipse indicates initial crack for the calculation.

PRECISE FORGING AND PROTO-TYPE PRODUCT

Formability of Cast Alloys

Deformation resistance at the temperature between 293 K and 693 K of the alloys casts was examined by direct extrusion, indirect extrusion and upset forging. Working rate was from 30% to 70% in reduction of area, respectively. Their formability was good to apply isothermal forging at above 600 K, while the deformation resistance was higher than that of A4032, especially at higher temperature.

Microstructural modification by RTMT was effective to improve tensile properties as mentioned in above.[3-4] To increase formability, drawing-annealing step was applied to the cast alloys. The working from 10 to 70% in reduction was employed and T7 heat-treated at 773 K for 3600 sec followed by water quenched and aged at 483 K for 18000 sec. As shown in Figure 9 the treated material over 20% reduction in Al-11.2Si-2.2Cu-0.27Mg-0.25Fe alloy exhibits a significantly enhanced ductility and strength as well as RTMT material.

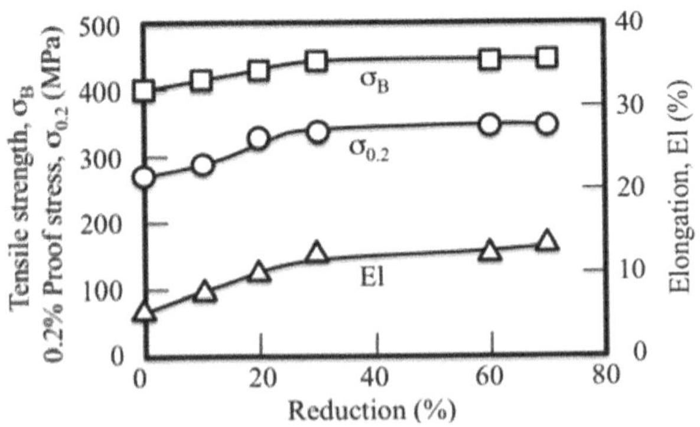

Figure 9. Effect of drawing reduction on the tensile properties of Al-11.2Si-2.2Cu-0.27Mg-0.25Fe alloy.

Precise Forging

The isothermal forging process with greener lubricant and precise mold technologies was developed. Test pieces of the alloys billets were solution treated at 783 K and peened their surface. Then greener lubricant which was water-soluble graphite coating on chemical oxidation of the test pieces was adopted. The isothermal forging with 400 ton press was done at sample temperatures of 633 K, 673 K and 693 K, respectively. The temperature of forging die was 543 K. All alloys were successfully forged into a motorbike engine piston to precise measure as shown in Figure 10. No defects were detected in the cross section of piston, and plastic flow indicated good formability. The product has merits of much less machining and lighter weight, since it gives a shape with fair accuracy. Further the combination of drawing-annealing step and isothermal forging on the alloy billets has an advantage to the manufacturing of high performance motorbike engine piston with lower cost and environmental load.

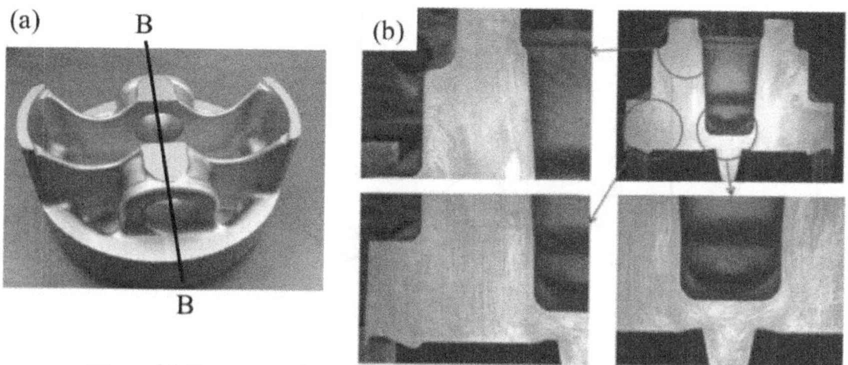

Figure 10. Prototype of motorbike engine piston (a) and macrostructure in the section B-B of UTM401 forged at 693 K (b).

Mechanical Properties of Forged Parts

Specimens were machined from the top part of forged pistons for all alloys as shown in Figure 11. The top part was controlled to keep its thickness for the specimen. The specimens were heat treated of T7 at 778 K for 4800 sec followed by water quenched and aged at 483 K for 21600 sec.

The test alloys showed higher strength such as tensile strength, high-cycle fatigue, creep and thermal fatigue than A4032 forged material and cast one at high temperature. Figure 12 shows S-N data at 523 K and 573 K of UTM407 compared with A4032 forged material and Al-12.7Si5Cu-1Mg-1Ni-0.15Ti (mass%) alloy cast. UTM407 showed the highest 10^7 cycles fatigue strength among test materials.

Through experiments mentioned in above the alloy, Al-(11~13)Si-2.5Cu-1Mg-1Ni-0.5Fe-0.1Mn-0.1Ti-0.1Zr (in mass%), was chosen as the candidate material of a motorbike engine piston. The combination of developed new semi-continuous casting process and precise forging process is promised to provide an up-grade solution to the recycle of secondary aluminum alloy.

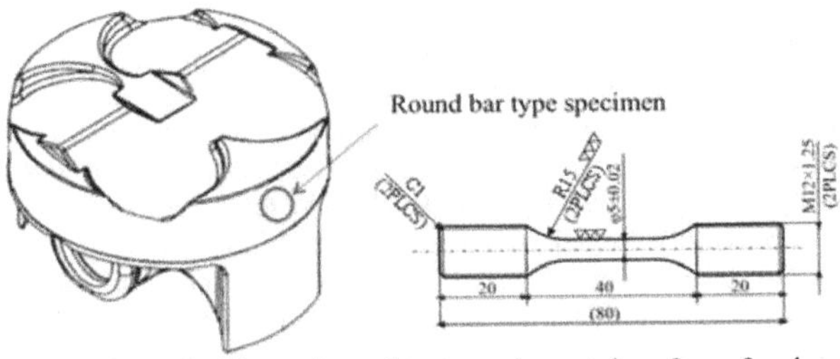

Figure 11. Schematic illustration of test specimen taken from forging parts.

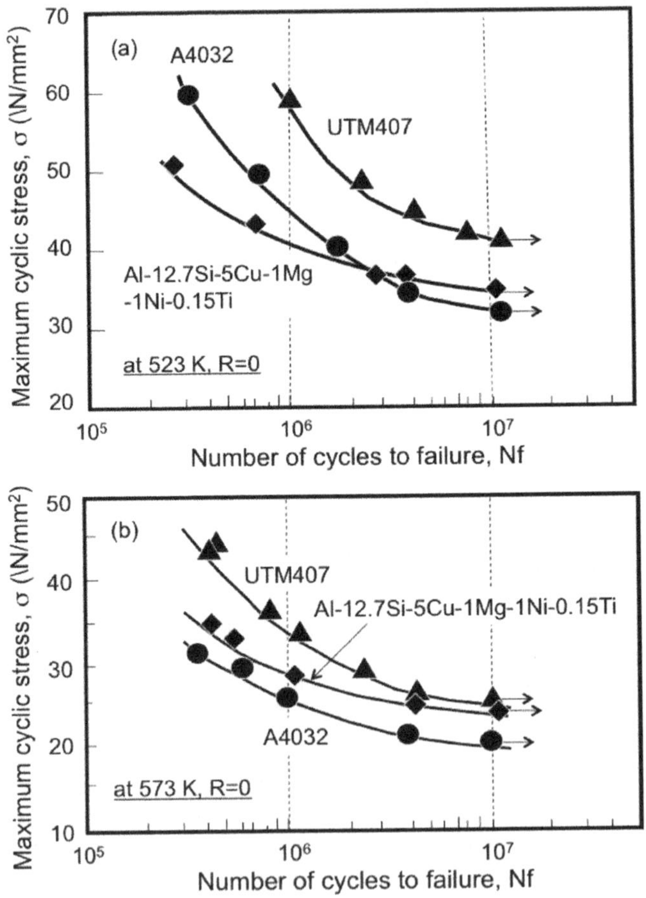

Figure 12. S-N data of the forged UTM407 at 523 K (a) and 573 K (b).

CONCLUSIONS

Novel processes have been successful in achieving microstructural refinement of Al-Si-Cu-Mg-(Ni, Fe, Mn) cast materials and producing a prototype of forged engine piston. Major results were as follows:

(1) A new vertical semi-continuous casting process was adopted an adiabatic and rapid cooling mold to produce a billet with smooth surface. Without peeling and hot-extrusion processes the cast billet was applied to warm forging.
(2) Coalesced Si crystals or compounds were still remained in the center part of the billets for hyper-eutectic alloys. Those gave a fatigue crack initiation site and decreased high-cycle fatigue strength.
(3) The isothermal forging process with greener lubricant and precise mold technologies was developed. All test alloys were applicable to the forging. The test alloys showed higher strength such as high-cycle fatigue and creep than A4032 forged material and cast one at high temperature.
(4) A combination of drawing-annealing step and isothermal warm-forging on the Al-Si alloys billets has an advantage to the manufacturing of high performance motorbike engine piston with lower cost and environmental load. The product has merits of much less machining and lighter weight, since it gives a shape with fair accuracy.
(5) The alloy, Al-(11~13)Si-2.5Cu-1Mg-1Ni-0.5Fe-0.1Mn-0.1Ti-0.1Zr (in mass%), was chosen as the candidate material of a motorbike engine piston. This process is promised to provide an up-grade solution to the recycle of secondary aluminum alloy.

ACKNOWLEDGMENTS

This study has carried out by the project "Novel process and product development with precise forging of Al-Si hyper-eutectic alloys towards high performance and low environmental load" in the regional research and development resources utilization program. Financial support from Japan Science and Technology Agency is gratefully acknowledged.

REFERENCES

[1]K. Halada, Ecomaterial –New Step of Material Towards the 21st Century-, *Bulletin Jpn. Inst. Metals*, **31**, 505-511 (1992).

[2]O. Umezawa, K. Halada and K. Shinohara, Ecomaterials in the Global Eco-society: Present Situation and Future Prospects, *Materials Science Forum*, **555**, 1-7 (2007).

[3]O. Umezawa and K. Nagai, Microstructural Refinement of As Cast Al-12.6wt%Si Alloy by Repeated Thermomechanical Treatment to Produce a Heavily Deformable Material, *Metall. Mater. Trans. A*, **30A**, 2221-2228 (1999).

[4]O. Umezawa, Mechanical Properties of Thermomechanical Treated Hyper-eutectic Al-Si-(Fe,Mn,Cu) Materials, *Materials Transactions*, **46**, 2616-2623 (2005).

[5]T. Yamashita, T. Watanabe, Y. Kondou, H. Anada, S. Ikeno, S. Tada and K. Nakahira, Solidification Position and Surface Condition in Aluminum Billets Produced by a Vertical Semi-continuous Casting Process Using a Heat Insulating Mold, *J. Jpn. Inst. Light Metals*, **46**, p 494-499 (1996).

[6]H. Takagi, M. Dohi, Y. Uetani, T. Watanabe, T. Yamashita and S. Ikeno, Microstructure of Eutectic Al-Si Alloy Billet Produced by Continuous Casting Process Used Heat Insulating and

Rapid Cooling Mold, *Proc. 11th Inter. Conf. on Aluminum Alloys*, WILEY-VCH Verlag, in the press.
[7]R. Mitamura, Continuous Casting of Aluminum, *J. Jpn. Inst. Light Metals*, **30**, 227-236 (1980).
[8]M. Matsuo, Continuous Casting of Aluminum, *J. Jpn. Inst. Light Metals*, **44**, 510-525 (1994).
[9]H. Takagi, T. Watanabe, M. Dohi, T. Yamashita, T. Kawabata, K. Matsuda, S. Ikeno, Characteristics of Hypereutectic Al-Si Alloy Billet Fabricated by Continuous Casting Process Used a Heat Insulating and Rapid Cooling Mold, *Proc.114th Conference of Jpn. Inst. of Light Metals*, 233-234 (2008).
[10]M. Shiratori, Assessment of Residual Fatigue Lives for Surface-Cracked Structures by an Influence Function Method, *Current Topics in Computational Mechanics*, eds. J.F. Cory, Jr. and J.L. Gordon, ASME, 357-364 (1995).
[11]T. Kobayashi, Strength and Fracture of Aluminum Alloys, *J. Jpn. Inst. Light Metals*, **54**, 333-347 (2004).

PILOT TESTING OF A GREEN, NO-WASTE PROCESS TO MAXIMIZE VALUE FROM HOT ALUMINUM DROSS

O. R. "Duke" Singleton
Singleton Technology, Inc.
Richmond, Virginia, USA

ABSTRACT

The Dross-Spin process was tested on a 150-kg scale in five plants. Hot dross was processed at the furnace (or remotely) with an immediate primary recovery of metal. Primary recovery varied from <20% to 80% of and with available metal content depending on now established furnace and operating conditions. Product remelt skim was 6% or less.

The residue (cooled and friable) was easily crushed, ground, and screened which caused some dust. Residue +1-cm yielded about 40% RSI on induction remelt. All the process residuals were able to be used to make foamed cement. An optimum value-recovery practice was found based on material balances and economics. At commercial scale, a suspended spinner drive would be used for the Dross Spin Process. Such a drive has been used in tonnage galvanizing operations for many years. Spin speeds would be less than 200-rpm.

INTRODUCTION

The Dross Spin Process[1] is a simple, fast process for the recovery of metal from reverberatory, melting-furnace dross. The process requires no pollution control. The essential equipment is inexpensive and portable. Immediate return of the reclaimed metal alloy to the furnace from which it came maximizes recovered metal value. The uncontaminated residue is readily sold or processed in house.

The pilot testing was in five operating plants. The process steps, equipment and procedures are described. Tests are described by plant with an introduction followed by results and discussion. The unique residue of the Dross Spin Process is illustrated and residue processing discussed. Molten-Salt-Flux Stirring of hot dross, studied at two plants is compared to Dross Spin results at those plants. Scale-up using galvanizing industry equipment will be discussed. The paper ends with conclusions.

What is Dross?

Dross begins with the reverberatory, melting-furnace charge. All solid input has a hydrated, surface-oxide envelope. Molten metal input contributes oxide films on transfer. Oxide films form bag-like networks with attached, dispersed metal globules and occluded gas as discussed by Ford and Christ[2] and Van Linden[3]. Dross (density < 1.3 g/cm^3) floats on the metal bath[4]. Dross usually will comprise about 1% to 3% of the charge[5, 6].

The dross layer is an insulator. The longer a thick dross layer ages on the bath the more fuel is wasted; dross temperatures rise and the dross tends to "dry out". The metal globules in dross do not return to the melt; instead they react with the furnace atmosphere to form oxides, carbides and nitrides[2, 3]. As dross ages the dispersed metal burns vigorously (thermiting) which generates specks of white light. Van Linden showed that metal can be wicked up from the bath to be burned in the dross layer[7]. Metal burned to dry up dross is permanently lost.

When Should Dross Be Processed?

On the basis of the reports from many workers[2, 5, 7, 8, 9], it is conservative to consider fresh dross as having a metal content of over 80% when the melt is first flat and the burners turn off. That is the best time to remove dross for reclamation by Dross Spin.

The enterprising cast house manager will make sure that fresh, shiny-bright, metallic-looking dross will be taken from the furnace and processed immediately. Now, to do that takes courage. However, that practice is not new! As Binczewski[10] stated: "The good cast shop practices of the 1930's which kept melt loss to under 1.0% [included] meticulous skimming procedures and detailed attention to immediate reclamation of hot skim."

Dross Process Comparisons:

Dross metal content is a function of furnace charge, furnace operation and design, as well as its age; that is, the time the dross layer has been on the melt. A measure of process efficiency, metal recovery as a per cent of available free metal in the dross (at the time the dross was removed from the furnace) is the best means for process comparisons.

PROCESS DESCRIPTION:

The "Dross Spin" process fits the recommended 1930's practice. The process steps are:

1) First-formed, hot, wet dross is skimmed into a preheated bowl at the furnace door.
2) As soon as possible the bowl is spun inside a portable deflector-collector at the furnace.
3) Liquid metal is quickly extracted by spinning at low speeds and collected.
4) The recovered metal is immediately returned to the furnace (for pilot: cooled & weighed).
5) A residue with unique and important characteristics remains in the bowl.
6) The bowl is removed from the spinner, covered and taken to a cooling station.
7) If needed, more preheated bowls are loaded and spun.
8) The spin unit and collector are then moved to the next scheduled furnace.

Pilot Equipment:

The pilot equipment consisted of a bottom driven spinner-plate surrounded by a circular, steel, product collector trough mounted in a frame. The drive was belt driven by an attached 2.2-KW, variable speed motor. Bowls of dross were placed on the spinner. A steel deflector shield placed on the collector trough deflected the extracted liquid metal into the trough.

Bottom Discharge Bowl: The three part bottom discharge bowl (BDB) shown in Figure 1 was used for all but the test at plant 5.

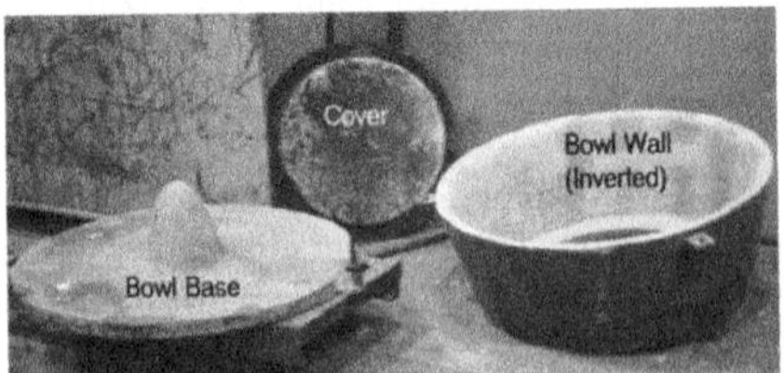

Figure 1. Refractory lined bottom discharge bowl parts

The bowl was top loaded and had a bottom discharge gap of about 1-cm. between the bowl wall and the bowl base Figure 2.

Figure 2. BDB mounted on spinner

A high-density castable refractory lining about 2.5-cm. thick was used for the bowl parts. After plant 1 tests a more insulating castable was used to reduce heat loss. The bowl internal diameter was 86-cm and height 33-cm. Bowl capacity was about 200-kg.

Spin Forces: The approximate "g's" (centrifugal force, gravity ratio) at the discharge gap for several speeds are: 200-rpm = ~19-g's, 300-rpm = ~ 43-g's, and 375-rpm = ~ 68-g's. The "dead-zone" cone in the center of the base at 200-rpm sees a force of about 4.5-g's

Top Discharge Bowl: A top discharge bowl (TDB) has advantages; it is loaded at its maximum diameter and, when moved or lifted, molten metal will not leak out. The TDB bowl is shown in Figures 3 and 4.

Figure 3. The TDB shown mounted on the spinner and above the collector trough.

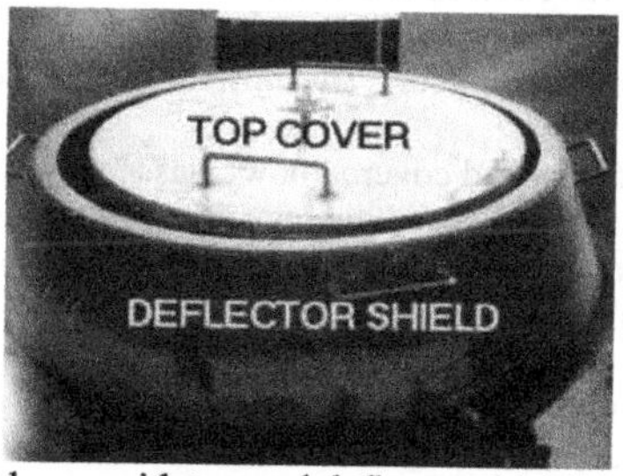

Figure 4. The TDB shown with top and deflector shield in place, ready to spin.

The inner top radius was 66-cm. and bowl depth was ~25-cm. The refractory, bowl lining was rigidized refractory fiber ~2.5-cm thick (for improved heat retention) and was topped with phosphate bonded alumina for durability.

The light-weight, castable-refractory-lined top had three protruding steel spacers which set the discharge gap between the bowl and top to about 1-cm. The centering shaft had a lock-pin to limit the upward movement of the cover.

PILOT TEST PROCEDURE:

We operated in four actively producing plants (another plant used the equipment for independent tests). We took dross when two conditions coincided: our equipment was ready and plant workers and their equipment were available. As it turned out, we never took dross when the melt was first flat and the burners had just turned off (the optimal time).

"At Furnace" Spin*:* Using readily portable equipment, Dross Spin is specifically designed for "at furnace", speedy processing of hot, fresh dross. Since only liquid metal can be extracted, in-bowl cooling-time must be minimized. "At furnace" recovery and return of the metal to that furnace maximizes the amount and value of the recovered alloy.

Remote-Spin*:* Many plants wanted spinning to be performed near their usual, central dross processing location. Sometimes only one remote-spin test was required "for comparative purposes. Remote spin incurred "delays to spin" of from 5 to about 30-minutes.

<u>Remote spinning of dross is very poor practice</u>. Recovery is always reduced by either cooling or by thermiting. Recovered alloy identity is complicated, possibly reducing value.

Dross Temperature*:* Several thermocouples in a sample spoon of dross typically gave reading which varied by over 60°C. Dross is heterogeneous both thermally and analytically. Hence, dross temperature was qualitatively estimated from very cool to hot by eye-judgments.

Bowl Preheating: For "at furnace" tests by researchers at plants 1 and 2 the BDB bowl was spun slowly while being preheated with a portable gas torch. Preheat temperatures were about 200 to 400°C (higher temperatures, would have been preferred).

For remote-spin tests the stationary bowl was preheated with one or more gas torches.

Loading Dross: The standard, dross-taking procedure was for a truck-mounted ram to pull dross up a sloping hearth to drain for about two minutes with the furnace door closed. Sometimes the burners were fired for that time. With burners off, the door was raised and the hearth-drained dross was manually raked down a steel chute into the preheated bowl.

For "at furnace" tests the BDB bowl was already mounted on the spinner and inside the collector-deflector. The bowl was covered and spun.

For "remote" tests, the loaded BDB bowl, on a steel pallet, or the loaded TDB was taken by fork truck to the spinner and mounted. The bowl cover and deflector shield were added before spinning.

Dross Spinning: The loaded and covered bowl was spun as soon as possible to maximize liquid alloy extraction. Bowls reached speed in about 15-seconds and were spun under power for a total of two minutes. Extraction seemed completed in about one minute. There was no dynamic braking; the bowl took about two minutes to slow to a stop.

Spin speed at Plant 1 was limited to 200-rpm; above that speed the apparatus rocked excessively. The frame was strengthened and weighted to permit higher speeds Bowl

Demounting and Product Recovery: When another bowl was to be spun, the deflector shield was removed and the bowl was demounted (for later removal to a cooling area). The hot, solidified product was removed and later weighed (gross recovery). The next bowl was spun.

If another bowl was not to be spun, the entire unit was picked up and taken to a cooling area for processing when cool.

Net Remelted Product: Recovered product was sawed to fit in a 20-kg induction furnace. The melt was stirred vigorously, and thoroughly skimmed with the dross quenched on a steel plate. Input weight less dross gave the per cent net remelted product weight.

Residue: The cooled bowl with its residue was weighed. The BDB wall and cover were released from the bowl base and lifted free of the coherent, friable residue. The TDB had its cover removed. The residue was either sampled or dumped. An empty bowl tare weight for each bowl was established at every plant. Residue weight was calculated as loaded weight less tare.

Input Dross Weight: Input dross was calculated as residue plus gross recovered metal.

Free-Metal Content of Dross and Residue: At plant 1, dross was sampled during bowl loading. The sample was quenched on a steel plate. Salt melt analysis (50 – 50 NaCl - KCl + 5% Cryolite) of duplicate sample sections determined free metal content. Residues were sampled by crushing and milling a representative "pie" section of the residue. Duplicate, riffled samples were subjected to salt melt analysis to determine metal content.

The sum of gross product plus residue metal content also was used for dross free metal content.

TESTS AT PLANT 1

Introduction:

Dross was from 6000-series alloy melts of about 30% scrap. The dross had aged for at least an hour and the furnace roof was cool for all tests. The dross was mildly thermiting when taken. Five tests were spun at the furnace. The one, remote-spin test required a delay of about 5-minutes. The maximum spin speed (200-rpm) was used with a discharge gap of ~1-cm.

Dross was sampled during loading and analyzed. Residues were also sampled and analyzed. Gross recovery as a per cent of available free metal in the input dross was calculated.

Results and Discussion:

During spinning here and at all the plants there was nothing to see: no smoke, no dust. The results are shown in Table I.

Table I. Results of Dross Spin Tests at Plant 1

Dross Temp.	Input Dross	Gross Recovered Metal		Avg. Available Metal in Dross		Gross Recovery of Available
*	kg	kg	%	kg	%	%
v. cool	131	47.6	36.5	86.9	66.5	54.8
v. cool	149	57.2	38.4	96.7	65.0	59.1
remote	171	70.8	41.3	107	62.6	65.9
cool	169	78.9	46.8	115	68.0	68.8
warm	142	64.4	45.2	85.2	59.8	75.6
warm	132	61.7	46.6	81.7	61.7	75.5

* Qualitative temperature; the remote dross was very warm at bowl loading.

Gross Recovery & Comparisons: The gross recovery varied from 37% to 47%. Those results compared favorably with stirring of hot dross which averaged 39% in a multi-plant study[11] and, at plant 1, averaged 35% gross recovery within a range of 21% to 58% [12].

The process was robust; it operated with very cool dross but with reduced recovery.

Net Product: Three products were remelted. The remelt recoveries were 95.9%, 97% and 97.1%. The process recovered metal with a minimum of non-metallic contamination.

After the deflector and bowl were removed a segment of recovered product was lifted up from the collector trough and is shown in Figure 5. With cooler dross considerable metal was solidified also by the uninsulated, deflector shield and metal was left in the discharge gap.

Figure 5. A segment of product lifted from the collector trough.

Free-Metal Content of Dross: The metal content of the dross ranged from 60% to 68%. Such low metal content is typical of dross that has lost significant free metal by in-furnace aging.

Material Balances: The average free metal content of the dross samples was only 2.8% less than the average sum of gross products and residue metal content. That is good agreement and indicated that either method would be acceptable for determining dross free metal content.

Recovery of Available Free Metal: From 55% to 76% of the average of the available metal in the input dross was recovered. Comparatively, of the available free metal, warm dross recovery was ~76%; whereas, very cool dross recovery was only ~57%. The process needs both hot dross and heat conservation. The low recovery of the remote test confirmed those needs. The metal globules in dross must be liquid to be extracted.

TESTS AT PLANT 2

Introduction:

The plant 2 furnaces (alloy 1350) operated more as holders than melting furnaces. Dross was formed from scrap charged (~20%) and from liquid metal input. All dross taken was unfluxed but had been aged for many hours and had been mildly thermiting. All bowls were spun at the furnace door with a fixed discharge gap of about 1.2-cm. Only once were two bowls spun taking all the dross available. Researchers operated the spinner for the first six tests and plant workers performed four tests (at slightly lower speeds) several months later.

Results and Discussion

Plant workers operated the process proficiently indicating both able workers and a robust process. Twice they overflowed the bowl and could not spin. When cooled, the dross was safely and easily cleared. However, that problem indicated a need for an improved bowl design.

The data obtained were maximum spin speeds and weights of recovered metal and residue. The latter were used to calculate input dross weight. The test results are in Table II.

Table II. Results of Dross Spin Tests at Plant 2

Comments	Spin Speed Maximum	Input Dross	Gross Recovered Metal	
	rpm	kg	kg	%
plant workers alone	235	120	70.8	59.1
plant workers alone	250	56.7	22.7	40.0
plant workers alone	275	123	60.3	49.1
plant workers alone	275	105	57.6	55.0
actively thermiting & longer aged	280	72.6	27.2	37.5
fast load, 1st bowl, f'ce. A	280	125.	66.7	53.5
fast load, 2nd bowl from f'ce. A	300	113	71.7	63.7
actively thermiting & longer aged	325	102	40.4	39.6
v. cool dross & longer aged	350	86.2	26.8	31.1
fast load	350	108	64.4	59.9
Average overall =				**48.8**

Gross Recovery: Gross recovery varied from 31% to 63.7%, average 48.8%. Researchers averaged 47.5% recovery; plant workers 50.7%. Five of the ten tests recovered over 50% metal from the dross (maximum 63.7%). That was very good, overall recovery.

The three lowest yields were from longer aged dross which was either very cool or had been actively thermiting. Active thermiting obviously lowers dross metal content.

Spin Speed: Spin speed from 235 to 350-rpm.did not show any benefit for higher rpm. That result would imply spin speed may be higher than needed.

Recovery of Available Free Metal: Dross Spin average recovery of available metal was 76%. based on a plant value of 64% free metal in unfluxed dross. Considering plant 2's long-aged dross 76% is good recovery and is the same as the value for the two warm tests at Plant 1.

TESTS AT PLANT 3

Introduction:

Researchers provided instruction and observed some of the first tests. Plant personnel spun 42-bowls of dross at a central, remote location. Three bowls were used so that all the dross from each of 15 furnaces could be treated..

Alloys, Charge and Fluxing: Most of the dross was from 3003 alloy with some from 3005, 5052 and 5252 alloy. The 5252 dross was an exception in that the charge was mostly pigs and the furnace was fluxed with the dross strongly thermiting when taken. The other tests used unfluxed dross from charges of 60% or more scrap.

Multiple Bowl Procedure: Usually three bowls were loaded; one was transported, loaded on the spinner, covered and spun. After the first bowl was demounted and the product removed for weighing, the other uncovered bowls were at the spinner and were processed in the same manner. The demounted, covered bowls were weighed and cooled (about a shift) before reuse.

Spinner residues in seven groups of either three or six bowls were gathered and processed through a rotary, salt melt furnace. The amount of recovered metal was recorded for each group.

Results and Discussion Reported by Plant 3:

The delay between bowl loading and spinning was "usually about 30-minutes". Specific delay times were recorded for 3-tests with 5252 and 3-tests of 3003 dross.

The plant reported that the Dross Spin Process returned overall 25.1% of the charged dross as metal. The metal was "clean, free of oxide and strictly aluminum". [Recovered metal was collected into bins by alloy and later charged to the furnaces.]

Grouped Tests: The seven, grouped, tests recovered 23.9% by spinning and 40.3% by rotary salt furnace. [Calculates to 54.5% free metal input; assume no allowance for black dross].

Regular Practice: For comparison, the "dross metal" from floor-cooled dross amounted to 37.4%. The salt furnaces yielded 41% metal. [Assuming as above 54.5% free metal in dross, then the "dross metal" returned to the furnace was just under 80% metal.]

Results and Discussion:

Fluxed Dross: Three bowls of fluxed, thermiting, 5252 dross were spun. One bowl delayed 12-minutes recovered 20.6%. The other two bowls, delayed 30-minutes, recovered 4.7% and 5.1% metal. In this case, during the longer delay in open bowls, metal content continued to actively burn. When spun, the dross had much less metal available for recovery.

Unfluxed Dross: Omitting the fluxed 5252, the plant 3 recovery ranged 11.5 to 47.7%; the average was 25.7%. Compare that recovery to plant 2's, at furnace, spin average of 48.8%. The almost two to one difference is attributed to dross cooling in the bowls delayed by transport.

Delay to Spin Effect: Three bowls of 3003 alloy, delayed a recorded 30-minutes, recovered an average of 14.4% (range: 11.5% to 16.8%). These recoveries were among the six lowest of all the 3003 recoveries. Recall that the delay to spin was "usually about 30-minutes". . A "Delay to Spin" effect may have overshadowed all the following effects.

Alloy Effect: Three bowls of 3005 alloy dross were spun: gross recovery average 23.8% with a range of 16.6% to 30.7%. Three bowls of 5052 alloy dross were spun: gross recovery average 21.5% with a range of 14.3 to 28.8%. There is no definite alloy effect as these values are within the 3003 alloy gross recovery (33-bowls): average 26.0%, range 11.5% to 47.7%.

Gap Effect: The 39 bowls spun with a nominal 1.2-cm gap averaged 26.5% recovery. Two bowls spun with a 1.9-cm gap averaged 15.5% recovery; two bowls spun with a 2.5-cm gap averaged 22.3% recovery. The reason for the increased gaps was not given and there is not a trend in recovery with these few tests. Product remelt recovery was never determined.

Spin Speed Effect: The average recovery for spin maximum speeds of: range 250-325-rpm (6 tests), 335-rpm (6 tests), 350-rpm (19 tests), and 375-rpm (8 tests) were, respectively, 27.2%, 19.5%, 24.7%, and 28.9%. There does not seem to be any clear trend within these rpm's.

TESTS AT PLANT 4

Plant 4 used the Dross Spin equipment when it was idle at plant 2. Plant 4 reported: "Yields from hot, thermiting, heavily fluxed dross ranged from 25% to 30% even though the amount of charge was only about half of the bowl's capacity."

Those at furnace results (no delay) and plant 3's 12-minute and 30-minute delayed recovery of fluxed dross form a straight line (negative slope) of recovery vs delay time. The results also speak to the process's robustness in that it was operated independently by the plant and, also, was able to effectively treat thermiting dross.

TEST AT PLANT 5

Introduction:

The top discharge bowl (TDB) was tested at plant 5. A very important TDB question to be answered was the effect of top-discharge on both product and residue. The dynamics of movement during residue formation and metal extraction are different from that of a bottom discharge bowl. Because at that time there was no good scale-up design to replace the bottom driven "washing machine" system which was not suitable for scale-up, only one test was conducted before the project was terminated.

Spinning was remote from the furnace, about a 5-minute delay. The furnace was a combination side-well-melter for can scrap and direct furnace input for larger scrap and sows. The furnace had been on intermittent low fire for about an hour; the dross was sparse and cool..

The bowl and top were only pre-warmed to about 150°C. The bowl, about half full, was transported to and loaded on the spinner. The cover and deflector shield were mounted and the bowl spun. Spinning was about 2-minutes at 300-rpm producing about 33 "g's" at the bowl periphery. After spinning the unit was trucked to a lab for analysis.

Results and Discussion:

The centrifugal force on the bowl contents raised the bowl cover up to the lock pin which increased the discharge gap to 1.6-cm. Metallic, horizontal "icicles" protruded from the bowl discharge gap. They were the result of metal solidified before it was extracted. The uncovered, bowl is shown in Figure 6.

Figure 6. The residue appeared well extracted except for the rim of metallic "icicles".

Gross Recovery*:* Gross product was 32.3-kg and residue was 22.3-kg.; thus, the gross recovery was 59.2% of the input dross. That is excellent recovery, particularly for remote spinning of cool dross.

Net Recovery: The remelt dross was 4.6% for a 95.4% remelt recovery. This is acceptable net recovery indicating a low, non-metallic content returned to the furnace.

Material Balance: The residue included the not fully extracted metallic rim material. The residue contained 31% metal (6.9-kg); thus the input dross had 71.8% available metal.

Recovery of Available free Metal*:* The recovery of available metal was a high 82%. The TDB dynamics and the improved bowl insulation contributed to this outstanding recovery under the adverse processing conditions of a smal,l cool load of dross delayed before spinning.

Residue Additional Recovery: The residue was processed. The plus 1-cm screened fraction (11.8%) was heel melted without flux. The remelt recovery was an additional 1.18-kg of metal for an additional 2.16% metal recovery (RSI) based on the input dross weight.

THE UNIQUE AND VALUABLE DROSS SPIN RESIDUE

The Dross Spin residue is unique, is fully useable and adds significant value to the process. Remarkably, the Dross Spin residue does not consolidate into a solid mass during spinning. The residue forms a porous structure. This natural, coherent structure provides flow channels for the easy extraction of coalescing liquid metal droplets into streams and the passage of these streams out of the bowl due to the force of rotation. Cooled, the residue is safe and easy to unload. The residue has a lumpy, ash-like appearance shown in Figure 7.

Figure 7. A normal, "dry" well-extracted residue except for a few "icicles" at the discharge gap.

The porous residue structure is very friable. Hence, the coherent residue is in an ideal form for crushing and milling. Cooling in covered bowls preserves metal values, generally 20% to 40%, so the residue has value. The high metal content is not unexpected from both material balances and from work of other investigators[13]. The process was unaffected by dross which contained tramp steel, chunks of bath or sintered dross, nor bales of burned foil.

Plant 1 residues were processed. Plus 1-cm samples varied from 40 to 60% metal content. Minus 4-Mesh samples varied from 10% to 30% metal content.

The form of metal in the two fractions was different. In the larger fraction, much of the visually identifiable metal consisted of ductile agglomerates which were recovered by induction melting in a heel of molten metal. No salt flux was needed. The dross from that remelting was grey to black and ash-like. It may be sent to crushing and milling to fines.

In the finer fraction, most of such metal was as thin films. Salt melt analysis of fines was difficult even with four weights of the salt per weight of fines because of very poor coalescence.

Uses for Fine Fraction Residue

Aluminum and magnesium metallic powders react with $Ca(OH)_2$ added to cement to produce aerated (foamed or cellular) cement which is a very "green" building material. Used in Europe for over 60-years, it has seen world wide use since the 1990's[14]. When residue fines replace expensive metallic powders, value is optimized [15].

When foaming cement, the carbides and nitrides as well as the fine free-metal can nucleate and produce the needed small gas bubbles in cement. The finely divided oxides are not impurities in cement. The fines may also act as foaming agents and extenders in plastics. We produced foamed cement with minus 4-Mesh residue In another application, H. E. Phipps [16] explored the use of pulverized dross fines in slurry explosives in tests with Thiokol. The use of the pulverized mixtures of dross fines and scrap containing 60 to 75% free metal produced desirable, high order, detonation speeds.

COMPARISON TO MOLTEN-SALT-FLUX STIRRING

Salt Stirring Procedure:

Molten-Salt-Flux Stirring stirs hot dross + salt at temperatures above 800°C [17]. Plants 1 and 2 made material balance studies of Molten-Salt-Flux stirring for dross reclamation[18]. A crucible of dross (~ 450-kg) was either decanted or tapped twice during a cycle of about 15-minutes. Plant 1 deliberately took fairly fresh dross (note the free metal content in Table III, below); Plant 2 also took unfluxed dross but after it had been on the melt for some hours.

Process Comparison:

Results from the molten flux salt stirring tests are compared in Table III with the average of the two "warm dross" tests at plant 1 and the average of all ten tests at plant 2. The best comparison is the recovery of the available metal in the dross. That comparison is shown in the fifth line of Table III. The two processes give equivalent metal recovery of available free metal.

Table III Comparison of Molten Salt Flux Stirring and Dross Spin at Two Plants

Reclamation Process	SALT	SPIN		SALT	SPIN
Test Location	Plant 1	Plant 1		Plant 2	Plant 2
Gross Metal Recovery %	56	46		46	49
Free Metal in Input Dross %	**76**	61		64	64 per plant
Recovery of Available Metal %	74	76		72	76
Salt Added % of Input Dross	27	0		33	0

The added economic advantages of the Dross Spin Process are: no salt, no pollution, reduced process time, no recovered alloy identification errors, and, importantly, the residue has real value versus a cost of disposal or reclamation for salt contaminated residue.

PROCESS SCALE-UP TO DROSS LOADS OF 500-KG AND LARGER

Spinning a somewhat unbalanced, total mass over 300-kg rocked the pilot machine even at the low rpm used. The pilot design was not suitable for scale up to commercial size.

Later it was learned that a galvanizing company had developed[19] a portable machine (a suspended spinner) for spinning loads of 1,000-kg and more shown in operation in Figure 8

Figure 8. A hook suspended air motor drive unit is driving a hook suspended basket loaded with hot zinc coated parts to spin off excess zinc. A 4000 series "Spin-a-Batch" is rated for a total suspended load of 2,000-kg. An April 2008 price was listed as $34,000 U.S. for the drive unit.

Over 150 units are in commercial operation. A piston driven, air motor is used for the drive and is reversed for braking. This suspended spinner seems directly applicable as the means to spin heavy loads of aluminum dross. Hence, a way is now open to scale-up the Dross Spin Process.

CONCLUSIONS

The Dross Spin Process recovered competitive amounts of metal (76 to 82% of the available metal) from hot, reverberatory, melting-furnace dross.

The equipment needed for scale up is simple, proven, easily portable and inexpensive.

The entire residue from Dross Spin's primary metal recovery is salable. Alternately it can be mechanically processed to metal and a useful by-product in house. Thus, the process is "Green"; it produces no fumes, uses no salts, and has no waste.

The Dross Spin Process is a very robust process. It was operated in Plants 2, 3, and 4 by plant workers alone with no researchers present. It can reclaim thermiting dross and cool dross.

ACKNOWLEDGEMENTS

Many coworkers at Reynolds Metals Co., Metallurgical Research Laboratories (MRD) gave invaluable help and assistance as did those in the Plants where tests were run and those in administration gave encouragement; it is impossible to name them all. Particular MRD people must include G. C. Robinson, Jr., S. B. Kessler, Jr., O. H. Holcomb, W. W. Berkey, G. R. Johnson, and G. T. Culbreth. Plant people include John Patton, Harrison Monk, Ray Crocker, A. Biscaro, G.N. Chaffin, and T. McKnight. Administrators include J. H. Dedrick, C. L. Brooks, G. E. Spangler, J. W. Korcz, J. L. Kerins and especially A. L. Moxon. Ms. M.H. Jordan was a major help in editing this paper.

REFERENCES

[1].) Ogle R. Singleton (Jr.), U.S. Patent 4,137,073, Jan. 30, 1979; U.S. Patent 4,938,457, Jul. 3, 1990; U.S. Patent 5,916,084, Jun 29, 1999.

[2].) Ford & N. Christ, A Study of the Formation and Behavior of Aluminum Skim, TMS Paper Selection A74-61, pgs. 639 – 648.

[3].) Jan H. L. Van Linden Van Den Heuvell. "Melt Loss Reduction in Recycling Processes and Oxidation of Molten Al-Mg Alloys", Doctorial Thesis, Katholieke Universiteit Leuven, Mei (May) 1988, pgs. 58, 61.

[4].) *ibid*, pg. 29.

[5].) S. Freti et al, Metallurgy of Dross Formation on Aluminum Melts, *Light Metals 1982*, Ed. J. E. Andersen, TMS, 1982, pg. 1003 - 1016.

[6].) Reference 3, pgs. 61, 75.

[7].) *ibid*, pgs. 61, 75.

[8].) I. Alfaro, "Report of Dross Study Prepared for Reymetal SA", Reymetal, SA, Bilbao, Spain, Private Communication.

[9] .) Carson L. Brooks, "Basic Principles of Aluminum Melting, Metal Preparation, and Molten Metal Handling", Reynolds Metals Co., 1970, pgs.23-26.

[10].) George J. Binczewski, What Our Cast Shop Predecessors Were Telling Us! [Why Didn't (Did) We Listen?], *Light Metals 1991*, Ed.: Edwin L. Roy, TMS, 1991, pg. 794.

[11].) H. K. Herr, "Dross Reclamation and Disposal", Internal Reynolds Metals Co. Report, May 17, 1965.

[12].) John Patton, "Bellwood Hot Skim Processing", Plant Tri-Annual Report, August 1970.

[13].) Reference 3, pg. 61.

[14].) William H. Lurz, German Aerated Concrete Block In First U.S. Applications, *Professional Builder and Remodeler*, March, 1993, pg138.

[15].)Jiann-Yang Hwang and X. Michael Song, Replacing Al Powder with Al Slag or Recycled Foil in Cellular Concrete, *J. of Metals*, August 1997, pgs. 29-30.
[16].) H. E. Phipps, "Reclamation of Aluminum Dross by Pulverization", Internal Reynolds Metals Co. Report, Sept. 12,1966, Private Communication
[17].) Olivio Sivilotti, Economical Aluminum Dross Recovery with In-House Stirring Plant, *Light Metal Age,* December, 1984, pg. 9
[18].) J. Patton and C. L. Brooks, "Molten Flux Salt Stirring Skim Reclamation", Reynolds MRD Report No. 72-16, Private Communication.
[19].) William H. Garrison, U.S. Patent 3,699,918, Oct. 24, 1972, Galvanizing Apparatus

EFFECT OF BISMUTH ON THE TENSILE PROPERTIES AND DRY MACHINING PERFORMANCE OF AL-12.7 WT% SI ALLOY

P. Chen, H. Hu, A. T. Alpas
Department of Mechanical, Automotive & Materials Engineering, University of Windsor
Windsor, Ontario, Canada.

ABSTRACT

Aluminum adhesion to cutting tools during machining of cast Al-Si alloys, should be avoided for the success of dry and near-dry machining processes. The machinability of Al-Si alloys can be improved by adding low melting point elements during casting. In this work, the effect of bismuth addition on the mechanical properties and dry machining performance of an Al-12.7 wt. % Si alloy were investigated. Different Sr/Bi ratios (0.05 - 0.40) were used to modify the silicon morphology. The reduction in alloy's strength could be minimized by the proper selection of the Sr/Bi ratio and using 0.5 wt% Bi. With increasing the Sr/Bi ratio, the fracture surfaces exhibited features consistent with higher ductility. Dry turning tests indicated that 0.5 wt% Bi addition promoted the discontinuous chip formation.

INTRODUCTION

Al-Si cast alloys are used extensively by the automotive industry to reduce fuel consumption and emissions [1]. However, engine blocks cast from hypoeutectic Al-Si alloys must be used with cast iron cylinder liners [2]. The use of iron-based liners adds extra steps to the manufacturing cycle, increases the cost of production, compromises mass savings and complicates their recycling. The requirements for an aluminum alloy intended for the mass production of an all-aluminum cylinder block (without cylinder liners) for an automotive engine are very demanding [3]. The use of hypereutectic alloys for this application presents other challenges, including poor casting process reliability, and difficulties encountered in machining processes, primarily due to large primary silicon particles. Eutectic Al-Si alloys, due to their excellent castability, good mechanical properties and wear resistance, are currently receiving attention [4-6] for the engine block applications.

Dry machining of eutectic Al-Si alloys is an environmentally sustainable alternative to the conventional flooded cutting of castings. However, dry machining of aluminum alloys is a challenging task since in the absence of cutting fluids aluminum chips adhere to the tools surface [7-9] and cause premature tool failure. The most commonly used criteria for assessing machinability consist of determination of tool life, chip forms and tendency to produce burr [10]. Wrought aluminum alloys such as 2011 and 6262 contain bismuth, which is added specifically to improve machinability [10]. It was shown that the dry drilling performance of a cast aluminum alloy B319 was improved when low melting-point elements, namely bismuth and lead were added [11]. Improvement in machinablility occurred for bismuth concentrations exceeding 0.1% [12]. However, there is almost no information on the dry machining of eutectic Al-Si alloys containing Bi. In this work, the effect of bismuth content on the mechanical properties of an eutectic Al-Si alloy was investigated and the dry machining performance of the alloys with and without bismuth was examined.

EXPERIMENTAL PROCEDURE

Melts were prepared using a commercial 319.2 aluminum alloy. An Al-50%Si master alloy was used to increase the silicon concentration to 12.7 wt% and an Al-20%Ni was alloyed to obtain 1.0 wt % Ni. Four different bismuth concentrations of 0.0, 0.4, 0.5, and 1.0 wt% were employed. Strontium within a composition range of 0.05 and 0.1 wt % was added using an Al-10% Sr master alloy. Each batch consisting of 2.2 kg alloy was melted in an alumina crucible under an Ar gas environment. The silicon modifier (Al-10% Sr) and the grain refiner (5Ti1BAl) were added to the fully molten alloy. Bismuth was then introduced in the bath after 15 minutes of melting and hand mixed. Molten alloy was poured into a mold at 750°C after a further holding for 20 minutes. The permanent mold (120mm×120mm×25mm) was preheated at 400°C. The composition of the alloys, which differ only in Sr and Bi concentrations, hence providing Sr/Bi ratios of 0.05, 0.20, 0.4 are shown in Table 1 with their designations. The alloy without the Bi addition (with 0.05%Sr) is designated as 0Bi-0.05Sr. Castings were subjected to T6 heat treatment prior to mechanical testing.

The Brinell hardness (HB) of the alloy was measured using a 10 mm diameter ball and a 500 kg load. The tensile tests were performed using an Instron 8500 testing system using a strain rate of 2.2×10^{-4} s^{-1}. The reported values are the averages of four tensile tests. Samples for the orthogonal cutting were machined in the form of 25.4 mm outer diameter hollow tubes with a wall thickness of 3.0 mm. Cutting tests were performed on a lathe equipped with a rapid action brake with electrical disengagement. The cutting speed and feeding rate were 370 rev/min (0.42 m/s) and 0.25 mm/rev. A diamond tip cutter (SANDVIC Coromant 332 FPC 010) with a positive 2° rake angle was used as the cutting tool insert. No cutting fluid was introduced into the system during machining. During the cutting process, Al-Si casting generated discontinuous chip morphology without the formation of a built-up edge at the tool tip. Compositional EDS maps on the cross-sections of chips were obtained to determine the bismuth distribution.

Table 1 Composition of Al-12.7% Si Castings With Different Bi And Sr Additions, in Wt%

Designation	Si	Cu	Fe	Mg	Mn	Ni	Zn	Ti	Sr	Bi	Sr/Bi
1Bi-0.05Sr	12.7	3.0	0.26	0.08	0.42	1.0	0.01	0.12	0.05	1.0	0.05
0.5Bi-0.10Sr	12.7	3.0	0.26	0.08	0.42	1.0	0.01	0.12	0.10	0.50	0.2
0.4Bi-0.16Sr	12.7	3.0	0.26	0.08	0.42	1.0	0.01	0.12	0.16	0.40	0.4
0Bi-0.05Sr	12.7	3.0	0.26	0.08	0.42	1.0	0.01	0.12	0.05	0.0	∞

RESULTS AND DISCUSSION

Microstructures

The microstructures of the Al-12.7 wt% Si base alloy with different bismuth concentrations are shown in Fig. 1. Primary aluminum dendrites and eutectic silicon particles present in the microstructure of 0Bi-0.05Sr are seen in Fig. 1a. This is a hypoeutectic microstructure indicating that strontium shifted the eutectic to a higher silicon concentration.

The microstructures of Al-Si alloys with bismuth additions (Fig. 1b, c and d) exhibit eutectic aluminum and silicon and also have a few primary silicon particles that are evenly distributed in the matrix. The bismuth phase is in the form of spherical particles, with a diameter of about 8μm (Fig. 1e). The average diameter, length and width of the eutectic Si in Al-Si alloys with different Bi concentrations after the heat treatment are shown in Fig. 2. The average silicon particle length is 11.2 ± 8.2 μm in the Al-Si alloy with 1 wt% Bi, 7.1 ± 6.5 μm in the alloy with 0.5 wt% Bi, and 2.3 ± 6.5 μm in the alloy without the bismuth.

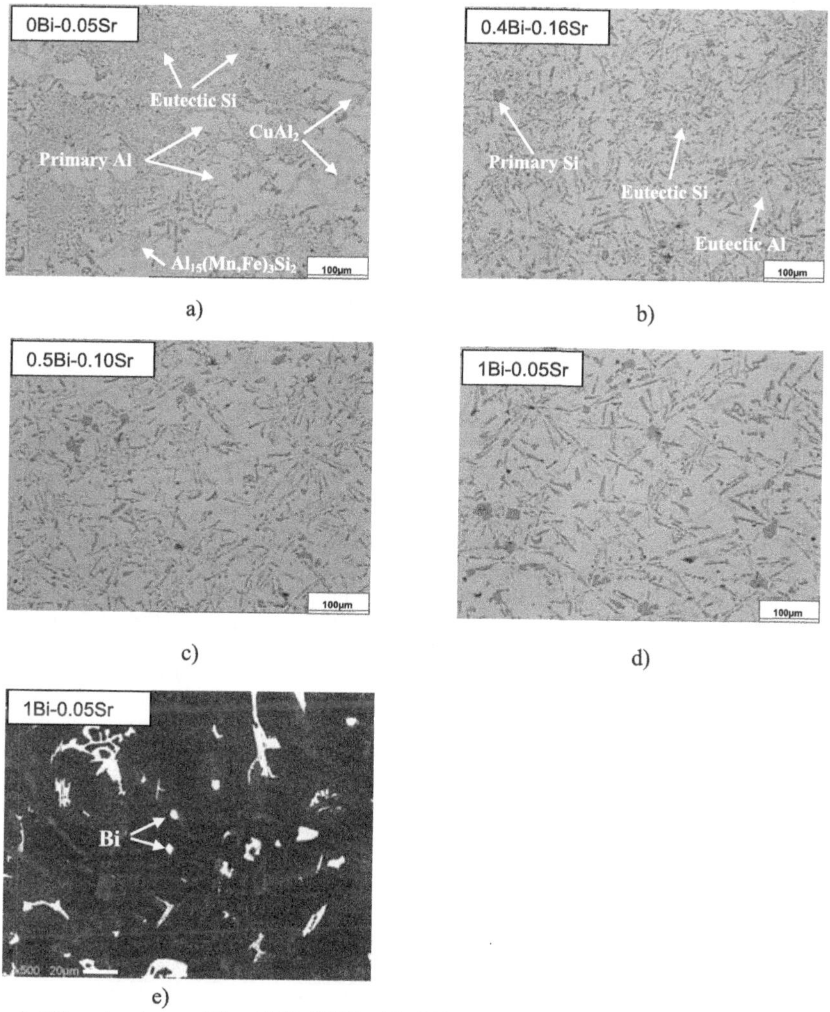

Fig. 1 Microstructures of the Al-12.7%Si with different Bi concentrations. a) 0Bi0.05Sr, b) 0.4Bi0.16Sr, c) 0.5Bi0.10Sr, d) 1Bi-0.05Sr, e) BEI-SEM of 1Bi-0.05Sr, Bi is the light spherical phase.

The strontium's modification effect is attributed to the absorption of strontium atoms onto the growth steps at the silicon solid-liquid interface [13]. Sr modification of the Si phase was effective as shown in (Fig. 1a). The refining effect of Sr on the eutectic silicon morphology was lessened by the increase in Bi as the silicon morphology became coarser with an increase in Bi content (Fig. 2). In order to describe the extent of Si modification, the Sr/Bi ratio should be also considered. A high Sr/Bi ratio, for a similar Bi content appeared to be more effective in maintaining the refined microstructure.

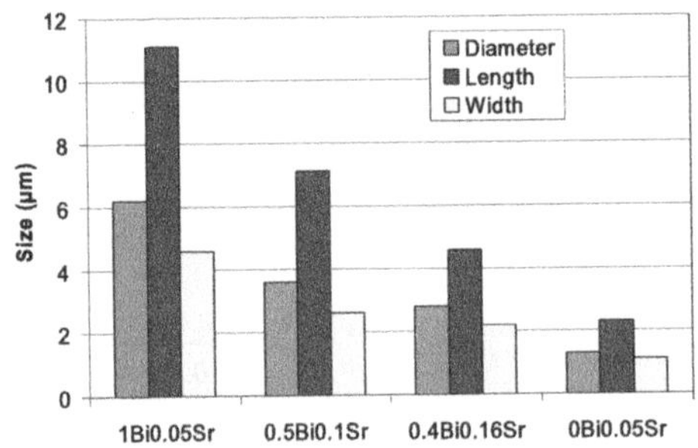

Fig. 2 Morphological parameters of eutectic Si phases in Al-12 % Si with different content of Bi.

Mechanical Properties

The tensile properties of T6 heat treated Al-12.7%Si with different bismuth amounts are listed in Table 2. The addition of bismuth reduced alloys' mechanical properties as expected. The extent of the tensile strength decrease was larger in the range of 0.5 ~ 1.0 wt% Bi compared to the 0.0 ~ 0.5 wt% Bi. The yield strength remained almost constant in the range of 0 ~ 0.5 wt% Bi. The same trend became also evident when the Brinell hardness of the Al-Si alloys with different Bi content was considered (Fig. 3). The average Brinell hardness of the as cast alloy with 0.5 wt% Bi was 93 ± 6.1, and decreased to 82 ± 4.1 as a result of 1.0 wt% Bi addition. T6 heat treatment increased the Brinell hardness in such a way that 0.5 wt% Bi alloy had the same hardness as the alloy without the bismuth addition. This result is in agreement with the tensile strength, which also remained approximately constant within the same range. It can be concluded that it is necessary to limit the Bi content to <1 wt% to avoid significant reduction of mechanical properties. The alloy with 0.5 wt% Bi and also with Sr/Bi = 0.2 provided reasonably good combination of hardness and tensile properties.

Table 2 Tensile Properties of T6 Treated Al-12.7% Si With Different Bi Content

Alloy (%Bi, %Sr)	Yield Strength (MPa)	Tensile Strength (MPa)	Elongation %
1Bi-0.05Sr	155.3 ± 4.4	195.9 ± 11.6	1.13 ± 0.16
0.5Bi-0.1Sr	197.3 ± 2.4	258.1 ± 3.2	1.22 ± 0.16
0Bi-0.05Sr	195.3 ± 1.4	267.2 ± 12.8	1.89 ± 0.46

Bismuth content influenced fracture surface morphology as shown in Fig. 4. The tensile fracture surface of the alloy with 1 wt% Bi exhibited a higher percentage of fractured silicon phases compared to the alloy without the Bi (Fig. 4b). Also, 1Bi-0.05Sr showed a more facetted fractured surface indicative of drop in ductility compared to the fibrous morphology of the Al-12% Si without the bismuth addition.

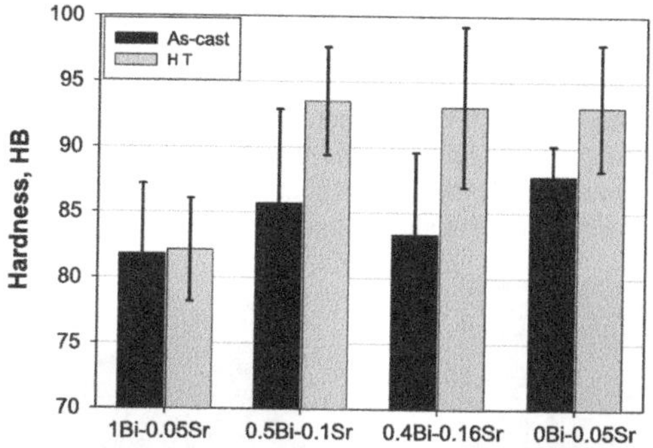

Fig. 3 Brinell hardness of the T6 treated Al-12%Si with different wt% Bi.

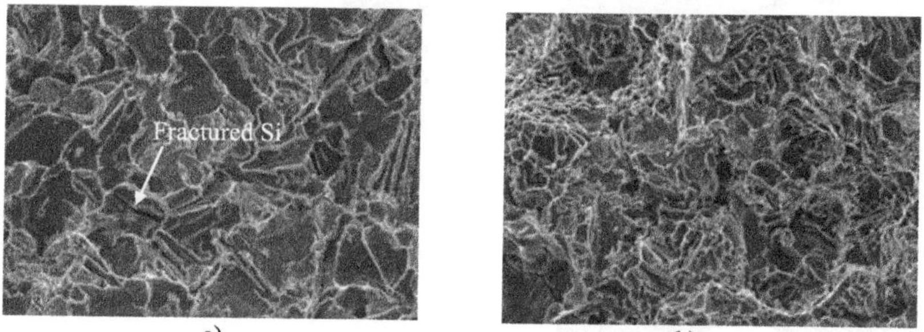

a) b)

Fig. 4 SEM tensile fracture surfaces of Al-12.7% Si alloys with different Bi content after heat treatment.
a) 1Bi-0.05Sr, b) 0Bi-0.05Sr.

Dry Machining

Al-Si alloy samples with different bismuth additions were subjected to dry orthogonal cutting tests using the parameters listed in the experimental section. The cross-sectional optical micrographs of chip roots, which were taken by interrupting the cutting process of 0Bi-0.05Sr and 0.5Bi-0.1Sr alloys, are shown in Fig. 5. Both alloys generated discontinuous type chips; where short chip segments were loosely attached to each other. The 0.5Bi-0.1Sr alloy produced narrow and short chip segments (Fig. 5b), which are smaller than those in 0.0Bi-0.1Sr alloy (Fig. 5a). The average length and thickness of chip segments of 0.5Bi-0.1Sr alloy was 320 ± 10 μm and 550 ± 25 μm compared to 420 ± 17 μm and 665 ± 30 μm for the alloy without the bismuth. Consequently, a slightly larger shear angle of about 30° in 0.5Bi-0.1Sr, compared to 25° in 0.0Bi-0.1Sr resulted. This may infer that the tool face friction was reduced in the alloys with bismuth [14]. Fig. 5(c) is taken from the same chip root of 0.5Bi-0.1Sr shown in Fig. 5b at higher magnification, and indicates presence of fractured eutectic silicon particles in the primary deformation zone as well as fracture in the secondary deformation zone.

Fig. 6a shows the free edge of the chip of 0.16Sr-0.4Bi alloy where the shear type fracture is evident. Severe deformation occurred inside the region outlined by dotted lines. Fig. 6b) a backscattered electron image (BEI) of the area within this region. It is clear that this area is subjected to large compressive stresses and the material was displaced in a direction perpendicular to the cutting direction.

Fig. 7 is the EDS Map Of An Area Between Two Segments From The Chip Separated By A Shear Crack From 0.5Bi-0.1Sr Alloy (See The Insert "A" In Fig. 5b). The area between the two parallel lines at each side of the crack shows that Bi smeared and elongated along the crack resulting from the shear deformation and sliding between the two segments.

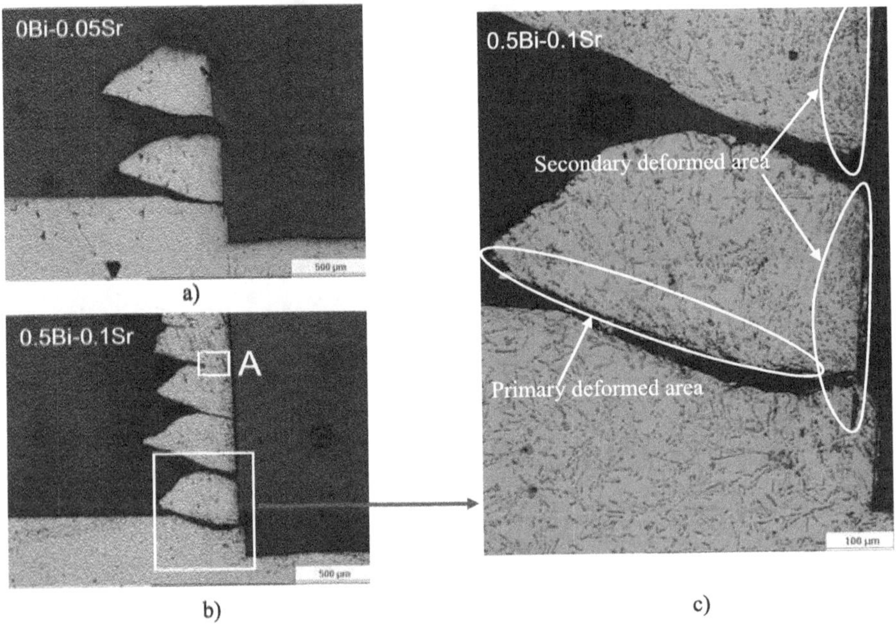

Fig. 5 Cross-sectional optical micrograph of chip roots cutting at 370rpm, and 0.25mm/rev a) 0Bi-0.05Sr, b) 0.5Bi-0.1Sr, see composition of area marked "A" in Fig. 7, c) 0.5Bi-0.1Sr at higher magnification.

The bismuth morphology changed from a spherical shape to elongated stringers as a result of temperature generated during shear deformation (Fig.7). Localized unstable flow at high strain rates can cause generation of high temperatures [15]. Plastic deformation increased the temperature of 1100 Al up to 244 °C [16]. This led to the formation of low-melting point bismuth layers running parallel to the tool face.

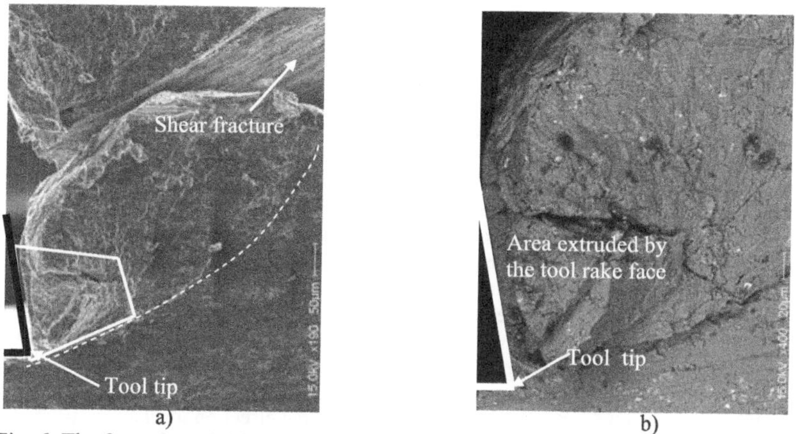

Fig. 6 The free edge of chip from 0.16Sr-0.4Bi alloy cutting at 370rpm, and 0.25mm/rev a) The free edge of chip, b) BEI image of the insert in a).

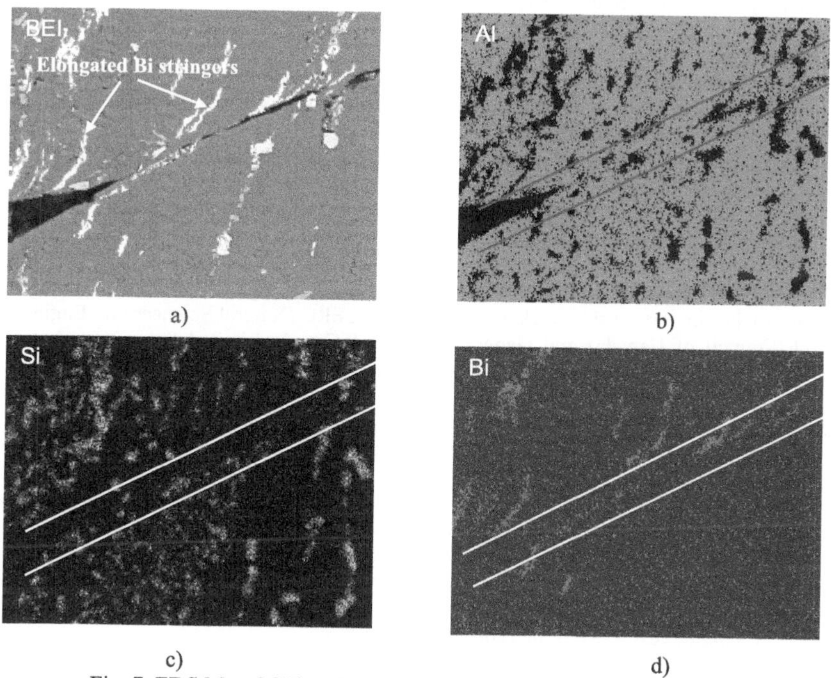

Fig. 7 EDS Map Of The Chip From The Chip Marked As A In Fig. 5b

a) Back-scattered electron image; b), c) and d) the EDS maps of element aluminum, silicon and bismuth.

The temperature increase in the shear deformation zone of chip enhances the effect of bismuth on facilitating the dry machinability. The energy absorbed during high temperature impact testing of binary Al-Bi alloy has been found to drop suddenly in a narrow temperature range of 222~238°C [17]. The embrittlement was the most severe in alloys containing liquid Bi inclusions. The reduction in the fracture surface energy brought about by inclusion melting facilitates crack initiation and propagation [17]. The elongated stringers of bismuth shown in Fig. 7a) indicate the possibility of localized melting during machining.

In summary, the dry machining performance of Al-12.7%Si alloy was improved with bismuth addition. The average chip segment thickness and length decreased, and consequently the shear angle increased. Discontinuous chips in bismuth-containing Al-Si alloys produced smaller segments. The bismuth morphology changed from spherical particulates to elongated stringers due to high temperature generation during the shear deformation process. However, bismuth additions up to ~0.5 wt% have no significant adverse effect on mechanical properties of Al-12.7%Si alloys. To maximize the benefit of bismuth on machining without sacrificing the mechanical properties, it appears that the acceptable level of Bi for strontium modified Al-12.7%Si alloys should be set at 0.5 wt%.

CONCLUSIONS

The role of bismuth addition on the morphology of eutectic silicon phase in an Al-12.7% Si alloy has been studied. Mechanical properties and chip formation characteristics of the resulting alloys are determined. The main results are follows:

1. Bi forms a spherical morphology in Al-Si alloys. The appropriate amount of Bi is 0.5% by weight for acceptable mechanical properties.
2. 0.5 wt% Bi addition results in the thinner and shorter chip segments and promotes the discontinuous chip formation.

ACKNOWLEDGEMENTS

Financial support for this work is provided by NSERC (Natural Sciences and Engineering Research Council of Canada) and General Motors of Canada Limited through an Industrial Research Chair Program entitled Tribology of Lightweight Materials.

REFERENCES

[1] W.S. Miller, L. Zhuang, J. Bottema, A.J. Wittebrood and P. De Smet, Recent development in aluminium alloys for the automotive industry, *Mater. Sci. Eng.*, **A280**, 37–49 (2000)

[2] M. Elmadagli, T. Perry, A. T. Alpas, A parametric study of the relationship between microstructure and wear resistance of Al-Si alloys, *Wear*, **262**, No. 1-2, 79-92 (2007)

[3] Herbert W. Doty, Aluminum alloy for engine blocks, *US Patent* 6921512B2, Jul. 26 (2005)

[4] K.G. Basavakumar, P.G. Mukunda, M. Chakraborty. Influence of melt treatments and turning inserts on cutting force and surface integrity in turning of Al–12Si and Al–12Si–3Cu cast alloys, *Surf. Coat. Technol.*, **201**, 4757–4766 (2007)

[5] M. Chen, T. Perry, A.T. Alpas. Ultra-mild wear in eutectic Al–Si alloys. *Wear*, **263**, 552–561 (2007)

[6] A. Ma, K. Suzuki, Y. Nishida, N. Saito et al., Impact Toughness Of An Ultrafine-Grained Al-11mass%Si Alloy Processed By Rotary-Die Equal-Channel Angular Pressing, *Acta Mater.*, **53**, No. 1, 211-20 (2005)
[7] Konca Erkan, Adhesion and material transfer between aluminum and surfaces coated with diamond-like carbon and other coatings, *Ph.D. thesis*, University of Windsor, Canada (2005)
[8] S. Bhowmick, A.T. Alpas. The performance of hydrogenated and non-hydrogenated diamond-like carbon tool coatings during the dry drilling of 319 Al. *Int. J. Mach. Tools Manuf.*, **48**, 802–814 (2008)
[9] M. Nouari, G. List, F. Girot, and D. Coupard. Experimental analysis and optimisation of tool wear in dry machining of aluminium alloys, *Wear*, **255**, 1359-1368 (2003)
[10] D. Stephenson, and J. Agapiou, *Metal cutting theory and practice*, second edition, Boca Raton, FL : CRC Taylor & Francis, 578-595 (2006)
[11] C. Ang, J. Dasch, Y. Cheng et al., Dry Machinable Aluminum Castings, *US Patent Application* Pub. (2006)
[12] J. G. Kaufman, Elwin L. Rooy. *Aluminum Alloy Castings: Properties, Processes, and Applications*, ASM International, 15 (2004)
[13] J. E. Gruzleski, B. M. Closset, *The treatment of liquid aluminum-silicon alloys*, American Foundrymen's Society, Inc. 27 (1990)
[14] Milton C. Shaw, *Metal cutting principles*, New York : Oxford University Press, 361- 365 (2005)
[15] H. Zhang, A.T. Alpas, Quantitative evaluation of plastic strain gradients generated during orthogonal cutting of an aluminum alloy, *Mater. Sci. Eng.*, **A332**, 249–254 (2002)
[16] H. Ni, M. Elmadagli, A.T. Alpas, Mechanical properties and microstructures of 1100 aluminum subjected to dry machining, *Mater. Sci. Eng.*, **A385**, 267–278 (2004)
[17] M. C. Roth, G. C. Weatherly And W. A. Miller, The Temperature Dependence Of The Mechanical Properties Of Aluminum Alloys Containing Low-Melting-Point Inclusions, *Acta Metall.*, **28**, 841-53 (1980)

ENGINEERING DECISIONS TO GREEN THE AUTOMOBILE SUPPLY CHAIN

J. A. Isaacs[*], B. Dolinsky[†], D. M. Qualters[§] and J. T. Laird[**]
[*] College of Engineering, Northeastern University
[†] School of Arts and Sciences, Endicott College
[§] Center for Teaching Excellence, Suffolk University
[**] Founder, Metaversal Studios

ABSTRACT

Designed for in-class play by undergraduate, graduate engineering or business students, a game is developed based on team competition of companies in the automobile supply chain. With objectives to achieve the highest profit and lowest environmental detriment, students playing *Shortfall* take on roles for each of the three companies in the supply chain: materials, parts, and cars. Within ten rounds of the game, each team takes its turn to invest and select among different technologies in three areas: production, storage and waste disposal. There are tradeoffs in investment costs and green values for each technology option, and there is a hierarchy to the innovation options available for each turn. Successful game strategy requires both cooperation and competition for players to succeed. Knowledge is gained along multiple dimensions: environmental policies and legislation that influence manufacturing in the global economy; current strategies used to address environmental burdens; tradeoffs and team-based decision-making. This research explores the extent to which today's students increase their understanding of complex relationships and whether learning objectives are better met through repeated play of a cooperative cross-disciplinary computer game focused on the auto industry compared to typical lecture-based case studies.

INTRODUCTION

Developments in digital technology allow new opportunities to engage students in collaborative and active learning. The desire to address complex technological and social issues in an engaged manner inspired the development of a game called Shortfall – with a goal to raise the awareness of environmental issues in engineering. Designed for in-class play by undergraduate and graduate engineering students as well as business students, the computer game structure is based on team competition of companies in the automobile supply chain. The game objectives are to achieve the highest profit and to achieve the lowest environmental detriment.

To fully develop a game that accurately measures cross disciplinary learning, it is necessary to include contributions from experts from the major content areas as well as experts in interactive and game design. This project brings together faculty and staff with expertise in two engineering disciplines, as well as graphic design, game design, and educational assessment. Metaversal Studios, a company specializing in the design of educational games and simulations, provides the required game design and software development expertise.

While the programming team worked on creating a more open-ended game system, the game design team worked on a phased transition plan for bringing the game online. This plan involved first developing and formally assessing a second version of the game (Shortfall Digital 2.0), making revisions based on this assessment (Shortfall Digital 2.5), and then releasing the game on the Internet for other universities to test. At this point, there are essentially two tracks of development for Shortfall: Shortfall Digital 2.5, the game that is designed to be played in a single classroom by student teams that each share a computer, and Shortfall Online 3.0, the game that is under development and will be designed to be played by individual students against other students on the Internet.

In the Shortfall Digital 2.5 computer game, six students create a team of three suppliers: materials, parts, and cars. Within this team, two students take on roles for each of the three companies in the supply chain. During each round in the game, each company within the supply chain takes its turn to invest and select among different technologies in three areas for each company: production, storage and waste disposal. There are tradeoffs in investment costs and green values for each technology option, and there is a hierarchy to the innovation options available for each turn. During each round, a current event was imposed on the players to provide players with a connection to actual events that students might have heard about on the news. The students work within their team and budget (and within ten rounds of the game) to try to create the most profitable and green supply chain. Students compete with other supply chain teams. Although this structure does not entirely represent the reality of business (in which individual tiers of supply chains often directly compete against one another), it allowed development of a prototype to test the game prior to development of the networked platform.

Successful game strategy requires both cooperation and competition for teams to succeed. The auto industry manufacturing supply chain allows exploration of relationships among design considerations, supply chain management, environmental issues, research and development, and profitability. Instead of using a lecture format for green manufacturing case studies, the concept of participative group game play is being tested and evaluated for successful learning outcomes. Although the supply chain is simplified, students can experience the ramification of technology selection and processing decisions through a unique educational format. The game is not designed to faithfully simulate reality, but rather to put the participants into situations in which they will be required to manage a multifaceted system that requires them to perform tradeoffs and to assess the impact of their decision making processes. Details associated with the specific technological innovations, the design of the game and its instructions, and the revisions adopted after student assessment are reported elsewhere [1-3].

ASSESSMENT OF SHORTFALL DIGITAL 2.5

The major learning objective for the Shortfall board game is for students to develop a better understanding of how to manage materials, products and waste in the multifaceted system of the automobile supply chain. A secondary learning objective is retention of facts about manufacturing processes, tested through a Knowledge Survey (albeit at lower levels of Bloom's Taxonomy). Some of these areas were tested in the first Knowledge Survey for the prototype board game [4], and similar questions were asked to a group of student volunteers who played the game in July 2008.

Of the nine students who participated, seven were males. Seven of the nine students were majoring in Mechanical Engineering Technology, and two were majoring in Mechanical Engineering. Seven of the nine students were first semester juniors. Of the eight students who reported their grade point average, the average GPA was 3.20 ±0.45. Students were first given a survey that collected: 1) demographic information; 2) likelihood of using individual decision making, study groups and external resources when completing engineering homework and team projects; 3) willingness to cooperate with competitors for long term gain; 4) concern about the environment; 5) perceived knowledge of the automotive supply chain, supply chain management, manufacturing practices; and environmentally benign manufacturing practices;

6) how well they work on teams when in a leadership role and when they are in a subordinate role; 7) their preferences for working individually. Responses to a pre-test knowledge survey was also requested with six content questions designed to measure their knowledge of supply chains, environmental issues effecting automotive supply chains, government regulations influencing automotive manufacturing; technology upgrades that enhance operations, and external events that impact business practices. Students were asked to rate their level of confidence as to the accuracy of each of their answers to the knowledge survey using an "A" (feels confident they could answer the questions sufficiently if it was a graded test), "B" (feels could answer 50% of the question or would know precisely where to get the information needed and could return in less than 15 minutes if it were a graded test), or "C" (does not feel confident that they could adequately answer the questions if it were a graded test).

Immediately after playing and completing the *Shortfall* game, students were asked to take a post-test knowledge survey with had the same questions as the pre-test. The purpose of the post-test was to obtain feedback on how participation in the game influenced: 1) their knowledge of the automotive supply chain, supply chain management, manufacturing practices, and environmentally benign manufacturing practices; 2) the confidence in leading others on a team; 3) their understanding of working collaboratively in a group; and 4) their preference to work individually. Participants were also asked if playing *Shortfall* increased their knowledge of environmentally benign automotive automobile manufacturing and their ability to work with a team. Students were also asked if they would enjoy playing *Shortfall* in tandem with class lectures about environmentally benign manufacturing and/or supply chain theory. Responses to these questions were rated on a seven point Likert scale. Using two open ended questions, students were asked to list three benefits of playing the game as well as list three suggestions for change.

Two weeks after the play test, a follow up discussion was conducted with seven of the nine students returning to participate in a two hour debrief of their experience. Students were asked to repeat the post-test to determine retention of information. A 60 minute focus group discussion followed, as well as time for participants to take a short survey to reflect on their experiences with the game.

PRE-TEST RESULTS

When asked if they liked to play computer games, four students indicated yes and five indicated sometimes. On average they reported playing games 4.33 ±2.74 hours per week, and reported using computers an average of 24 ± 15.9 hours.

Seven of the nine students stated they prefer to learn new topics by learning on the job. Students were asked to rate the likelihood of making decisions themselves, working in a study group and checking external resources to achieve the best results on engineering homework. On a scale from 1 (indicating never) to 10 (indicating always), students were likely to use all three strategies when completing homework as shown in Table 1. When completing engineering homework, no one strategy was rated significantly more often than another ($F\ (2,16) = 0.31$, $p=0.74$). A strong negative correlation that approached significance was found between subjects' tendency to prefer making decisions themselves with the tendency to using a study group when completing homework ($r(9) = -0.61$, $p = 0.08$). When completing homework, participants who tended to prefer making decisions themselves preferred to use study groups less. A similar strong negative correlation between individual decision making and use of external resources was also found ($r(9) = -0.64$, $p=0.06$. When completing homework, participants who tended to prefer making decisions themselves preferred to use external resources less.

In contrast, to achieve the best results when working on team projects, students were significantly more likely to consult with their project team and use external resources ($F(2,16) = 3.60$, $p = 005$). There was a strong negative correlation between preferences for individual work as compared to a preference for using external resources. The greater the preference for working individually, there was a predictable decrease in a preference to use external resources ($r(9) = -0.80$, $p=0.01$). There was a strong positive correlation approaching significance between preference for consulting with the project team and the use of external resources ($r(9) = 0.63$, $p=0.07$).

Students also indicated a willingness to cooperate with 'competitors' to achieve long term benefits even though it might mean short-term detriment as shown in Table I. However, seven out of the nine students stated that their willingness to cooperate depended upon the situation. When asked to elaborate the majority of answers focused on the nature of the long term benefits, three students mentioned their tendency to be cooperative depended upon the final positive benefits implying financial positive benefits. Two students mentioned their increased willingness to cooperate if the information resulting from the collaboration improved decision making.

Using a Likert scale from 1 (indicating strong disagreement), 4 (indicating no opinions) and 7 (indicating strong agreement), students were asked how well they work on teams in which they play leadership ($m = 5.44 \pm 1.00$) and subordinate roles ($m = 4.78 \pm 1.48$). Five out of nine students agreed or strongly agreed they worked well on teams when in a leadership role. Only two out of nine students agreed or strongly agreed they worked well on team when in a subordinate role. Six out of nine students indicated some level of agreement regarding their preference to working individually when possible ($m = 4.67 \pm 1.73$).

Table I. Descriptive Statistics Summarizing Learning Preferences

	Minimum	Maximum	Mean	Std. Deviation
Q4a: Homework: Prefers to make decision themselves	4.00	10.00	6.5556	2.00693
Q4b: Homework: Prefers to consult with a study group	2.00	10.00	6.7778	3.15348
Q4c: Homework: Prefers to check external resources	2.00	10.00	5.6667	3.53553
Q5a: Engineering team projects: Prefers to make decision themselves	2.00	9.00	5.3333	2.73861
Q5b: Team projects: Prefers to consult with project team	6.00	10.00	8.5556	1.42400
Q5c: Team projects: Prefers to check external resources	3.00	10.00	7.1111	2.42097
Q6: Willing to cooperate with competitors to benefit in long term despite possible short term detriment	5.00	10.00	6.7778	1.56347

On a scale from 1 (lowest) to 10 (highest), students were asked to rate their level of concern regarding the importance of environmental issues in general. Six out of the nine students rated their concern as 8 out of 10. The mean response was 7.56 ± 0.73. In general, students showed a moderately strong concern for the environment. Students concern rating did not predict their rating of their knowledge of environmentally benign manufacturing practices ($r(9) = 0.03$, $p > 0.05$) or their knowledge of two federal regulations regarding automobile manufacturing environmental regulations. However, there was a strong positive correlation between a students expressed concern prior to their game and their confidence in identifying federal government regulations following game play ($r(7) = 0.84$, $p = 0.02$).

Table II summarizes students perceptions regarding their knowledge of the automotive supply chain, supply chain management, manufacturing practices and environmentally benign manufacturing practices. Students perceived content knowledge of automotive supply chain theory was rated as the strongest. The majority of students agreed they were knowledgeable of supply chain management (n=6) and manufacturing processes (n=6) with the two female students who had not yet completed co-cops strongly disagreeing that they were knowledgeable. Only three students 'somewhat agreed' that they were knowledgeable about environmentally benign manufacturing practices, while four out of nine disagreed with this statement.

Table II. Descriptive Statistics Summarizing Perceived Content Knowledge

	Minimum	Maximum	Mean	Std. Dev.
Q9a: Knowledgeable about automotive supply chain theory	1.00	7.00	4.7778	2.22361
Q9b: Knowledgeable about supply chain management	1.00	6.00	4.1111	1.83333
Q9c: Knowledgeable about manufacturing practices	1.00	6.00	4.2222	1.92209
Q9d: Knowledgeable about environmentally benign manufacturing practices	1.00	5.00	3.5556	1.42400

POST-TEST RESULTS

Table III summarizes students opinions regarding the impact playing *Shortfall* had on their knowledge of the automotive supply chain, supply chain management, manufacturing practices and environmentally benign manufacturing practices on a scale with 1 (indicating strong disagreement), 4 (indicating no opinion) and 7 (indicating strong agreement). In general, students somewhat agreed that the game increased their knowledge of supply chain theory, management and the incorporation of environmentally benign manufacturing practices both immediately following game play and two weeks later when asked to reflect on the game again.

Table III. Post Test Statistics Summarizing Impact of ShortFall Play on Content Knowledge

	Minimum	Maximum	Mean	Std. Dev.
Q1a: Shortfall increased my knowledge of the automotive supply chain	4.00	6.00	5.1111	0.78174
Q1b: Shortfall increased my knowledge of supply chain management	2.00	6.00	4.6667	1.11803
Q1c: Shortfall increased my knowledge of manufacturing practices	4.00	6.00	5.1111	0.60093
Q1d: Shortfall increased my knowledge of environmentally benign manufacturing practices	4.00	6.00	5.1111	0.78174
Q1h: Playing Shortfall again would further increase my knowledge of environmentally benign automotive mfg	1.00	7.00	4.6667	1.73205

These answers were further supported when students identified the benefits of playing *Shortfall* immediately following game play in an open ended question. The primary benefit of playing the *Shortfall* game as perceived by 14 of the 20 student comments focused on an increase in content knowledge pertaining to automobile manufacturing, supply chain theory and how each of these impacts the financial market as well as the environment. These attributes were once again listed as the primary benefits of playing the game when students reflected on their experiences two weeks later.

Table IV summarizes the confidence ratings of students when answering content based open ended questions following gameplay with a score of 1 (indicating confidence), 2 (indicating 50% confidence) or 3 (indicating a lack of confidence). When comparing their confidence scores prior to playing the game with after, gameplay significantly increased students confidence when answering questions regarding knowing the characteristics of successful supply chains ($T(8) = 2.29$, $p=0.05$). Playing *Shortfall* also resulted in an increased confidence in answering questions regarding knowing two environmental factors associated with automotive manufacturing practices that approached statistical significance ($t(87) = 2.65$, $p=0.08$). All other confidence ratings moved in a more confident direction although not significant, with the exception of identifying Federal regulations regarding environmental regulations for auto manufacturing where gameplay resulted in a slight decrease in confidence.

Table IV. Post Test Confidence Ratings of Content Knowledge

	Minimum	Maximum	Mean	Std. Deviation
Post Confidence - successful supply chain	1.00	2.00	1.4444	0.52705
Post Confidence - environmental issues	1.00	2.00	1.5556	0.52705
Post Confidence - federal government regulations	1.00	3.00	2.1429	0.89974
Post Confidence - practices for reclamation	1.00	2.00	1.7143	0.48795
Post Confidence - technology upgrades	1.00	2.00	1.5000	0.53452
Post Confidence - events impacting business practices	1.00	2.00	1.5000	0.53452

Two weeks later while following up in the focus groups, students again reported their greatest confidence in being able to identify the characteristics of a successful supply chain theory (m=1.57, m-0.53). Their increased confidence in answering questions regarding knowing two environmental factors associated with automotive manufacturing practices remained two weeks later ($m = 1.83 \pm 0.41$). Students reported a decreased confidence in their ability to name two practices for the reclamation of waster or pollution reduction in manufacturing processes that approached statistical significance ($t(4) = -2.449$, $p=0.07$) and a decreased confidence in identifying two technology upgrades in a manufacturing facility that could make operation more successful ($t(5) = -3.66$, $p = 0.03$).

Student s agreed that they would enjoy playing *Shortfall* in tandem with class lectures. They also stated that playing *Shortfall* helped them understand the importance of working collaboratively with a team as well as increasing their ability to work with a team. There was no significant change in opinion on these dimensions two weeks later.

Immediately following game play, only three out of the nine students stated that playing *Shortfall* helped them to feel more confident when leading others in a team. Two weeks later, four out of seven stated the playing the game led to more confidence in leading times, however the change was not significantly significant ($t(6) = -0.225$, $p=0.83$). Five out of the nine students indicated they had no opinion as to whether *Shortfall* influenced their preference to work individually when possible. This remained consistent two weeks later with four out of seven stating no opinion.

The positive influence of developing teamwork skills was confirmed when students were asked in an open ended question to identify the benefits of playing *Shortfall*. Of the 20 comments made, six comments reflected benefits in enhancing meta-cognitive skills including an opportunity to work with teams, an increased knowledge of and ability to work in a team as well as enhancing reasoning skills. These positive attributes were again reflected in the focus group two weeks later. An additional positive benefit identified in the focus group was that the game was fun and enjoyable.

RESULTS FROM FOCUS GROUP DISCUSSION

Focus group questions were designed by the assessment members of the team with input from the other researchers. One researcher conducted the initial portion with preset questions, at the end the other researchers were invited to add questions that had been raised in their minds during the discussion. The data was voice recorded and transcribed. The transcript was coded by the qualitative researcher producing 29 original codes; these were reviewed and collapsed into 11 primary codes. Further analysis identified four emerging themes of environment, learning, gaming, and goals. A concept map, shown in the Figure 1, was developed from the thematic data to show the interrelationship of the elements.

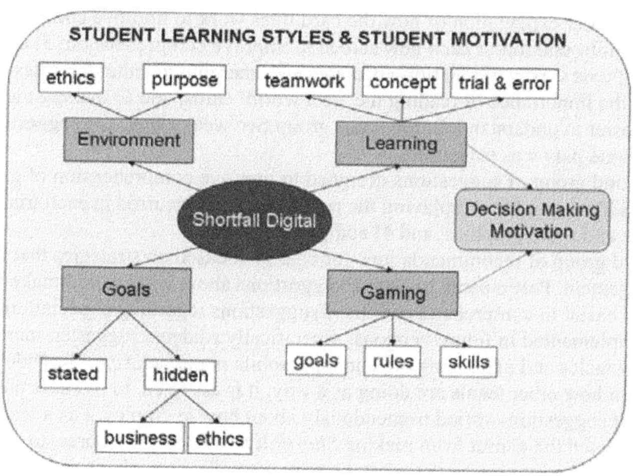

Figure 1. Concept Map of Focus Group Discussion

Concept Mapping is a qualitative research strategy to deal with the methodological biases often inherent in data analysis. "A concept map is a schematic device for representing a set of concept meanings embedded in a framework of propositions" [5]. Because of the small number of participants and their varied responses, this strategy helped to place the individual data responses into a larger and more definable context. The overarching themes of the student discussion focused on student learning styles and student motivation. In Figure 1, these themes are shown as a yellow area that encompasses the ideas generated in the discussion. The green rectangles represent the four major categories that the data was reduced to after multiple analysis using

Miles and Huberman's check code methodology. These four categories identify the major thrusts of the student conversation: gaming, goals of the game, environment of the learning game, and learning itself. Under each category, there are the subsets (white boxes) of codes that referred to these themes. Finally, the blue rectangle summarizes the students' feelings about the game and its role in their learning. It should be noted that it is difficult to make any strong claims from this data primarily due to the small number, the different learning styles, and the different assumptions about gaming and learning each student expressed.

While students expressed seemingly contradictory opinions and ideas around the *Shortfall* game, careful analysis of the data revealed that a lot of their assumptions and opinions were based on their personal definition of what it means it learn, how they best learn, and whether the game is seen as a game or a learning device and whether or not they would be held accountable (tested) in some form for the information.

SUGGESTIONS FOR GAME REVISIONS

The clear recommendation for changes in the *Shortfall* game focused on improved instructions (14 out of 27 comments) to enhance the comprehension of game play and strategy. Students recommended more and clearer instructions at the beginning. Specific suggestions included: 1) a better explanation of how the card trees work to improve comprehension; 2) improving the introduction of each new screen to improve comprehension; 3) adding an explanation of cost versus production differences; 4) merging together the rules of the tutorial; 5) emphasizing the importance of reading the 'real world' cards; and 6) making the green score calculation easier to understand. In the focus group two weeks later, the suggestion of adding an example of game play was suggested.

A second group of suggestions designed to improve comprehension of game play included: 1) adding graphics displaying the profits and cost incurred in each trial; 2) increase text size; 3) better and more graphics; and 4) adding more trials.

A third group of recommendations for change focused on strategies that might enhance student engagement. Participants had great suggestions about what would make the game more engaging and easier to comprehend and those suggestions regarding expectations, rules, and pop ups can be implemented in future versions. Specifically students suggested increasing the number of obstacles and expanding beyond automobile manufacturing. A student also suggested adding data on how other teams are doing as a way, it is assumed, to increase engagement in the game. Student suggestions varied tremendously about how to employ it as a learning tool. Their suggestions ranged the gamut from making *Shortfall* a "lab" with a course, to an introduction to a course to stimulate interest, to the central component of a class playing it over and over at different levels throughout a semester. Much of this mirrors the research around distance education. Researchers [6-9] have identified the challenges and opposing tensions inherent in the concept of learning styles and the reality of the pedagogy of on-line formats. These authors are showing that certain types of learners, i.e., those more self-directed and with more visual learning styles, are more likely to learn and persist in an on-line format.

CONCLUSIONS AND FUTURE WORK

A prototype computer game has been developed to increase environmental awareness of engineering students. The game was play tested on volunteers from several engineering degree programs. Prototype information surveys and knowledge surveys were developed and vetted.

Assessment of game play with these participants allowed improvements to be made to the assessment tools and to the game itself.

Students enjoyed playing the game and felt it helped increase their content knowledge of automobile manufacturing and supply chain theory as well as how these factors influence final financial outcomes. It increased their awareness of environmental considerations and how these interact with supply chain processes. A secondary benefit of the game was an increased awareness of the importance of teamwork and an increased knowledge of what it takes to be a successful team member. Students strongly recommended improving game play instructions to enhance the comprehension of game play and strategy.

The focus group feedback on Shortfall Digital 2.5 provided ideas for several incremental improvements to the game that help to increase playability. The team decided that in the interest of ongoing assessment of Shortfall Digital, one more revision would be released, this version including further user interface changes, context-sensitive help screens, and a new interactive "tutorial mode" to help players through the first round of play. Faculty and students have been commenting on this version (Shortfall Digital 2.6) as we continue to move towards 3.0. Feedback on Shortfall Digital 2.6 (available at http://www.coe.neu.edu/Groups/shortfall/) has offered addition changes to the game to improve assessment of student learning. However, the development goal is for Shortfall Online 3.0 to eventually supplant Shortfall Digital (1.0-2.x), these changes will be implemented towards release of Shortfall Online 3.0.

Shortfall Digital 2.6 will be play tested during the 2008-2009 academic year as development of the next version continues concurrently. Students in industrial engineering program at Northeastern University will be asked to play as a classroom activity. The game will also be play tested at several other universities. Assessment is expected to inform both the game development and the methods for its use to offer the highest impact for student learning.

REFERENCES

[1] J. A. Isaacs, J. Laird, S. Sivak, and M. Sivak, "Greening the Supply Chain: Development of a Computer Game to Teach Environmentally Benign Manufacturing", Proceedings of ASEE 2008 Annual Conference and Exposition, American Society for Engineering Education, Pittsburgh, PA, CD-ROM, June 22-25, 2008.

[2] J. A. Isaacs, J. T. Laird, and T. P. Seager, *Engineering Students Game to Green the Automobile Supply Chain,* Proceedings of the 2008 IEEE International Symposium on Electronics and the Environment, San Francisco, CA, May 19-21, 2008.

[3] D. M. Qualters, J. A. Isaacs, T. P. Cullinane, J. Laird, A. McDonald, and J. Corriere, "Shortfall: An Educational Game on Environmental Issues in Supply Chain Management", *International Journal of Scholarship and Teaching and Learning*, Vol. 2, No. 2, July, 2008.

[4] D. M. Qualters, J. A. Isaacs, T. P. Cullinane, A. McDonald, and J. Laird, *Assessment of Shortfall: A Board Game on Environmental Decisionmaking,* Proceedings of ASEE 2006 Annual Conference and Exposition, American Society for Engineering Education, Chicago, IL, CD-ROM, June 18-21, 2006.

[5] Novak, J. D., & Gowin, D. B., *Learning How to Learn,* New York and Cambridge, UK: Cambridge University Press (1984).

[6] James, W.B. & Gardner, D.L. (Fall, 1995). Learning Styles: Implications for Distance Learning. New Directions in Adult and Continuing Education. no. 67, pg. 19-31. San Francisco: Jossey- Bass.

[7] Sherry, L., Issues in Distance Learning, *International Journal of Educational Tele-Communications*, 1(4) pp. 337-365 (1996).

[8] Ouellette, R.P., The Challenges of Distance Education: A New Paradigm for Teaching and Learning, *Technology Management*, 5(1) (1999).

[9] Diaz, D.P. and Cartnal, R.B., Student Learning Styles in Two Classes: On-line Distance Learning and Equivalent on Campus, *College Teaching*, 47(4) pp. 130-135 (1999).

NOVEL CERAMIC FORMING METHODS WITH A REACTIVE ORGANIC BINDER

Kimiyasu Sato, Yuji Hotta, Takaaki Nagaoka, Koji Watari
National Institute of Advanced Industrial Science and Technology (AIST)
Shimoshidami, Moriyama-ku, Nagoya, 463-8560 Japan

Cihangir Duran
Gebze Institute of Technology
Gebze-Kocaeli, Turky

ABSTRACT

Most ceramics are produced as particle assemblages made dense by firing. Since the fine ceramic particles usually possess poor formability, the addition of considerable amount of organic binders in the forming stage is unavoidable at the present time. However, the pyrolysis of the organic binders can result in the emission of carbon dioxide or other hydrocarbon gases. A reduction in the emission gases is one aim of current ceramics industry. A novel ceramic-forming process was developed that involves immobilizing reactive molecules on the surfaces of the ceramic particles, which on subsequent external stimuli, form bridges that bind the entire particle assembly together. The chemical forces involved produce strong bonds, resulting in a reduction in the amount of organic binder that is required to maintain the shape of the ceramic green body. The obtained green body with covalent bondings is also available for debinding methods not relaying on pyrolysis. The methods will help to decrease emission gases produced from pyrolysis of the organic additives.

INTRODUCTION

In ceramic processing, organic substances are often used as binders, dispersants, plasticizers, or lubricating agents.[1,2] However, these must be removed before the sintering process by converting them into carbon dioxide and hydrocarbon gases, which are emitted into the environment. Because of serious global environmental problems, all manufacturing industries are being forced to pay more attention to reducing their emission of pollutants. The current trend in the ceramic industry is to minimize environmental problems by reducing the amounts of organic additives that are used.

A conventional organic polymer binder functions by absorption onto the surfaces of ceramic particles that otherwise have insufficient mutual binding forces; however, a poor affinity between the binder molecules and the ceramic particle surfaces will result in phase separation and nonhomogeneous partial segregation, which impair the function of the binder. Weak bonding by the binder and subsequent phase separation necessitate the use of disproportionately large amounts of organic binder. In a recent paper, we reported a novel ceramic-forming method in which a photoreactive thin film of a chemically bonded organic binder on the surfaces of the particles acts as a bridge for the binding of the whole particle assembly when the green sample is exposed to UV radiation.[3] The stronger bonding arising from chemisorption results in a reduction of the amount of organic binder that is required to retain the shape of the green body, and the green body contains only a minimal amount of the organic binder.

In the present study, the use of microwave irradiation and heating as reaction triggers was investigated. A macromolecule containing carbodiimide groups (–N=C=N–) and water-attracting (hydrophilic) segments in its structure was employed as a linking agent to form the particle assembly. The carbodiimide group can react with a variety of chemicals.[4–6] When green bodies containing the macromolecule within their structure are irradiated with microwaves, water molecules near the hydrophilic segments are dielectrically heated in a time-efficient manner. Subsequently, the increase in the internal temperature of the green bodies induces a reaction of the carbodiimide groups. Direct heating of the green bodies should also induce the chemical reaction. By these methods, we prepared green bodies composed of mutually connected ceramic particles.

Problems related with debinding such as carbon dioxide emission arise because the organic binders are removed by heat treatment. Debinding methods not relaying on pyrolysis could bypass the nuisances. The forming method described above is of advantage also in this scheme. The green bodies prepared by the method possess a feature that they are stable even in aqueous environment for a long time. When the green bodies contain some organic additives, they can be extracted from the particle compacts by dissolving them out into surrounding aqueous phase. The methodology can be recognized as a special mode of solvent debinding. For instance, in tape casting, it is necessary to confer adequate flexibility and adhesiveness for easy lamination of the green tapes by adding organic materials. In consequence, the green tapes contain well over 10 mass% of organic materials. The organic additives can be removed from the green tapes through solving out as is evidenced below.

MATERIALS AND METHODS

Spherical silica particles (Fuso Chemical Co. Ltd., Japan) were used as the ceramic phase. The number-average diameter of the particles was 270 nm, as observed by transmission electron microscopy. (3-Aminopropyl)triethoxysilane [3-APS; $H_2N(CH_2)_3Si(OCH_2CH_3)_3$] from Shin-Etsu Chemical Co. Ltd., Japan, was used as received. Silica particles were treated with 90 mM aqueous 3-APS solution under vigorous stirring for 4 h. The resulting powdery material was washed with copious amounts of water and dried overnight at 105 °C. The dried powder was washed with water again to remove any unreacted 3-APS. This procedure resulted in silica particles whose surfaces were covered with amino groups ($-NH_2$).[7–9]

3-APS-coated silica was dispersed in ultra-pure water by using intense agitation. A water-dispersible polycarbodiimide (WDC; MW 2,000; Nisshinbo Industries Inc., Japan) was added to the suspension to link the 3-APS-coated silica particles. On irradiation with the microwaves, the WDC forms covalent bonds with the amino-functionalized silica, binding the particles together: the carbodiimide group reacts with the amino group to form a guanidine structure when it is heated to ~80 °C (Fig. 1).[10] The WDC employed has reactive carbodiimide segments (–N=C=N–) and water-attracting (hydrophilic) oxyethylene segments ($-CH_2CH_2O-$): the molar ratio of carbodiimide to oxyethylene segments is approximately 1:10. The slurry was formed by slip casting into tablets of diameter 10 mm and height 2 mm and into rectangular 10 × 10 × 4 mm solids. Microwave irradiation was carried out in a tunable magnetron multimode microwave furnace (MW-Master, Mino Ceramics Co., Ltd., Japan). The green bodies were placed in a thermally insulated box and irradiated with microwaves at 2.45 GHz.

Figure 1. Binding mechanism of a polycarbodiimide reactive segment with an amino group.

A tape-casting slurry was prepared by first dissolving WDC in water. After that, 3-APS-coated silica was added and stirred for 15 min. The suspension was ultrasonicated for 10 min. Polyvinyl alcohol (PVA; MW 22,000; ICN Biomedicals Inc., USA) and glycerol (Wako Pure Chemical Industries, Japan) were added to the suspension and the slurry was stirred overnight. Tape casting was performed on a glass substrate at a casting speed of ~10 cm/s and with a blade gap of 300 μm. After drying, tapes were cut, stacked and laminated at room temperature by applying 1 MPa for 10 min. Some of the green samples were maintained at 100 °C for a certain time (e.g., 1 and 6 days). The green samples, both heat treated and not heat treated, were deaired in water for 10 min and the soaked for 1 week.

The existence of chemical bonding between the organic phases and particle surfaces was investigated by infrared (IR) spectroscopy. IR spectra were recorded with a Fourier-transform IR (FTIR) spectrometer (Spectrum GX, Perkin-Elmer, U.S.A.). The mechanical properties of the green bodies were evaluated by using a universal testing machine (AG-IS Autograph, Shimadzu Co., Japan). Thermal analyses were performed by thermogravimetry (TG) to determine the amounts of organic substance incorporated into the green bodies (RTG320, Seiko Instruments, Inc., Japan).

RESULTS AND DISCUSSION

Microwave-Driven Bonding Formation

A diagnostic absorption band from carbodiimide stretching can be found at 2120 cm^{-1} in IR spectra. IR diffuse reflectance spectra of the microwave-irradiated green bodies in the carbodiimide stretching region are shown in Fig. 2. When carbodiimide groups are heated to ~80 °C, they initiate bonding to neighboring amino groups on the silica surface, and this reactivity can be used to form covalent linkages between the particles. The intensity of the absorption band decreased with increasing time of microwave irradiation. The decrease in the band intensity indicated the formation of linkages between the silica particles.

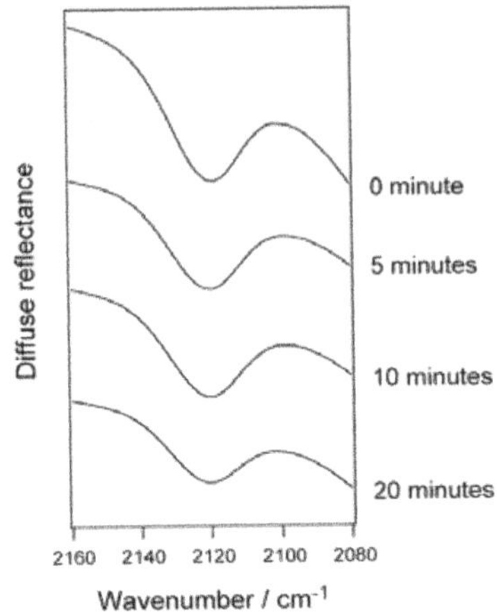

Figure 2. Spectral changes in the carbodiimide absorption band during microwave irradiation. The elapsed times are 0, 5, 10 and 20 minutes, from top to bottom.

We prepared two kinds of green body, one with and one without covalent linkages. The former was a green body containing WDC treated by microwave irradiation for 20 min, and the latter was a green body prepared in the same way, but using polyethylene glycol (PEG) instead of WDC. PEG is composed solely of oxyethylene segments and contains no carbodiimide segments. Because there is no chemical bonding between the silica particles, the PEG-containing green body can be used as a control specimen. The two kinds of specimen were soaked in water and kept for a long period. Remaining air in the green bodies was expelled by keeping the whole assembly under a reduced-pressure atmosphere. When the green bodies are soaked in water, capillary condensation forces between particles caused by adsorbed water should disappear. Green bodies with no attractive force other than capillary condensation acting between their constituent particles should not be able to maintain their shapes. Whereas the control specimen collapsed immediately on soaking in water, the green bodies with covalent linkages maintained their shape in water (Fig. 3).

We examined the influence of the chemical bonding between the particles on the mechanical properties of the green bodies. We employed a simplified breaking test using the universal testing machine. The total weight on the load cell was recorded when the push-pin broke the rectangular green bodies. We tested 18 specimens of green bodies containing WDC and a similar number of control specimens and recorded the weight values when the green bodies were broken. The mechanical strength of the control specimen was, naturally, poor. Disintegration began at a load of about 500 g. Disintegration of the green bodies with the chemical bonds occurred at loads of more than 1000 g, and the mechanical properties of these green bodies were significantly better than those of the control specimens.

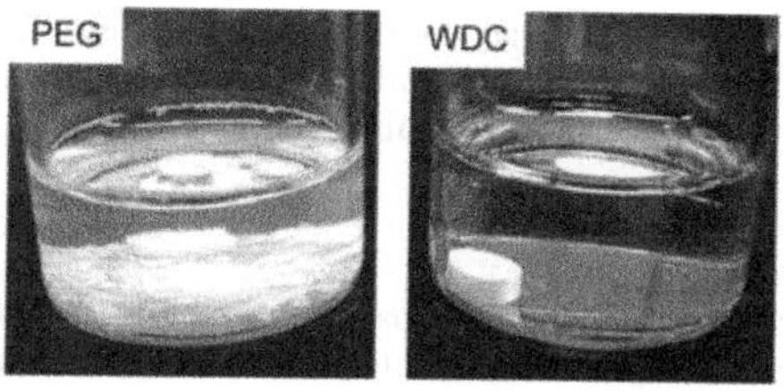

Figure 3. Photographs of the green bodies prepared by using PEG and WDC. The green bodies were soaked in water to evaluate the advantageous effect of interparticle bonds.

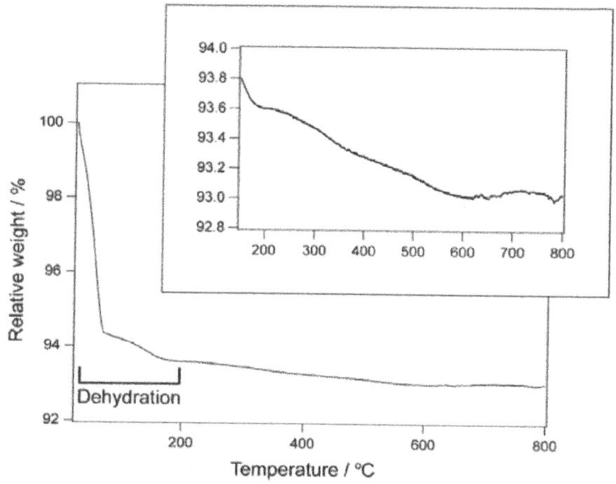

Figure 4. TG curve of the WDC-containing green body. The region relating to the burning out of organic substances is magnified.

Figure 4 shows typical burnout profiles of the WDC-containing green body, measured by TG analysis. The measurements were conducted in the atmosphere. Oxidation of the organic substances in green bodies starts at about 200 °C, and weight loss below this temperature is due to dehydration.[11] Because the oxyethylene segments in WDC are highly hydrophilic, the green body still contained significant amounts of water even after microwave irradiation. The majority of the organic substances in the green body were removed between 200 and 600 °C. The shapes of the green bodies were maintained in the presence of only 0.6 mass% of organic substances, which is much less than required for conventional polymer-based molding systems. With such a small amount of binder, green bodies are usually too fragile to handle. In conventional methods typically 3 mass% or more of binder is necessary for various operations.[1] Therefore, the present method, through linking of the particles induced by microwave irradiation, results in a marked reduction in the amount of organic binder that is required compared with that used in conventional methods.

Novel Debinding without Heating

We prepared green tapes whose constitutive particles are mutually connected with covalent bondings. A slurry consisting of 3-APS-coated silica, WDC, PVA and glycerol was prepared and used for tape casting. The obtained tape was flexible and easily stripped from the glass substrate after solvent evaporation. The green samples were first kept at 100 °C for 1 or 6 days to bind 3-APS-coated silica particles covalently to each other. The green samples which were subjected to heat treatment were then deaired for 10 minutes and soaked in water for 1 week. Figure 5 shows photos of the samples before and after the binding reaction and the result of subsequent water-soaking test. Because the green tapes contain only water-soluble substances, the sample without interparticle connection collapses immediately after immersion in water. On the other hand, the sample heated for 6 days kept its original shape even though it was held in water. There was no disintegration of the sample, neither by removal of particles from its sharp corners, nor by rounding off the edges. The sample heated at 100 °C for 1 day had some peeling off from the sharp edges and disintegration from the top layers after soaking due to an incomplete bonding.

Figure 6 shows the TG plots of the samples heated at 100 °C for 6 days. The sample without water-soaking has 14% relative weight lost up to 800 °C as compared to 11% for the water-soaked sample. This result indicates that polymer in the green sample can be removed by

Figure 5. Appearances of the green tapes and the result of water-soaking test. Green tape without heating (a), and that heated at 100 °C for 6 days (b). Water-soaked green tape without heating (c), and that heated at 100 °C beforehand (d).

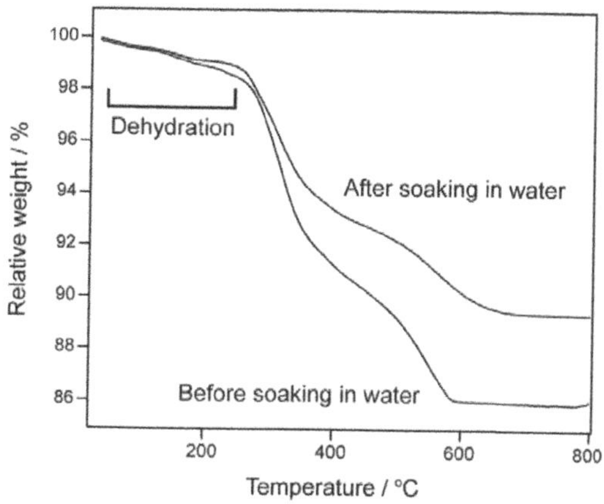

Figure 6. TG curves of the heat-treated green tapes.

water dissolution without distorting the green body shape, which, in turn, results in lesser amount organics to be removed during polymer pyrolysis.

Since the organic additives extracted from the green bodies by leaching are not deteriorated by thermolysis, they might be reused for ceramic forming via condensation. Although there remain problems to be solved, e.g., inhomogeneous distributions of residual organic additives in the greens and considerable durations required for leaching, it still is meaningful exemplar for eco-friendly forming process.

CONCLUSION

Reactive organic molecules were used to form bridges to bind the entire particle assembly. The resultant assembly can be considered as a ceramic green body. This unique methodology results in a significant reduction in the quantity of organic binder that is required to maintain shape, while providing a stronger binding as a result of the chemical forces that are involved. The green bodies obtained in this study showed improved mechanical properties in the presence of only 0.6 mass% of organic substances.

Green tapes with the mutual linkage were used for a debinding method not relaying on pyrolysis. When the green tapes are soaked in water, organic additives can be extracted from the particle compacts by dissolving them out into surrounding water, without distorting the green body shape. This results in lesser amount organic substances to be removed by pyrolysis.

The present methods should contribute to reduce the amounts of organic binder required or emission of pollutants during thermal debinding.

ACKNOWLEDGEMENT

The present work has proceeded as a part of the collaborative research project entitled “Research and Development of Low Environmental Load Processes” between AIST and NGK insulators Ltd.

REFERENCES
[1]M. Bengisu, Engineering Ceramics; pp. 85-207. Springer, Berlin, Germany, 2001.
[2]W. E. Lee and W. M. Rainforth, Ceramic Microstructures: Property Control by Processing; pp. 3-66. Chapman & Hall, London, UK, 1994.
[3]K. Sato, Y. Hotta, T. Nagaoka, K. Watari, M. Asai, and S. Kawasaki, Mutual Linkage of Particles in a Ceramic Green Body through Photoreactive Organic Binders, *J. Ceram. Soc. Japan*, **113**, 687-691 (2005).
[4]M. Mikołajczyk, and P. Kiełbasi´nski, Recent Developments in the Carbodiimide Chemistry, *Tetrahedron*, **37**, 233-284 (1981).
[5]A. Williams, and I. T. Ibrahim, Carbodiimide Chemistry: Recent Advances, *Chem. Rev.*, **81**, 589-636 (1981).
[6]L. C. J. Hesselmans, A. J. Derksen, and J. A. M. van den Goorbergh, Polycarbodiimide Crosslinkers, *Prog. Org. Coatings*, **55**, 142-148 (2006).
[7]B. Arkles, J. R. Steinmetz, J. Zazyczny, and P. Mehta, Factors Contributing to the Stability of Alkoxysilanes in Aqueous Solution, *J Adhesion Sci. Technol.*, **6**, 193-206 (1992).
[8]C. R. Suri, and G. C. Mishra, Activating Piezoelectric Crystal Surface by Silanization for Microgravimetric Immunobiosensor Application, *Biosens. Bioelectron.*, **11**, 1199-1205 (1996).
[9]M. Nisnevitch, M. Kolog-Gulco, D. Trombka, B. S. Green, and M. A. Firer, Immobilization of Antibodies onto Glass Wool, *J. Chromatogr. B*, **738**, 217-223 (2000).
[10]Y. Imashiro, I. Takahashi, N. Horie, T. Yamane, and S. Suzuki, E.U. Patent 0878496, J.P. Patent 10316930, U.S. Patent 6124398, Nisshinbo Industries Inc., May 15, 1998.
[11]A. C. Young, O. O. Omatete, M. A Janney, and P. A. Menchhofer, Gelcasting of Alumina, *J. Am. Ceram. Soc.*, **74**, 612-618 (1991).

ENVIRONMENTAL ASSESSMENT OF MANUFACTURING WITH CARBON NANOTUBES

L. J. Dahlben and J.A. Isaacs
Department of Mechanical and Industrial Engineering
Northeastern University
Boston, MA, USA

ABSTRACT

Current applications for carbon nanotubes include manufacture and assembly to create carbon nanotube (CNT) switches and CNT-polymer meshes. While these nanomanufacturing processes are still in the developmental stage, it is critical to understand the tradeoffs of scaled-up process economics with their associated environmental footprints. Two case studies that involve using single-walled nanotubes (SWNT) are considered: 1) a comparison of SWNT production alternatives and 2) a SWNT switch for non-volatile memory devices. Using life cycle assessment (LCA) – a methodology to explore the environmental impacts associated with a product's life cycle, inventories for these nanomanufacturing processes have been assembled through data collection of inputs (raw materials, energy, labor and equipment) and outputs (emissions). From this life cycle inventory data, select sensitivity analyses are performed to investigate parameters that drive economic and environmental impacts. From these analyses, recommendations can be offered to improve specific process parameters and practices that affect cost and the process environmental footprints prior to full-scale manufacturing.

INTRODUCTION

Nanotechnology offers great potential in a wide range of applications including electronics, materials, energy, and biomedicine. Because of its promising commercial viability, substantial investment has been afforded to nanotechnology research and development (R&D). United States federal R&D funding jumped from $464 million in 2001 to $1,425 million in 2007 [1]. In 2005, $9.6 billion was spent on nanotechnology R&D worldwide by governments, corporations, and venture capitalists, which is an increase of 10% from 2004 [2]. The global carbon nanotube (CNT) market is projected to exceed $1.9 billion by 2010, and the single-walled nanotube (SWNT) market has a forecast of over $5 billion by 2012 [2]. As investments continue to rise, many have stated that nanotechnology is the next great technological wave [3]. With any emerging technology, however, it is important to assess the associated economic and environmental attributes to determine its commercialization potential and sustainability.

Carbon nanotubes have unique electrical and mechanical properties that make them desirable in industrial applications. Such properties include superconductivity, sustainable high current densities high thermal conductivity and high tensile strength. Hence, CNTs have great potential to be used in a variety of applications such as structural polymers, energy conversion, batteries, sensors, and shielding. To date, however, there are a very limited number of publications regarding the environmental and economic impacts of nanotechnology processes or products [4-7]. There is also great uncertainty over the health effects and handling of nanomaterials. Some studies report that SWNT's appear to damage lung tissue in mice [8-10], whereas other studies state that there is little effect [11, 12]. Due to the high risk and unknown outcomes, many organizations have begun to adopt guidelines and protocols for the handling of engineered nanomaterials [13-16]. While this indicates proliferation of safe handling practices for nanomaterials, there are still no standards for handling this developing suite of engineered nanomaterials.

Given these concerns, it is prudent to assess the economic and environmental attributes of nanomanufacturing processes while they are in a developmental phase. This exploration can inform the commercialization of safe, economically competitive, and environmentally responsible nanotechnologies. Such work has already begun at the Center for High-rate Nanomanufacturing (CHN) at Northeastern University, focusing on two case studies: 1) a comparison of three SWNT production alternatives and 2) a SWNT switch for nonvolatile memory devices. Each case study uses a similar methodology, with development of process-based cost models and life cycle assessment (LCA) tools to investigate the economic and environmental impacts. These studies provide key information on the process flow paths, material and energy consumption, and cost drivers of each process step.

METHODOLOGY

Process-Based Cost Models

Economic assessment is carried out by development of process-based cost models, whereby costs are categorized into a comprehensive set of cost elements [17-20] to indicate the economic competitiveness of operational factors in a particular process and to compare alternative manufacturing processes for the same product. These models capture relationships among process variables to determine the fixed and variable costs of manufacturing operations. Unlike other cost systems prevalent in industry where the manufacturing overhead is allocated on the basis of direct labor or machine rent, process-based cost models break down manufacturing overhead into individual process steps, providing a more detailed cost assessment and the ability to explore how costs change as input parameters change. Typical cost elements include materials, labor, energy, capital equipment, tooling, building space, and maintenance, as illustrated in

Figure 1.

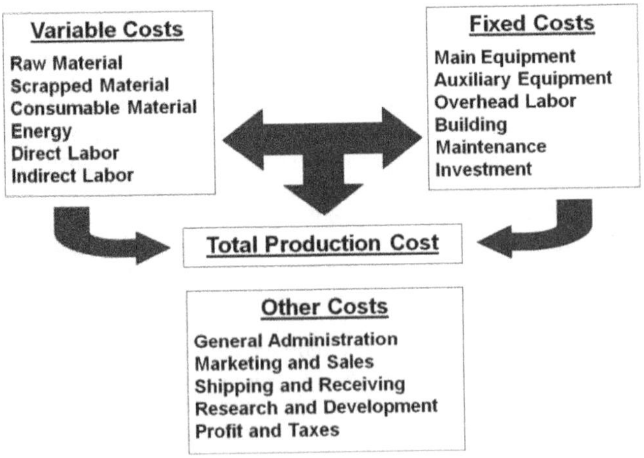

Figure 1. Costs included in process-based cost models.

Cost model development consists of four steps: definition of the process steps, construction of the model logic and framework, collection of information, and validation of the model. Results from process-based cost models can allow informed decision making before investing in a prototype or pilot production process. For developmental technologies, cost models can indicate the cost drivers and process parameters that may require additional development before the process would become economically viable.

Life Cycle Assessment

To assess the environmental impacts of products and processing technologies, LCA is a widely used technique that allows the investigation of the environmental impacts associated with a product's life cycle. LCA is comprised of 4 interdependent phases: goal definition and scope, inventory analysis, impact assessment, and interpretation, as illustrated in Figure 2.

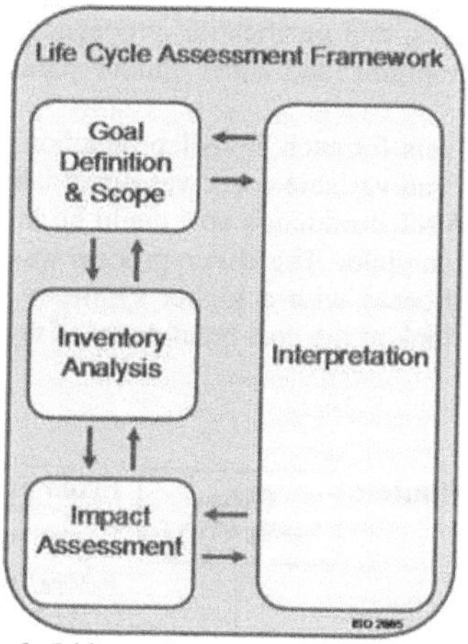

Figure 2. Life cycle assessment framework [21].

The goal definition and scope phase of LCA defines the objective of the study, the system boundaries, functional unit, and other LCA process parameters. For both case studies, the studies are limited to the manufacturing processes, and do not include use or disposal of materials or products. Inventory analysis consists of collecting input data (i.e., raw materials, energy, labor, and equipment used) and output data (i.e., emissions to land, water, and air), describing this data, and modeling the product system using functional unit processes. The nanomanufacturing inventory data is gathered by observation of laboratory scale processes and discussions with researchers. The process-based cost models developed to assess process economics are also used as for tracking the life cycle inventories.

Impact assessment involves characterizing the environmental impacts associated with the given process, evaluating the contributions to each category, and weighting to determine relative significance. The most widely used LCA software program, SimaPro™, is used to perform the impact assessment. SimaPro enables first order prediction of the environmental footprint of the processes, utilizing software databases to calculate the contributions to a set of recognized impact categories from the inventory data. By using Eco-Indicator 1999 methods, emissions are categorized into impact areas such as climate change, acidification, eutrophication, land use, mineral depletion, ecotoxicity, ozone layer depletion, and carcinogens. The significance of each impact is compared by normalizing with respect to reference values corresponding to the damage caused by one European person in one year.

CASE STUDY I: SWNT PRODUCTION ALTERNATIVES

Three SWNT processes were investigated to assess the economic and environmental attributes generated from production. These synthesis processes included: arc ablation (arc), chemical vapor deposition (CVD), and high pressure carbon monoxide (HiPco). Process-based cost models were developed for each process, which determined that the total manufacturing cost per 1 g SWNT for arc, CVD, and HiPco at $1,906, $1,706, and $485, respectively. Some of the key assumptions imposed on the models included a base production volume of 10,000 g SWNT/yr, electricity cost at $0.10/kWh, direct labor cost at $20/hr, appropriate synthesis product yields for each process, and purification proceeding synthesis. A more complete and descriptive list of the assumptions and other model parameters can be found in previous work [22-23].

The primary cost drivers for each SWNT production process included the synthesis and filtration steps, and for fixed and variable costs was direct labor. Through sensitivity analysis, it was determined that each SWNT production cost could be improved by over 14% by increasing the process's synthesis reaction yield. The HiPco process was found to be the most cost-efficient because it is a continuous process with a higher synthesis yield due to recycling the carbon monoxide. A more detailed look at the cost breakdown of the HiPco process is shown in Figure 3.

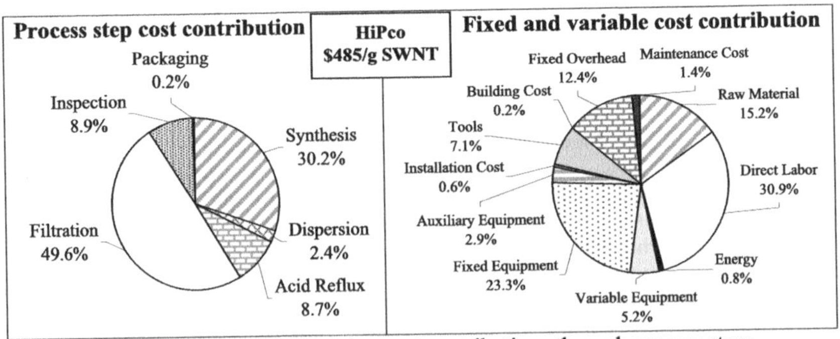

Figure 3. HiPco manufacturing cost contributions through process steps and fixed and variable costs.

Filtration, which is the last step of purification, is the process step that contributes the most to the HiPco production cost at 49.6%. The synthesis step is the second largest cost contributor at 30.2%. In terms of fixed and variable costs, direct labor was most significant at 30.9%, followed by fixed equipment costs at 23.3%.

The environmental footprint of each SWNT production process was determined by utilizing the cost model inventories and SimaPro software. The significance of each resulting impact was compared through normalization and yielded the identification of the dominate impact categories for this study: airborne inorganics, climate change, and acidification [23]. At present, there are no toxicological data for the health impacts involving SWNTs, hence outputs from SimaPro only reflect environmental implications of the materials and processes used to create CNTs, and not the implications of the CNTs themselves.

The environmental impact of the arc, CVD, and HiPco production processes were compared using three different Eco-Indicator 99 method valuation methodologies: Individualist, Egalitarian, and Hierarchist. The Individualist perspective assumes a short-term time perspective, includes substances only if there is complete proof regarding their effect, assumes damages to be recoverable by technological and economic development, and asserts fossil fuels cannot be depleted. The Egalitarian perspective is long-term and includes substances when there is any indication regarding their effect, assumes damages cannot be avoided and may lead to catastrophic effect, and assumes fossil fuels cannot be substituted. The Hierarchist perspective is also long-term and assumes substances are included when there is consensus regarding their effect, damages can be avoidable by good management, and fossil fuels are not easily substituted [24]. As shown in Figure 4, the HiPco process showed the lowest environmental impact for all impact categories, independent of the valuation method applied, followed by arc and then CVD. The y-axis label of "Pt" refers to a point scale for the normalized impact per gram of product. Figures 4b and 4c illustrate that each production process has a considerable fossil fuels environmental burden, with that of CVD being twice more than arc and over four times greater than HiPco.

Increasing the synthesis reaction yields of each SWNT production process to a best case scenario was compared with the initial base case. By improving the synthesis reaction yields of each process, the airborne inorganics, climate change, and acidification impact categories for are significantly reduced for arc (-78%) and CVD (-85%) [23]. This drop in environmental load is attributed to the reduction of batches in the arc and CVD processes and reduced power requirements. HiPco, however, experienced no changes in the airborne inorganics and acidification categories and a decrease of only 1% in climate change.

For these three SWNT production processes, the environmental impacts were dominated by energy use, particularly from electricity consumption in production [23]. Again, because there are no data on the health or environmental impacts of SWNTs, the results only indicate environmental attributes of known chemical and energy impacts. It is interesting to note that electricity consumption was 99% of the environmental impact for each process, however only approximately 1% of each process' total manufacturing cost. Such analysis demonstrates the environmental and cost trade-offs associated with each SWNT production process and provides guidance on how to improve the production processes.

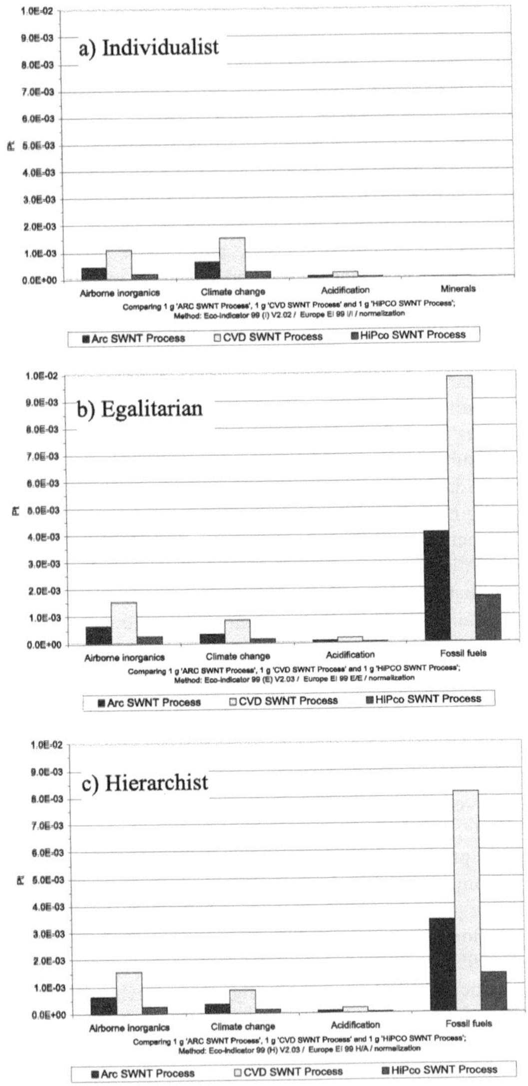

Figure 4. Environmental impacts of arc, CVD, and HiPco production process with varied Eco-Indicator 99 method perspectives.

CASE STUDY II: SWNT SWITCH FOR NONVOLATILE MEMORY DEVICES

Preliminary work on the economic and environmental assessment of the fabrication a SWNT switch has been undertaken. The memory device consists of a SWNT that crosses a trench in a silicon wafer and upon application of a voltage; the SWNT deflects to complete the circuit. The SWNT remains in the deflected position without the addition of more energy, and this non-volatile switch will use less energy over its lifetime and should reduce thermal loading. A scanning electron microscope (SEM) photo of this SWNT switch is provided in Figure 5.

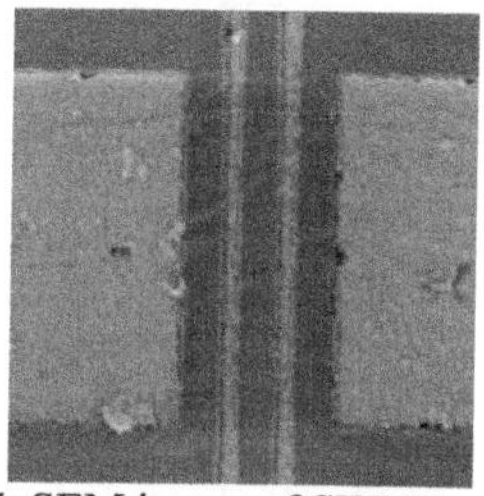

Figure 5. SEM image of SWNT switch [25].

The manufacture of this SWNT switch begins with cleaning the silicon wafer substrate and proceeds through the process flow path seen in Figure 6 until the SWNT is deposited via dispersion in pure solution or dielectrophoresis.

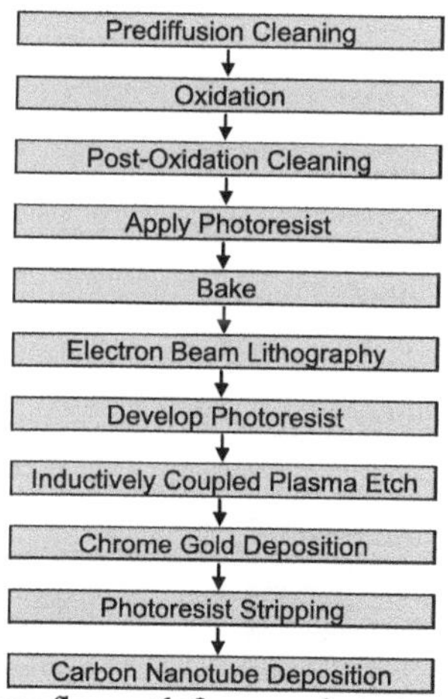

Figure 6. Process flow path for manufacturing SWNT switch

The process parameters and inventory data were collected by observation of laboratory practices, and included collection of the costs of the materials, equipment, and energy management. A sample inventory data spreadsheet for the prediffusion step of the SWNT switch process is shown in Table 1.

Table 1. Inventory data for prediffusion cleaning step of manufacturing SWNT switch

Process Step:	PREDIFFUSION CLEANING		Quantity	Units	Time			Power	Cost/Unit	Frequency changed	Waste Disposal
					In use	units	Idle on		cost/unit		
Equipment		Equipment Components									
	Wet bench*								≈ $25,000		
	Rinse Dryer*										
Materials											
	Silicon Wafer		25	wafers					$8-$25		
	Piranha etch				10	min				3 weeks	Waste water treatment
		Sulfuric acid	6000	mL					$23.92/gal		
		Hydrogen peroxide*	3000	mL					$22.42/gal		
	Hydrochloric acid				10	min				3 weeks	Waste water treatment
		Hydrochloric acid	1200	mL					$33.36/gal		
		Hydrogen peroxide*	1200	mL							
	Hydrofluoric acid				15	sec				2 months	Waste water treatment
		Hydrofluoric acid	180	mL					$33.36/gal		
		DI water	9000	mL							Drain
	Rinse				7.5	min					
		DI water**	1	L							
	Cascade rinse				15	min					
		DI water**	1	L							
		Nitrogen dioxide gas*			150	sec			$50/tank		
* = Not inputted into SimaPro											
** = Assumption											

The life cycle inventory data was input into SimaPro software to assess the environmental attributes of manufacturing a SWNT switch on the lab-scale. Eco-Indicator 99 methods were applied with the "Individualist" perspective selected for valuation. The resulting environmental assessment is shown in Figure 7.

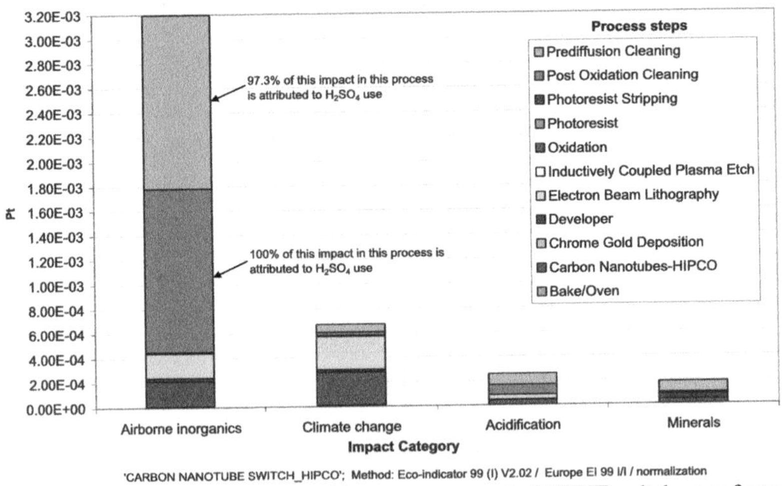

Figure 7. Preliminary environmental impact assessment of SWNT switch manufacture.

In this assessment the HiPCo production of SWNT was included as a feedstock to the memory device fabrication. The main impact categories for the manufacture of the SWNT switch include airborne inorganics, climate change, acidification, and minerals. In terms of airborne inorganics, the process steps that have the greatest environmental impact on manufacturing are prediffusion cleaning and post-oxidation cleaning. These large impacts are due to the amount of sulfuric acid used in these processes. This information highlights some of the drivers associated with environmental burden and can be used to explore whether processes can be adapted to reduce burdens.

CONCLUSIONS

The two case studies described above demonstrate the successful utilization of process-based cost models and LCA techniques to assess economic and environmental attributes of nanomanufacturing. This methodology can be used for numerous other applications of nanotechnology, including research currently underway at CHN. Because of the similarity in process steps, these same process-based cost models and LCA techniques could be applied to the research on processes to fabricate CNT-polymer meshes that would lead to SWNT batteries and electromagnetic inference (EMI) shielding. Preliminary LCA work on this CNT-polymer mesh has been initiated, including the process flow diagram as shown in Figure 8.

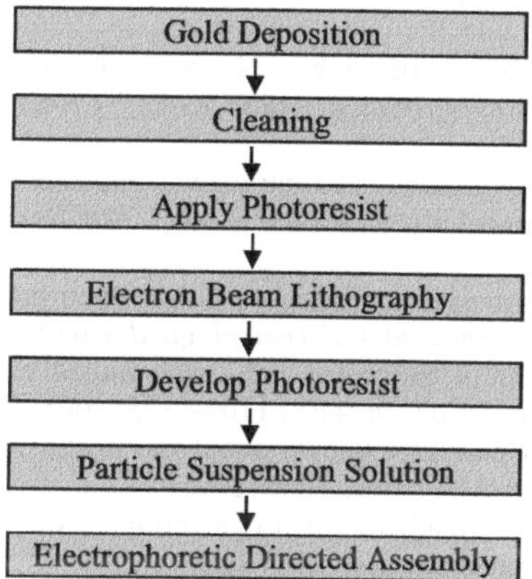

Figure 8. Process flow path for CNT-polymer mesh using in EMI shielding.

Assessment of the economic and environmental attributes of nanomanufacturing is critical in bridging nanoscale research with commercialization of nano-products. Through the use of process-based cost models and LCA techniques, lab-scale nanomanufacturing processes for alternative SWNT production processes have been scaled-up to provide detailed information on both the cost and environmental impacts. Preliminary life cycle inventory data for the manufacture of SWNT switches has been collected and utilized with SimaPro software to explore its process economics and environmental burden.

By evaluating the economic drivers of nanomanufacturing products, researchers and system developers can work towards optimizing production conditions and processes. Therefore, this work serves as a foundation for development of environmental footprints and assessment of cost trade-offs in process design involving CNTs and can be used to promote commercially viable and responsible nanomanufacturing.

REFERENCES
[1]NNI, "FY 2009 Budget & Highlights," 2008.
[2]L. Research, "The Nanotech Report, 4th Edition," pp. iii, 2006.
[3]J. Miller, K.J. Cho, and M.D. McGehee, "A Realistic Assessment of the Commercialization of Nanotechnology: A Primer for Lawyers and Investors," *Nanotechnology Law & Business*, 1[1] (2004).
[4]S.M. Lloyd and L.B. Lave, "Life Cycle Economic and Environmental Implications of Using Nanocomposits in Automobiles," *Environmental Science and Technology*, 37 3458-3466 (2003).
[5]M. Steinfeldt, et al., "Nanotechnology & Sustainability," Institute for Ecological Economy Research, Berlin, Germany, 2004.
[6]S.M. Lloyd, L.B. Lave, and H.S. Matthews, "Life Cycle Benefits of Using Nanotechnology To Stabilize Platinum-Group Metal Particles in Automotive Catalysts," *Environmental Science and Technology*, 39 1384-1392 (2005).
[7]S.I. Olsen and M.S. Jørgensen, "Environmental assessment of micro/nano production in a life cycle perspective," in Mater. Res. Soc. Symp. Proc., Vol. 895 Edited. Materials Research Society, 2006.
[8]C.-w. Lam, et al., "A Review of Carbon Nanotube Toxicity and Assessment of Potential Occupational and Environmental Health Risks," *Critical Reviews of Toxicology*, 36[3] 189-217 (2006).
[9]C.A. Poland, et al., "Carbon nanotubes introduced into the abdominal cavity of mice show asbestos-like pathogenicity in a pilot study," *Nat Nano*, advanced online publication (2008).
[10]K. Donaldson, et al., "Carbon nanotubes: A review of their properties in relation to pulmonary toxicology and workplace safety," *Toxicological Sciences*, 92[1] 5-22 (2006).
[11]S. Fiorito, et al., "Effects of fullerenes and single-walled carbon nanotubes on murine and human macrophages," *Carbon*, 44[6] 1100-1105 (2005).
[12]D.B. Warheit, "What is currently known about the health risks related to carbon nanotube exposure?," *Carbon*, 44[6] 1064-1069 (2005).
[13]Royal Society & Royal Academy of Engineering, "Nanoscience and nanotechnologies: Opportunities and uncertainties," 2004 July
[14]EPA, "Nanotechnology White Paper," pp. 134, 2005.
[15]T. Medley and S. Walsh, "Nano Risk Framework," pp. 104, Environmental Defense - Dupont Nano Partnership, 2007.
[16]NIOSH, "Progress towards safe nanotechnology in the workplace," National Institute for Occupational Safety and Health, U.S. Department of Health and Human Services, 2007.
[17]D.A. Chiango, J.A. Isaacs, and S.A. Miller, "Production of steel powder by rotating electrode processes: Economic analyses," *International Journal of Powder Metallurgy (Princeton, New Jersey)*, 36[4] 49-56 (2000).
[18]J.P. Clark, F.R. Field, and R. Roth, "Techno-Economic Issues in Materials Selection," *ASM Handbook Materials Selection and Design*, 20 256-265 (1997).
[19]R. Kirchain, "Cost Modeling of Materials and Manufacturing Processes," pp. 1718-1727 in Encyclopedia of Material Science and Engineering. Edited by E.S. Ltd, 2001.

[20]J. Busch, "Cost Modeling as a Technical Management Tool," *Research Technology Management*, 37[6] 50-56 (1994).

[21]ISO, "Environmental Management- Life Cycle Assessment- Principles and Framework ISO 14040," 2006.

[22]M. Healy, "Environmental and economic comparison of single-wall carbon nanotube production alternatives," pp. 124, Vol. MS, Northeastern University, Boston, 2006.

[23]M.L. Healy, L.J. Dahlben, and J.A. Isaacs, "Environmental Assessment of Single-Walled Carbon Nanotube Processes," *Journal of Industrial Ecology*, [Nanotechnology & the Environment] (2008).

[24]Pré, "SimaPro 7," Pre Consultants, Amersfoort, The Netherlands, 2004.

[25]CHN, "CHN Overview," 2008.

NITRATE FINING AND EMISSIONS DURING GLASS MANUFACTURING

Linda E. Jones and Shengchun Luo
Picker Engineering Program
Smith College
Northampton, MA 01063

ABSTRACT

The decomposition behaviour of KNO_3 is studied in Ultra High Purity (UHP) Ar using a facility for the analysis of chemical thermodynamics, FACT-sage 5.0 software package, Thermogravimetry (TG), and Differential Scanning Calorimetry (DSC) over a temperature range of 298-1573 K. The TG/DSC system is coupled with Fourier Transform Infrared (FTIR) for emissions identification and quantification. It is observed that A KNO_3 phase change occurs at 405 K. KNO_3 melts at 608 K before decomposition. Liquid KNO_3 decomposition initiates at 873 K. It decomposes into K_2O_2 (s), evolving NO (g) and NO_2 (g). NO_2 (g) evolves initially and disappears after 1073 K. Total NO (g) evolution is much higher than NO_2 (g) evolution. Two sets of reaction paths are put forward to explain the observed decomposition phenomena.

INTRODUCTION

Potassium nitrate plays an important role in glass batches as a fining agent, but its decomposition results in NOx emissions that are an environmental problem for the glass industry [1]. In order to further elucidate the glass melt and fining chemistries and provide some insight into NOx emission control for the glass industries, an exhaustive study of the nitrate decomposition behaviour is undertaken. Historically, these investigations of potassium nitrate decomposition primarily used Mass Spectroscopy and/or Gas Chromatography as the tools for evolved gas identification and quantification. The gas emission species identified from nitrate decomposition are O_2 (g), NO (g), and N_2 (g). NO_2 (g) is not considered a decomposition product in this earlier work. Although various reaction mechanisms have been put forward, disagreements exist and a few theories address the chemical reactions related to the generation of NO (g) and NO_2 (g) emissions [2-6]. The objective of this work is to study the reaction chemistry of nitrate decomposition based on accurate and sensitive measurements of the NOx emissions from nitrate decomposition using FTIR identification and quantification. In this work, NO_2 gas is observed to evolve directly from the decomposition product.

EXPERIMENTAL

The software package F*A*C*T (Facility for the Analysis of Chemical Thermodynamics) Sage 5.0 is developed jointly by Thermfact Ltd. (Montreal, Canada) and GTT-Technologies (Aachen, Germany) [7]. The module for equilibrium calculations primarily used in this work. This program uses Gibbs free energy minimization as the calculation method. Three inputs are required for equilibrium calculations: the quantities of reactants, a list of possible product phases and the chemical species in each phase, and desired thermodynamic conditions. In this work, the quantities of reactants are 1mol KNO3 and 1mol Ar. The products considered are all the liquid, solid and gaseous chemical species provided by the database of this program and the desired thermodynamic conditions are the increasing temperatures in a closed system under 93.3 kPa.

Potassium nitrate used in the study was 99.999% pure and from the Alfa Aesar (A Johnson Matthey Company, MA. USA). The sample weight of KNO_3 in the experiments was 10 mg. Thermal analysis is undertaken using a NETZSCH 409PC TG/DSC system. It consists of a silicon carbide heating element, an alumina tube sample chamber, and a top load electromagnetic compensation balance. The decomposition was performed in 93.3 kPa ultra high purity (UHP) Ar at a flow rate of 100 ml/min and a constant heating rate of 20 K/min. The temperature of the instrument was calibrated by the measurement of the melting point of a series of the pure metal under the same experimental conditions. The crucible material of this work is alumina. FTIR is used to undertake the decomposition gas analsis. The FTIR used in this work is a Nicolet Magna-IR 560 (ThermoNicolet, USA) FTIR spectrometer. A detailed description of experimental TG/DSC –FTIR instrumental set up and parameters is given in an earlier Ceramic Transactions[8].

RESULTS

Thermochemical calculations and TG/DSC decomposition analysis

Figures 1 and 2 are the analyses of the thermochemical calculations and TG/DSC experimental results, respectively. There is remarkable similarity between the calculated reaction chemistry and that observed experimentally. KNO_3 in Ar undergoes a crystal structure transition from KNO_3 (s) to KNO_3 (s2) at 405 K. This transition is calculated to occur at 403 K. KNO_3 in Ar melts at 608 K. The calculated solid to liquid phase transformation occurs at 610 K. KNO_3 (l) begins to decompose at 873 K and the peroxide, K_2O_2 (s) begins to evolve at 903 K. The consecutive endothermic peaks in the DSC curve indicate a slight difference between the predicted thermodynamic sequence and the experimental observations. This is because the calculations are conducted for a closed equilibrium system; whereas, the experiments are performed in an open system with a carrier gas flow at 100 ml/min. The evolved gases have are expected to have an influence on the decomposition. At 1073 K, the gas analysis indicates that NO_2 has completely disappeared: whereas, NO reaches its highest concentration at 1073 K. An unknown double peak was observed on FTIR spectra in the frequency range of 1000-1500 cm^{-1}. This unknown peak starts at about 973 K, increases to its maximum at 1173 K, and then decreases and subsequently disappears.

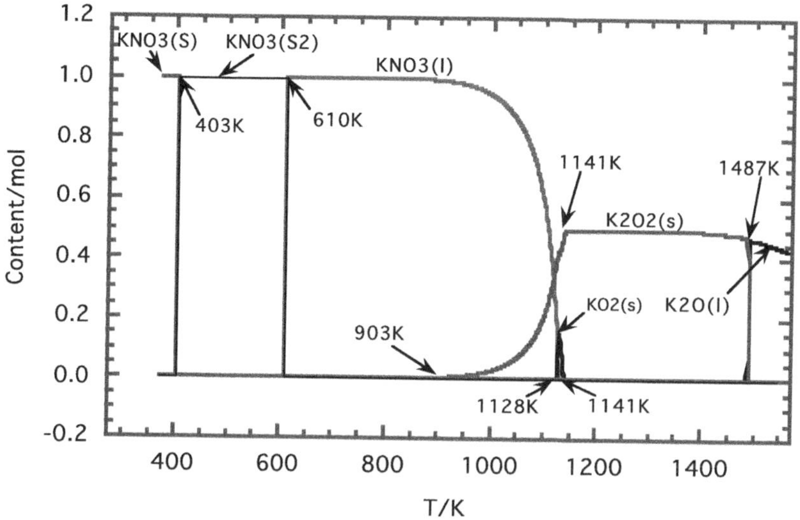

Figure 1. Equilibrium decomposition products for 1mol KNO_3 in 1mol Ar at 93.3 kPa using FACT- Sage 5.0 (solid and liquid phases).

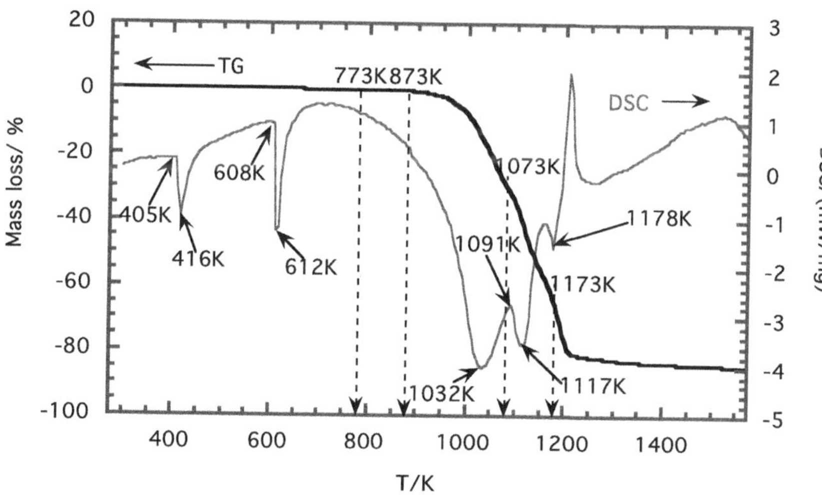

Figure 2 TG and DSC of pure KNO_3 in 93.3 kPa UHP Ar at a flow rate of 100 ml/min and a heating rate of 20 K /min.

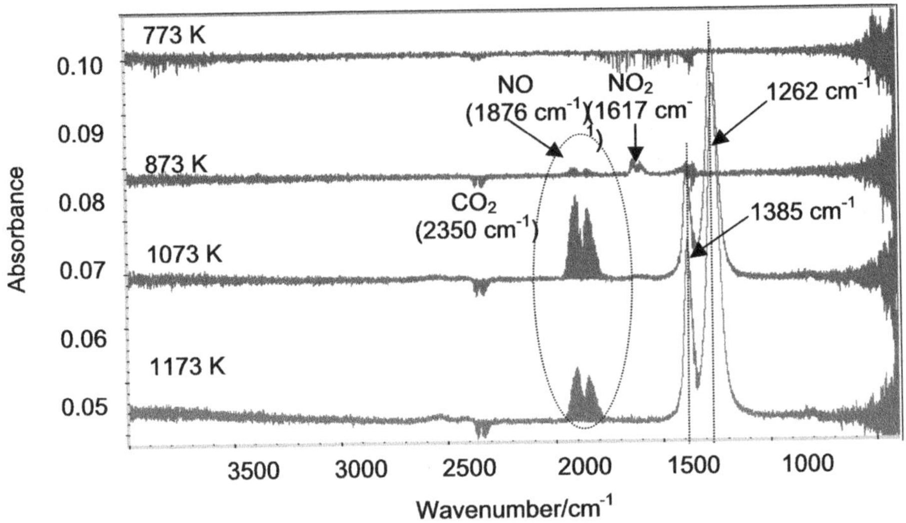

Figure 3. FTIR gas spectra taken from pure KNO_3 decomposition in 93.3 kPa UHP Ar at a flow rate of 100 ml/min and a heating rate of 20 K/min. (Window: KBr, gas cell temperature 413 K, resolution 0.5 cm^{-1}, number of scans 16, mirror velocity 0.4747 cm/s).

In order to identify if this unknown is N_2O gas emission, a N_2O FTIR gas spectrum is collected from a Ready-whip® canister using FTIR with the same resolution setting (0.5 cm^{-1}) and the same detection conditions. The spectrum of this N_2O gas is compared with the unknown spectrum that is collected from KNO_3 decomposition at 1173 K. Figure 4 is the comparison of these two spectra, which indicate that this unknown peak is not N_2O.

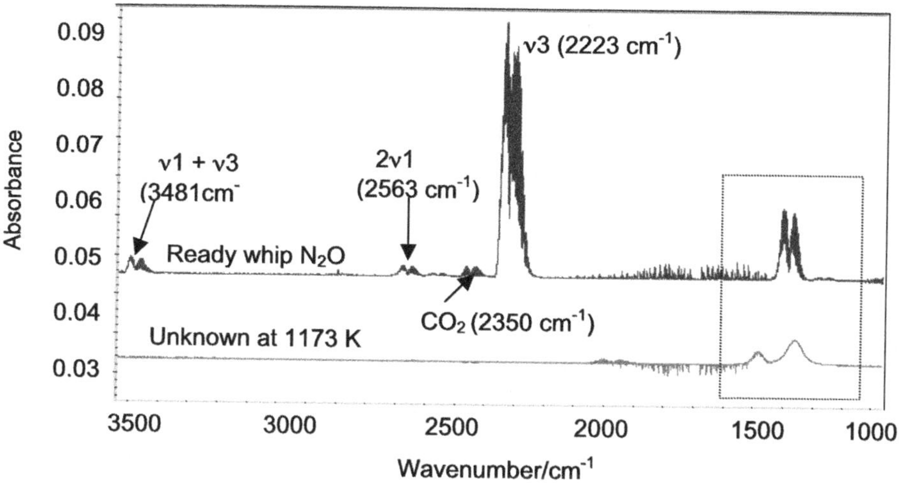

Figure 4. Comparison of the unknown peak from KNO_3 decomposition in UHP Ar at 1173 K with the spectrum of N_2O obtained from Ready-whip® (Window: KBr, gas cell temperature 413 K, resolution 0.5 cm^{-1}, number of scans 16, mirror velocity 0.4747 cm/s).

In order to determine whether this unknown is an intermediate potassium product, the vibration modes of KNO_3, KNO_2, and potassium oxides in the frequency range of 800-1500 cm^{-1} as reported in the literature are listed in Table I [9-15]. KNO_3 vibration modes are widely studied; therefore, the vibration modes of crystal, molten liquid, vapor, and aqueous solution of KNO_3 are obtained, but no exact match with the unknown exists. Therefore, this unknown peak is not KNO_3. The wavenumber of KNO_2 is 1250 cm^{-1}, which is also not consistent with the wavenumber of this unknown peak. The only vibration mode for potassium oxides obtained in this frequency range is that KO_2, but the wavenumber does not match the "unknown".

Table I Vibration Modes of Potassium Nitrate, Nitrite, and Oxides [9-15]

			v1 wavenumber (cm^{-1})	v2 wavenumber (cm^{-1})	v3a wavenumber (cm^{-1})	v3b wavenumber (cm^{-1})
KO_2			1108			
KNO_2			1250			
KNO_3	Aqueous solution		1050	820	1390	1383
	Molten liquid		1045	829	1338	1388
	Crystal		1049	829	1343	1361
	vapor	monomer	1031	830	1291	1462
		dimer	1047	835	1337	1420

Based upon this analysis a reaction between the nitrate and the with the KBr window is considered. To evaluate this further, a chemically stable CaF_2 window is used. KNO_3 is decomposed under the same TG experimental conditions and the same FTIR measuring conditions. This unknown peak is also be detected after changing the window. Figure 5 gives a comparison of the unknown peak in these two experiments at the detected highest absorbance. The peak shape in these two spectrum is similar. One of the unknown double peak position using CaF_2 window shifts 12 cm^{-1} compared to the peak using KBr window. The highest peak absorbance decreased from 0.065 to 0.021. These results indicate that the reaction with KBr window may occur but, it does not totally explain the formation of this peak. It is observed that the baselines of the spectra warp while the unknowns gradually arrive to their maximal intensity. Additionally, after the unknowns reach their maximal intensity, their peak intensities decrease and gradually disappear with the disappearance of the warped baselines. Based on these observations, it is proposed that this unknown peak is mainly some form of vapor species or tiny solid species of intermediate potassium compounds that dispersed in the gas stream and is brought to the FTIR gas cell by the carrier gas concentration as a function of the corresponding TG/DSC heating cycle.

Since the "unknown" species is not a gas phase constituent, quantification of the NO (g) and NO_2 (g) emissions is performed assuming this unknown peak does not affect the concentration data obtained for NO (g) and NO_2 (g). Based on the calibration and calculation protocol described in author's another paper [8], a profile of the gas emission. The mass evolutions of NO and NO_2 for 1mg KNO_3 decomposition is calculated. They are 0.071 mg and 0.001 mg for NO (g) and NO_2 (g), respectively. NO (g) evolution is much higher than NO_2 (g) evolution, the mass evolution ratio of NO/NO_2 are 71.

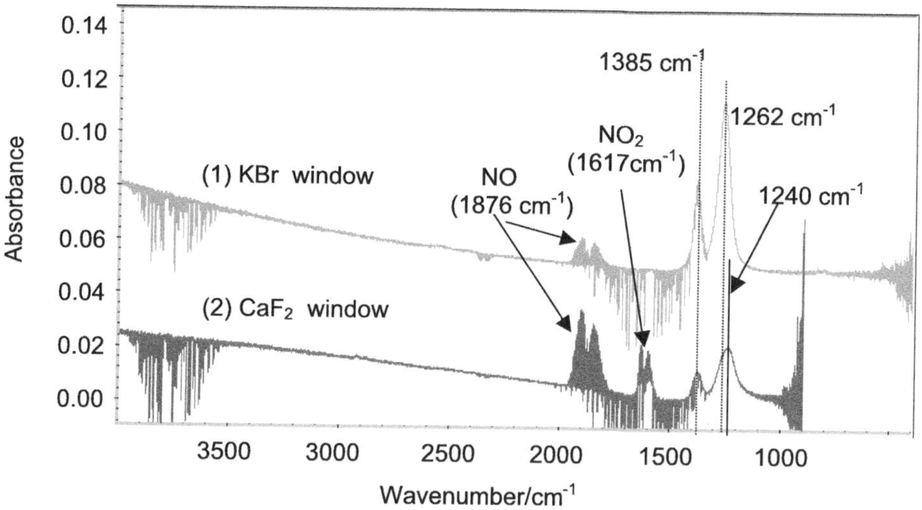

Figure 5. FTIR spectra of the unknown peak from KNO_3 decomposition in UHP Ar using KBr and CaF windows under the same experimental conditions.

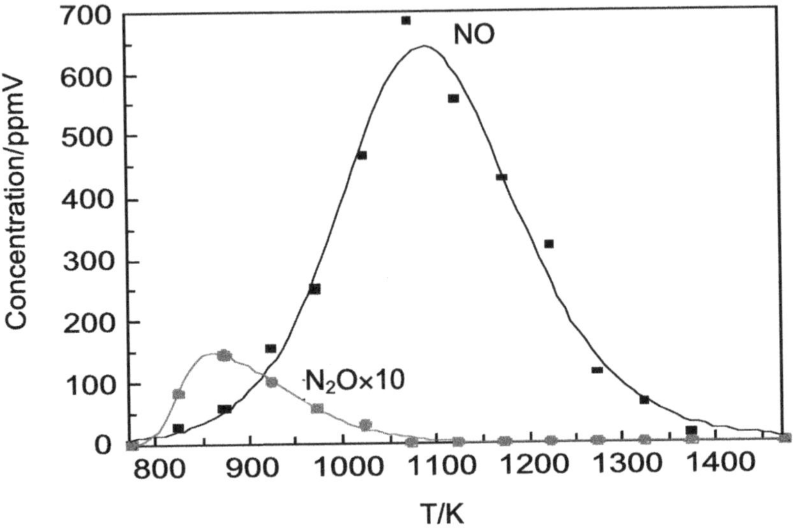

Fig.

Figure 6. NOx emission concentrations as a function of temperature for pure KNO_3 decomposition in 93.3 kPa UHP Ar at a flow rate of 100 ml/min and a heating rate of 20 K/min

DISCUSSION

Using the structure of KNO_3 as the template, the decomposition can start with two possible reactions [16]. This is supported by this work as two sets of reaction paths can be proposed to explain the decomposition process. The first reaction path is proposed is the following:

$$KNO_3\ (l) \leftrightarrows KNO_2\ (l) + 1/2\ O_2\ (g) \quad (1)$$

Figure 4 indicates that NO_2 (g) evolves initially, but decreases after 873 K. NO (g) concentration increases through 1073 K and then decreases after NO_2 (g) disappears. These two facts indicate that the second step can be the dissociation of KNO_2 (l) according to Equation 2, and then NO_2 (g) dissociates into NO (g) and O_2 (g) at higher temperatures according to Equation 3. This dissociation temperature for Equation 3 is higher than 893 K as described in the literature [6,17] and in this work is about 873 K as indicated by the gas profile of NO_2 in Figure 4.

$$KNO_2\ (l) \rightarrow K\ (g) + NO_2\ (g) \quad (2)$$
$$NO_2\ (g) \leftrightarrows NO\ (g) + 1/2\ O_2\ (g) \quad (3)$$

The consecutive reaction is that of K (g) reacting with O_2 (g) to form the peroxide,K_2O_2 (s) according to Equation 4, which is supported by thermodynamic calculations as indicated in Figure 2.

$$2K\ (g) + O_2\ (g) \rightarrow K_2O_2\ (s) \quad (4)$$

From this we are proposing that the unknown peak that starts at 973 K is the a precipitate of $K_2O_{2(s)}$ that is dispersed in the gas stream and brought to the FTIR gas cell by the carrier gas Ar. Since its phase is different from that of the gas stream, it causes the FTIR spectra baseline to warp.

The gas concentration profiles in Figure 4 indicate that the concentration of NO (g) is much higher than NO_2 (g) during decomposition. This finding demonstrates that the dissociation rate of NO_2 is much higher compared to the dissociation rate of KNO_2 (l) which, results in a rapid consumption of NO_2 (g). Therefore, Equation 2 is proposed to be the rate-determining step for this reaction path. A secondary reaction path is proposed that the decomposition starts from KNO_3 (l) dissociating into K(g) and NO_3 (g) according to Equation 5, and NO_3 (g) dissociates into NO_2 (g) and O_2 (g) according to Equation 6.

$$KNO_3\ (l) \rightarrow K\ (g) + NO_3\ (g) \quad (5)$$
$$NO_3\ (g) \leftrightarrows NO_2\ (g) + 1/2\ O_2\ (g) \quad (6)$$

The subsequent reactions are the dissociation of NO_2 (g) according to Equation 3 and the reaction of K (g) and O_2 (g) to form K_2O_2 (s) according to Equation 4, which also agrees with the experimental observations. In this reaction path, the rate-determining step is proposed to be Equation 5, which limits the rate of NO_2 (g) formation through NO_3 (g), resulting in a much lower concentration of NO_2 (g) than NO (g).

REFERENCES

[1].G.C. Beerkens,"The Role of Gases in Glass Melting Processes," Glastech. Ber., **68** (1995) 369-380.
[2] Y. Hoshino, T. Utsunomiya, and O. Abe, "The Thermal Decomposition of Sodium Nitrate and the Effects of Several Oxides on the Decomposition," Bull. Chem. Soc. Jpn., 54 (1980) 1385-1391.
[3].C.M. Kramer, Z.A. Munir, and K.H. Stern, "Evaporation of $NaNO_3$, KNO_3, and $NaNO_2$," High Temp. Sci., 16 (1983) 257-267.
[4]C.M. Kramer, Z.A. Munir, and J.V. Volponi, "Simultaneous Dynamic Thrmogravimetry and Mass Spectrometry of the Evaporation of Alkali Metal Nitrates and Nitrites," J. Therm. Anal., 27 (1983) 401-408.
[5]H.R. Bartos and J.L. Margrave, "Communicatioon to the Editor: Th Termal Decomposion of $NaNO_3$," J. Phys. Chem., 60 (1956) 256.
[6]V.F. Kramer,"Gas Profile Measurement as a Means of Determining Gas Evolution during Glass Melting," Glastech. Ber., 53 (1980) 177-188.
[7]C.W. Bale, A.D. Pelton, W.T. Thompson, P. Chartrand, S.A. Degterov, J. Melancon, G. Eriksson, K. Hack, S. Petersen, P. Coursol, E. Jak, I. Jung, R. Lacroix, M. Piche, C. Robelin, J.M. Sangster, P. Spencer, and M.W. Wadsley, FactSage 5.0 [Computer Program]. Thermfact Ltd., Montreal, Canada, 2001.
[8]S. Luo and L. Jones, "Analysis of Emissions from Glass," Environmental Issues and Waste Management Technlogies in the Ceramic and Nuclear Industries VIII, Ceramic Transactions, Vol. 143, 49-57 (2003).
[9]K. Nakamoto, Infrared and Raman Spectra of Inorganic and Coordination Compounds, 4th ed., John Wiley & Sons, New York, 1986.
[10]D. Smith, D.W. James, and J.P. Devlin, "Vibrational Spectra of Molecular Metal Nitrate Monomers and Dimers," J. Chem. Phys., 54 (1971) 4437-4442.
[11]I. Nakagawa and J.L. Walter,"Optically Active Crystal Vibrations of the Alkali-Metal Nitrates," J. Chem. Phys., 51 (1969) 1389-1397.
[12]R.K. Khanna, J. Lingscheid, and J.C. Decius, "Infrared Absorption Spectra of Potassium Nitrate in Phase I,II, and III," Spectrochim. Acta, 20 (1964) 1109-1115.
[13]S.C. Wait, A.T. Ward, and G.J. Janz, "Anion-Cation Interaction in Mlton Inorganic Nitrates: Vibrational l Analysis," J. Chem. Phys., 45 (1966) 133-137.
[14]K. Williamson, P. Li, and J.P. Devlin, "Vibrational Spectra and Structures of Ionic Liquids, II. The Pure Alkali Metal Nitrate," J. Chem. Phys., 48 (1968) 3891-3896.
[15]D.W. James and W.H. Leong, "Vibrational Spectra of Single Crystal Group I Nitrates," J. Chem. Phys., 49 (1968) 5089-5096.
[16]F.A. Cotton and G. Wilkinson, Basic Inorganic Chemistry, 3rd ed., John Wiley & Sons, New York, 1995.
[17]G.D. Parkes, Mellor's Modern Inorganic Chemistry, 6th ed., Longmans, London, 1967.

LOW-COST SOLID GEOPOLYMERIC MATERIAL FOR WATER PURIFICATION

M. Alshaaer[a], B. El-Eswed[b], R. I. Yousef[c], F. Khalili[a], H. Khoury[a]
[a]University of Jordan, Amman11942, Jordan
[b]Zarka University College, Al-Balqa Applied University, P.O. Box 313, Zarka, Jordan
[c]Petra University, P.O. Box 961343, Amman 11196, Jordan

ABSTRACT

This research aims to develop water purification geopolymers with high mechanical properties produced at low temperatures from natural raw materials. As a case study, the adsorption of heavy metal ions (Cd^{2+}) onto these geoploymers prepared by solid-state conversion was investigated. The adsorption isotherms were obtained and the Langmuir parameters were calculated in order to make a comparison between the adsorption capacities of geopolymers and their constituents. The results showed that geopolymerization in the presence of natural zeolitic tuff leads to an increase of adsorption towards Cd(II) ions relative to the precursors. This is due to the increase of surface area upon destruction of the precursor's structure, as shown by XRD patterns, and generation of new cation exchange sites upon geopolymerization. Thus, they could be used in water transportation and storage (pools and dams) and other construction applications.

INTRODUCTION

The important sources of accumulation of heavy metals in wastewater include input from municipal and industrial wastes and applications of fertilizers and pesticides on farmlands. The environmental impacts of metals on plants, health of humans, and wildlife have been studied [1,2]. Several methods have evolved over the years on the removal of these metal ions present in industrial wastewaters and soils. These are chemical precipitations, conventional coagulation, reverse osmosis, ion exchange, and adsorption on activated carbon [3]. Out of these methods, adsorption appears to be the most widely used for the removal of heavy metals. Natural substances like activated carbon and zeolites have assumed a wide application in this regard but their high cost limits their utilization. The possibility of using cheaper adsorbents like kaolin clay suffers from the problem that adsorption of metal ions onto kaolin is much less than that of zeolite [4-6]. Since both Jordanian zeolitic tuff and Jordanian kaolin are available, a new low-cost adsorbent based on geopolymerization (solid state conversion [7]) of these local materials is suggested.

Little work was found in the literature on the adsorption behaviour of geopolymers. Li et al. [8] studied the adsorption of methylene blue (MB) dye onto geopolymeric adsorbent based on fly ash. The synthesized geopolymer was found to have much higher adsorption capacity towards MB (0.12 mmol MB/g adsorbent) than fly ash itself (5.61×10^{-3} mmol MB/g adsorbent). Wang et al. [9] reported an amorphous aluminosilicate geopolymer resulted from solid-state conversion of fly ash. The synthesized geopolymer was found to have higher adsorption capacity towards Cu(II) ion (92 mg Cu/g adsorbent) than the fly ash itself (0.1 mg Cu/g adsorbent).

This research focuses on the influence of using natural Jordanian zeolitic tuff as filler on the adsorption properties of kaolin-based geopolymers products. This filler was characterized by its availability, high adsorption capacity, and stability. Silica sand filler was used as reference for zeolitic tuff filler.

EXPERIMENTAL

Fabrication of Geopolymer's Specimens

Four different series of geopolymers specimens with different compositions were prepared from kaolin (Jordanian kaolin-JK and reference kaolin-RK), filler (Jordanian zeolitic tuff-ZT and silica sand-SS), and alkali solution (NaOH) as shown in Table I.

Table I. Composition of the geopolymers[a]

ID	clay	filler	NaOH	water
M-RS	(RK, 28.1%)	(SS, 56.1%)	5.1%	10.7%
M-RZ	(RK, 48.3%)	(ZT, 24.2%)	8.7%	18.8%
M-JS	(JK, 29.8%)	(SS, 59.5%)	4.2%	6.5%
M-JZ	(JK, 41.7%)	(ZT, 41.7%)	5.8%	10.8%

[a] M-RS: geopolymers prepared using reference kaolin (RK) and Jordanian silica sand (SS), M-RZ: geopolymers prepared using reference kaolin (RK) and Jordanian zeolitic tuff (ZT), M-JS: geopolymers prepared using Jordanian kaolin (JK) and Jordanian silica sand (SS), and M-JZ: geopolymers prepared using Jordanian kaolin (JK) and Jordanian zeolitic tuff (ZT).

The first two series of geopolymers, M-RS and M-RZ, which are based on Ukrainian kaolin (reference kaolin), differ in the filler type. Silica sand (SS) was used as a filler material to prepare M-RS geopolymer while the zeolitic tuff (ZT) was used in M-RZ geopolymer. The other two series, M-JS and M-JZ, are based on local Jordanian kaolin, again the sand (SS) was used as filler in M-JS geopolymer while zeolitic tuff (ZT) was used in the preparation of M-JZ geopolymer. The weighed filler and kaolinitic soil were mixed (KitchenAid mixer-5K5SSS) first and then the sodium hydroxide solution (96%, Gainland Chemical Co.) was added. The amount of water used was based on the plastic limit of the powder component, kaolinitic soil and the filler, to ensure a workable mixture. Specimens (50 g each) of each mixture were molded immediately in a stainless steel cylinder (diameter of 25 mm and height of 45 mm) at a pressure of about 15 MPa (Carver hydraulic laboratory press). The molded specimens were cured by placing them in a ventilated oven (Binder-ED115) at a temperature of 80 °C for 24 hours. After this curing, the specimens were removed and allowed to cool down at room temperature before determining their bulk density, water absorption, mechanical properties, and adsorption behavior towards Cd(II) ions.

Mineralogical Composition

X-ray diffraction analysis using Philips PW1740 machine was carried out on powdered random preparations to identify the dominant crystalline phases. The samples were scanned from 2° to 60° 2-theta at a scan rate of 2°/min. The phases are identified by detecting and analyzing the d-spacing of the peaks.

Adsorption of Cd(II) Ions

About 100 g sample of each geopolymer M-RS, M-JS, M-RZ, and M-JZ were crushed using Retsch RM 100-crusher. The crushed samples as well as the filler (zeolitic tuff-ZT) and the clay materials (Jordanian kaolin-JK, reference kaolin-RK) were sieved (Fema-SAIZGITTER) at different meshes. The portion with a particle size of 250-500 micron was washed seven times with distilled water (100 mL each) then dried in the oven overnight at 105 °C and kept in a desiccator over anhydrous $CaCl_2$. Several Cd(II) standard solutions (10-100 ppm) were prepared from cadmium acetate (98%), Hopkin and Williams in 0.1 M $NaClO_4$ (98 %, Eurostar Scientific LTD) at pH = 6 (pH adjustment using $NaOH/HClO_4$). The absorbance of standard solutions were measured using atomic absorption spectrometer (Varian, AA-250 plus) in order to draw the calibration curve. In different sets of 100 mL-stoppered reagent bottles, a 0.05 g of geopolymer samples M-RS, M-JS, M-RZ, and M-JZ as well as the ZT-filler and clay materials (JK, RK) were placed. To each set of samples, 50.0 mL of standard Cd(II) solutions were independently added. A 10.0 mL portions from each reagent bottle were withdrawn after 24 h of shaking (Clifton-11268), then filtered by microfilters (0.45 um Nylon), and centrifuged (ANNITA.III, 6500 rpm, 10 minutes). The metal concentrations were determined using the atomic absorption spectrometer.

The effect of ionic strength on the adsorption properties of zeolite containing samples was evaluated by repeating the adsorption experiments of ZT, M-RZ, and M-JZ at pH = 6 without employing supporting electrolyte.

RESULTS AND DISCUSSION

The physical and mechanical properties of the geopolymers were reported elsewhere [10]. Geopolymers based on zeolitic tuff filler showed lower bulk density (1.65 g/cm^3 in M-RZ and 1.80 g/cm^3 in M-JZ) than those based on silica sand (2.16 g/cm^3 in M-RS and 1.93 g/cm^3 in M-JS). Thus, using Jordanian zeolitic tuff as a filler material in the geopolymers decreases the bulk density of the end products. This is due to the fact that the Jordanian zeolitic tuff is composed from porous medium with low bulk density. It was also observed that water absorption is higher in the geopolymers based on zeolitic tuff filler (23.2 % in M-RZ and 17.6 % in M-JZ) than those based on silica sand filler (11.0 % in M-RS and 13.0 % in M-JS). Again this behavior is due to the fact that incorporation of zeolitic tuff into the geopolymer increases the porosity of the product, which increases the water absorption capability. In spite of their low density and high water absorption, it was observed that M-JZ specimen (used Jordanian kaolin as a precursor and Jordanian zeolitic tuff) expresses a high strength, i.e. 20 MPa, compared with the rest of geopolymers series, 5 – 11 MPa, [10].

Geopolymerization-Induced Phase Changes

The x-ray diffraction (XRD) patterns of the powdered geopolymer M-JZ, Jordanian kaolin (JK), and Jordanian zeolitic tuff (ZT) samples [Figure 1] show peaks that correspond to the crystalline phases. XRD peaks confirm the presence of significant amounts of phillipsite (P) in the Jordanian zeolitic tuff (ZT). Figure 1 shows the major characteristic peaks corresponding to phillipsite (P), Kaolinite (K) and calcite (Ca). The peaks corresponding to phillipsite (P) and calcite (Ca) have diminished completely because of geopolymerisation as indicated by the XRD pattern corresponding to M-JZ. The same XRD pattern of M-JZ indicates a significant reduction in the intensity of quartz (Q) and kaolin (K) major peaks at the expense of forming feldspar (F). These results of XRD confirm the destruction of the structure of Jordanian kaolin (JK) as well as the Jordanian zeolitic tuff (ZT) because of geopolymerization. The constituents of the resultant geopolymers are dominant by amorphous phases due the lack of strong peaks corresponding to crystalline phases.

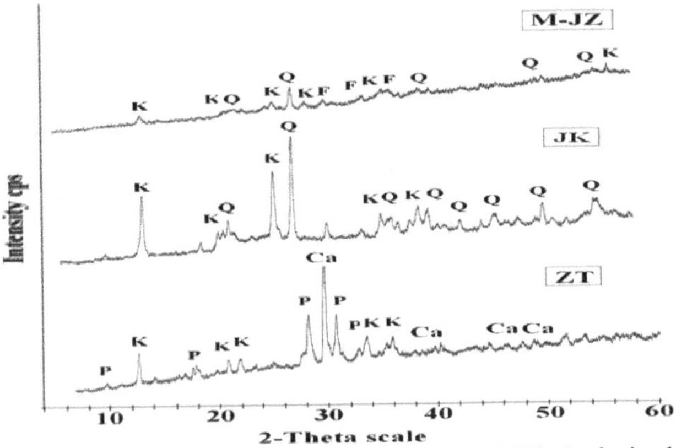

Figure 1. Qualitative XRD patterns of powdered geopolymer (M-JZ), Jordanian kaolin (JK), and Jordanian zeolitic tuff (ZT), [K: kaolinite, P: phillipsite, Q: quartz, F: feldspar, Ca: calcite].

Adsorption of Cd(II) Ions

The adsorption isotherms of Cd(II) ions onto geopolymers (M-RS, M-JS, M-RZ, and M-JZ) were investigated. In order to compare the adsorption properties of the geopolymers with the minerals from which they are made, the adsorption isotherms of Cd(II) on Jordanian zeolitic tuff (ZT), Jordanian kaolin (JK), and reference kaolin (RK) were also investigated. The Langmuir parameters Q_m and K were determined by nonlinear regression of the Langmuir equation (Eq. 1) as shown in Table II.

$$Q = Q_m KC/(1+KC) \tag{1}$$

Where, Q is the amount of Cd(II) adsorbed (mg Cd(II)/g adsorbent), Q_m is the monolayer adsorption capacity (mg Cd(II)/g adsorbent), K is the affinity constant (L/mg Cd(II)), and C is the equilibrium concentration of Cd(II) (mmol/L). The resultant plots of nonlinear fitting of the adsorption isotherms of Cd(II) on M-RS, M-RZ, M-JS, and M-JZ geopolymers at ionic strength = 0.1 M ($NaClO_4$) and pH = 6.0 are shown in Figure 2.

Table II. Langmuir adsorption parameters for the adsorption of Cd(II) ion onto geopolymers

ID	Ionic strength = 0.1 M		Ionic strength = 0 M	
	Q_m (mg Cd/g adsorbent)	K (L/mg)	Q_m (mg Cd/g adsorbent)	K (L/mg)
RK	a	a	b	
JK	a	a		
ZT	9.98	0.263	28.2	0.244
M-RS	13.6	0.949	b	
M-RZ	28.9	5.49	54.6	0.230
M-JS	7.40	2.44	b	
M-JZ	19.7	2.67	28.5	1.71

[a] No detectable adsorption was observed. [b] Not determined.

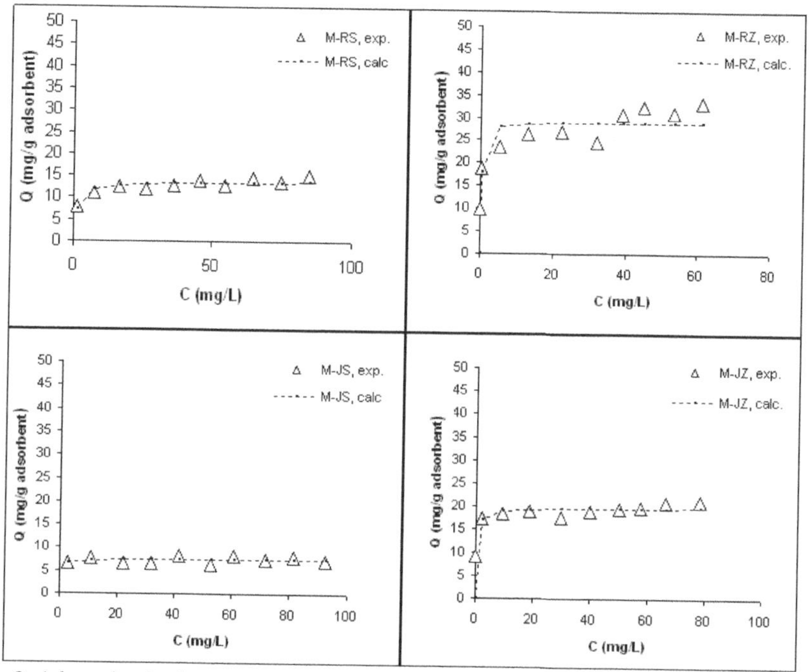

Figure 2. Adsorption isotherms of Cd(II) onto M-RS, M-RZ, M-JS, and M-JZ at pH = 6.0 and 0.1 M ionic strength.

Adsorption of Cd(II) onto geopolymers based on reference kaolin (M-RS and M-RZ)

In spite of their large silica sand content, reference kaolin based geopolymer (M-RS) has higher adsorption capacity towards Cd(II) ions than kaolin itself (Table II). This indicates that geopolymerization process generates new adsorption sites for Cd(II) ions. The zeolite containing geopolymer which is based on reference kaolin (M-RZ) has much higher adsorption capacity than that containing silica sand (M-RS) (Figure 2 and Table II).

Adsorption of Cd(II) onto geopolymers based on Jordanian kaolin (M-JS and (M-JZ)

Geopolymers based on Jordanian kaolin (M-JS and M-JZ) revealed lower adsorption capacity than their corresponding geopolymers based on reference kaolin (M-RS and M-RZ) (Figure 2 and Table II). This is due to that Jordanian kaolin contains less amount of kaolin (60%) than reference kaolin (95%) [10]. The zeolite containing geopolymer (M-JZ) exhibits higher adsorption capacity than silica sand containing geopolymers (M-JS) (Figure 2 and Table II), which is similar to the trend observed in the case of reference kaolin geopolymers. This is due to the increase of surface area upon destruction of the precursor's crystal structure, as shown by XRD (Figure 1), and the generation of new cation-exchange sites upon geopolymerization.

Effect of ionic strength on the adsorption of Cd(II) ions onto geopolymers (M-RZ and M-JZ)

The adsorption of Cd(II) ions onto zeolite-containing geopolymers (M-RZ and M-JZ) was decreased at higher ionic strength (Figure 3 and Table II). This is due to the competition between the Cd(II) ions and Na^+ ions of the background electrolyte.

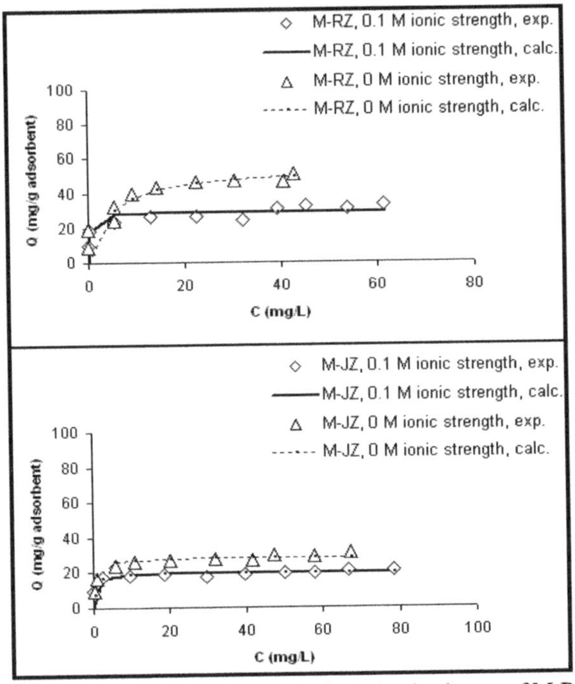

Figure 3. The effect of ionic strength on the adsorption isotherms of M-RZ and M-JZ geopolymers.

CONCLUSIONS

The geopolymerization of kaolin in the presence of zeolitic tuff leads to an increase in the adsorption of Cd(II) ions relative to zeolitic tuff and kaolin. This can be attributed to increase of surface area as a result of destruction of zeolite and kaolinite structures during geopolymerization. The observations were also confirmed that the local materials (kaolin and zeolitic tuff) satisfied the criteria to be used as precursors of inexpensive, stable, and high quality construction materials.

ACKNOWLEDGMENT

The financial support of the University of Jordan and the project "Chemical stabilization of natural geomaterials for construction and industrial applications" funded under the contract number (ZEIN2006PR33) by the Flemish (Belgium) Vlaamse Interuniversitaire Raad (VLIR) within the "eigen initiatieven' programme is gratefully acknowledged.

REFERENCES

[1]R. K. Sharma, M. Agrawala, and , F. Marshall, Heavy metal contamination of soil and vegetables in suburban areas of Varanasi, India, *Ecotoxicology and Environmental Safety*, **66**, Issue 2, 258-266 (2007)
[2]O. Rimawi, E. Salameh, Chemical changes of yeast industry wastewater during infiltration in Ruseifa area, Jordan, *Hydrological Sciences—Journal—des Sciences Hydrologiques*, **44**, 5, (1999)
[3]R. A. Meyer, D. K. Dittrick, *Environmental Pollution and Cleanup*, John Wiley and Sons, **2**, 1798 (1999)
[4]I. Heidmann, I. Christ, C. Leu, R. Kretzschmar, *J. Colloid. Interface Sci*, **282**, 270 (2005)
[5]J. Hizal, R. Apak, *J. Colloid. Interface Sci.*, **295**, 1, (2006)
[6]P. Srivatava, B. Singh, M. Angove, J. *Colloid. Interface Sci.*, **290**, 28, (2005)
[7]M. Alshaaer, H. Cuypers, J. Wastiels, Stabilization of kaolinitic soil for construction purposes by using mineral polymerisation technique, *6ICCT*, **3**, 1085-1092, (2002)
[8]L. Li , S. Wang , Z. Zhu , *J. Colloid Interface Sci*, **300**, 52, (2006)
[9]S. Wang, L. Li, Z. H. Zhu, *J. Hazrd. Mater.* Vol. **139**, 254, (2007)
[10]R. I. Yousef, B. El-Eswed , M. Alshaaer, F. Khalili, H. Khoury, The influence of using Jordanian zeolitic tuff on the adsorption, physical, and mechanical properties of geopolymers products, *J. Hazrd. Mater*, accepted in October (2008).

EXERGY ANALYSIS ON LIFE CYCLE OF CERAMIC PARTS

Hideki Kita, Hideki Hyuga, Naoki Kondo, and Tatsuki Ohji
National Institute of Advanced Industrial Science and Technology (AIST)
Nagoya 463-8560, Japan

ABSTRACT

It is critically important to consider the value of a product in terms of natural resource consumption and environmental problems throughout its life cycle. In order to achieve this, we need a method for quantitative evaluation of natural resource consumption that is similar to measurements of economic cost. In this regard, exergy analysis is an effective measurement approach. Based on these considerations, we carried out an exergy analysis for ceramics in order to clarify the value of environmental impact reduction throughout their life cycles and to develop a design policy for a highly efficient ceramics production process. Analysis was conducted based on large ceramic heat-tubes used in aluminum casting, and the amount of exergy used in the manufacturing process and the degree of effectiveness were evaluated. In addition, we estimated the reduction in environmental impact and in the use of natural resources in the production and usage stages when ceramic heat-tubes were used, compared to conventional steel heat-tubes.

INTRODUCTION

Although structural ceramics have excellent and unique properties compared with metals, they are not in widespread use because of the high initial costs and others. However, in addition to the financial costs, it is also important to consider the value of reducing the impact of a product in terms of natural resource consumption and environmental problems throughout its life cycle. In order to critically assess this issue, we need a method for the quantitative evaluation of natural resource consumption that is similar to measurements of economic cost.

The modes of the energy and resources coming in and going out of the ''systems'' during each process in a product's life cycle are always varying, making it difficult to evaluate the overall consumption and rate of effective utilization of natural resources. In order to design products that are highly energy efficient throughout their life cycles, a measurement index that can quantitatively and integrally express natural material and energy resources is important. Exergy could be an effective measurement index that can be used for this purpose.

Examples of exergy analysis that have been carried out to date for artificial products include research into exergy consumption in the usage stage of electrical lighting, in the production and consumption of reinforced concrete, and in the life cycle of aluminum and steel cans; among natural products, analysis of exergy consumption in plants during photosynthesis has been carried out.[1–6] Studies using exergy analysis on ceramic related issues have not been found.

Based on these considerations, we carried out exergy analysis for ceramics in order to clarify the value of environmental impact reduction throughout their life cycles and to develop a design policy for a highly efficient ceramics production process. In our previous report, analysis was conducted based on the manufacturing process of ceramic parts, and the process efficiency was evaluated.[7] In this report, we calculated the exergies for natural resources and energies in the production and usage stages when ceramic heater protection tubes were used, compared with conventional steel heater protection tubes, to estimate the reduction in environmental impact.

EVALUATTION PROCEDURE

Object for Analysis

In the melting and casting production of aluminum, various silicon nitride ceramics are now being used as components exposed to the molten aluminum because of their superior corrosion and thermal shock resistance. One of these applications is a heater protection tube, which contains heat sources such as gas burners or electrically heated wires inside to keep the temperature of molten aluminum constant. It is said that use of a highly durable silicon nitride heater protection tube provides great improvement in thermal efficiency.[8]

Figure 1 shows the structure of an aluminum melting furnace and the mounting location of heater protection tubes. Stainless steel tubes have been mainly used; however, the high corrosion rate due to the reaction between molten aluminum requires frequent exchanges, and then, this inevitably leads to the exchangeable vertical soaking type heater.

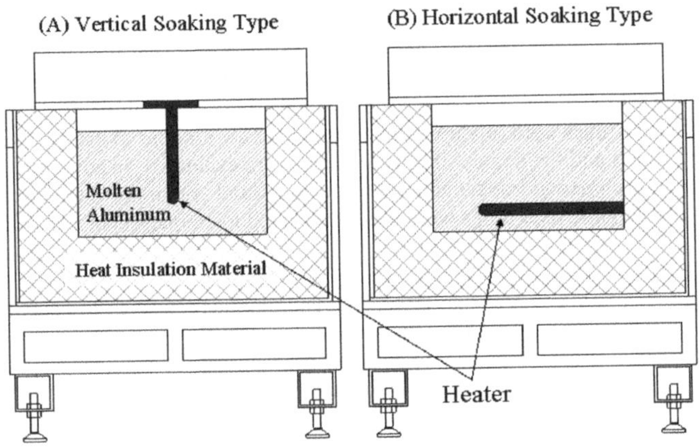

Fig. 1 Heater mounting location in aluminum melting furnace [8, 9]

On the contrary, silicon nitride, which is extremely stable in molten aluminum, changed the furnace structure. A horizontal soaking-type heater mounted at the bottom of the furnace was designed, which reduced heat loss drastically[8] and provided higher efficiency when compared with the vertical soaking type.

Figure 2 shows the schematic of heater protection tubes studied in this report. The tube measures 1346.5 mm in length, 155 mm in outer diameter, and 19 kg in weight for both types of steel and ceramic. Initial thicknesses of the tubes are 3 and 8 mm for steel and ceramic, respectively.

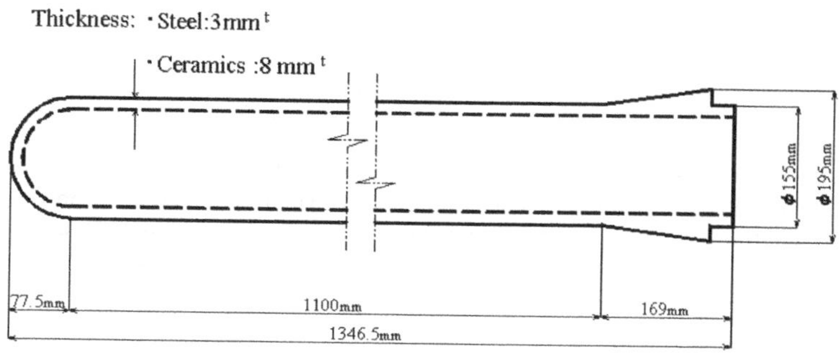

Fig. 2 Dimension of heater protection tube [8, 9]

We conducted exergy analyses on the production of heater protection tubes. Data for its production (quantities of raw material, fuel, waste, waste gas, water, etc.), which were necessary for analyzing exergy, were obtained from companies.[8] Some data were unavailable, and were supplemented with the authors' knowledge of ceramic processing.

Exergy Calculations

Chemical Exergy Calculations For Metal And Inorganic Compounds[9]*:* Referential species of inorganic compounds are denoted as *Xx*, *Aa*, *Bb*, and so on. They are formed according to the following reaction formula:

$$xX+aA+bB+ \ldots \rightarrow Xx\,Aa\,Bb \ldots \qquad (1)$$

Using ΔG^0 for the Gibbs free energy change, the chemical exergy of the inorganic compound can be calculated using the following formula:

$$E_x^{\,0} = 1/x\,(-\Delta G^0 - aE_x^{\,0}(A) - bE_x^{\,0}(B) - \ldots) \qquad (2)$$

As a reference, the exergy value of a substance is defined as zero if it does not react in a temperature environment of 25°C (298.15 K). References for some artificial materials are published in Japanese Industrial Standards,[10] but for those not mentioned in JIS, the reference is the smallest free energy value.[11]

Chemical Exergy Calculation for Organic Compounds: The chemical exergy equation for organic compounds is derived based on chemical exergy values for the individual hydrocarbon compounds that act as the building blocks of petroleum. Although the formulas of Rant[12] and Szargut and Styrylska [13] are known to be statistically dependent on elemental composition, in this report we revised the Rant formula for solid fuels, and Nobusawa's equation[14] was used for practical application:

$$E_x = mH_l\,(1.0064 + 0.1519\,\Phi_H/\,\Phi_C + 0.0616\,\Phi_O/\,\Phi_C + 0.0429\,\Phi_N/\,\Phi_C) \qquad (3)$$

Here, E_x is chemical exergy of organic compounds, m and H_l are dry weight (kg) and lower heating value (kJ/mol), respectively, and Φ_C, Φ_H, Φ_O, and Φ_N are the weight fractions of carbon, hydrogen, oxygen, and nitrogen contained in the compounds.

Exergy Calculation for Input Energy: The input energy used in this study was from electric power and LPG (liquefied petroleum gas). The exergy value for electrical power was the same as the electrical energy value, because electrical power has extremely low entropy. The calculation for LPG exergy was based on Nobusawa's equation,[14] as used in the previous section. The distribution was obtained for a mixed mole rate of the exergy values of C_3H_8 and C_4H_{10} (0.2 and 0.8, respectively), and the mixed entropy was calculated based on the difference between LPG exergy and the lower heating value:

$$E_x(\text{mix}) = H(\text{mix}) + T_0 \text{ x } \Delta S(\text{mix}) \quad (4)$$

$$H(\text{mix}) = \Sigma(X_i \text{ x } H_i) + R \text{ x } T_0 \text{ x } \Sigma(X_i \text{ x } \ln(X_i)) \quad (5)$$

$$\Delta S(\text{mix}) = (E_x(\text{mix}) - H(\text{mix}))/\ T_0 \quad (6)$$

Here, E_x(mix) is the chemical exergy of mixed gas (kJ/ mol), H(mix) the lower heating value of mixed gas (kJ/mol), T_0 environmental temperature (K), R the gas constant (58.314 J/mol), ΔS(mix) entropy of mixed gas (J/K·mol), X_i the volume fraction of each component, and H_i the lower heating value of each component (kJ/mol).

Production Subsystems: For the calculation of exergy input, consumption, and waste, the production system was assumed to be an assembly of subsystems, and quantification was performed for the input/output exergy of each production subsystem. The purpose of performing the analysis in sub-process units was not only for the sake of orderliness, but also for clarifying the location of problems with a view for improving the process efficiency and production method; an example is shown in Fig. 3. For each subsystem, raw materials, fossil fuels (including electric power and steam), *etc.* containing exergy were inserted (input), and waste, waste heat, *etc.* as well as a product or an intermediate product containing exergy were generated (output). The obtained intermediate product was used as the starting material for the next subsystem.

The final product was obtained by going through these subsystems and calculating the exergy values that enter or exit the respective subsystems. In principle, processes and items such as mining for the raw materials, construction, dismantlement, transportation, facilities, packing material for transportation, and palleting should also be included. The focal point of this study, however, is the production of a heater protection tube; thus, the above-mentioned processes were beyond the scope of the present study.

Process Efficiency: In this report, the ratio between the exergy fixed in the product and the total input exergy was determined by the above calculation (rate of exergy fixation), and this rate was regarded as the process efficiency:

$$\eta = E_x\,(\text{p}) \,/\, E_x\,(\text{in}) \quad (7)$$

Here, η is the rate of exergy fixation, E_x (p) is the exergy fixed in the product, and E_x (in) is the actual input exergy.

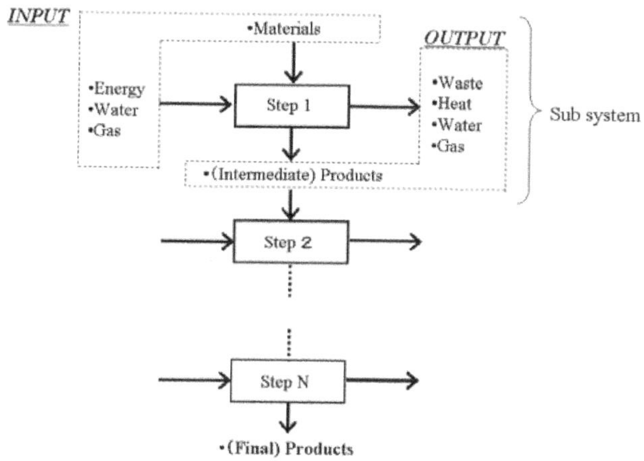

Fig. 3 Assembly of subsystems for the process

RESULTS AND DISCUSSION

Exergy in Production

The exergy balance for the production of a steel protection tube is shown in Table I. The production process of the steel protection tube consists of sintering, reduction, transfer, decarbonization, and metal rolling.[5] Based on the incoming and outgoing materials and energies for the respective processes, the exergy input for one tube was calculated to be 621.0 MJ. Of this exergy input, 126.0 MJ was fixed in the product; thus, the remainder of 495.0 MJ is considered to be disposed of as waste or heat. Then, the process efficiency of steel was approximately 20%.

Table I Materials and Exergy Balance for the Production of Steel Protection Tube.[5,9] Each exergy value was calculated.

	Raw materials			***Intermediate products***			***Disposal***		
		Weight / kg	Exergy / MJ		Weight / kg	Exergy / MJ		Weight / kg	Exergy / MJ
Sintering	Ironstone	27.1	0.0	Sintered ore	26.8	0.0	Wastage	4.3	0.0
	Limestone	4.0	0.0						
	Fuel		48.5				Waste exergy		48.5
	Sum	*31.1*	*48.5*	*Sum*	*26.8*	*0.0*	*Sum*	*4.3*	*48.5*
Reduction	Sintered ore	26.8	0.0	Pig iron	19.9	163.8	Wastage	17.4	30.4
	Ironstone	6.0	0.0						
	Coke	4.5	150.9						
	Fuel		211.1				Waste exergy		193.8
	Water/m3	0.3	26.0				Water vapor/m3	0.3	0.0
	Sum	*37.2*	*388.0*	*Sum*	*19.9*	*163.8*	*Sum*	*17.4*	*224.2*
Transfer	Pig iron	19.9	163.8	Pig iron	19.9	160.4	Heat waste		3.4
De-carbon	Pig iron	19.9	160.4	Steel	25.5	173.6	Wastage	6.1	28.3
	Scrap	9.7	64.4						
	Oxygen	2.1	3.8						
	Fuel		58.1				Waste exergy		93.9
	Water/m3	0.0	9.2				Water vapor/m3	0.0	0.0
	Sum	*31.6*	*295.8*	*Sum*	*25.5*	*173.6*	*Sum*	*6.1*	*122.2*
Transfer	Steel	25.5	173.6	Steel	25.5	169.7	Heat waste		4.0
Metal rolling	Steel	25.5	169.7	Steel	19.0	126.0	Wastage	6.5	42.8
	Fuel		43.3				Waste exergy		50.0
	Water/m3	0.0	5.7				Water vapor/m3	0.0	0.0
	Sum	*25.5*	*218.7*	*Sum*	*19.0*	*126.0*	*Sum*	*6.5*	*92.7*
Total process	Materials	53.3	260.0	Steel	19.0	126.0	Wastage	34.3	101.4
	Fuel		361.0				Waste exergy		393.6
Sum			***621.0***			***126.0***			***495.0***

Table II Materials and Exergy Balance for the Production of Ceramic Protection Tube.[9] Each exergy value was calculated.

	Input resouces					***Products***		***disposal***		
Process	Electric power	LPG		Raw materials		Products		Materials		Exergy
	MJ	Weight / kg	Exergy / MJ	Weight / kg	Exergy / MJ	Weight / kg	Exergy / MJ	Weight / kg	Exergy / MJ	MJ
Mixing	266.8			30.9	291.1	30.9	291.1	0.0	0.0	266.8
Spray dryng	300.1	46.4	2246.5	197.4	311.7	20.3	291.1	223.6	103.9	2463.2
Forming	175.0			20.3	291.1	20.3	291.1	0.0	0.0	175.0
Dewaxing	98.4			20.3	291.1	19.0	229.4	1.4	61.8	98.4
Sintering	776.3			42.7	229.9	19.0	229.4	23.7	0.6	776.3
Sum	***1616.6***	***46.4***	***2246.5***	***202.1***	***312.2***	***19.0***	***229.4***	***248.7***	***166.2***	***3779.7***

Table II shows the exergy balance for the production of a ceramic heater protection tube.[7] The process involves mixing, granulation, cold isostatic pressing (CIPing) and green machining, dewaxing, and sintering. As the input energy, LPG was used for drying in granulation, and electric power was used in dewaxing and sintering. Most of the material removed by green machining is supposed to be reused; thus, the raw material powder is not wasted. When the same weight of silicon nitride as the above was produced, the total input exergy of 4175.3 MJ, which is the sum for raw materials, electric power, and gas (312.2 MJ, 1616.6 MJ, and 2246.5 MJ, respectively), was necessary. The exergy fixed in the product was 229.4 MJ, and the remaining 3779.7 MJ was discharged into the environment as waste or heat. The process efficiency of ceramics was calculated to be approximately 5.5%. The details were studied in a previous report.[7]

Exergy Consumption During Operation

Dissolution of Heater Protection Tube and Its Disposal: The ceramic protection tube is chemically stable in molten aluminum. In the case of the steel protection tube, corrosion and dissolution occur because of the chemical reaction with molten aluminum. Therefore, once a predefined thickness has been reached, it is necessary to replace it, and its lifespan is about 6 months. The loss due to dissolution is assumed to proceed according to the following equation:

$$D = D_0 \times (2 - \exp(kt)) \qquad (8)$$

Here, D is the thickness (mm) of the heater protection tube, D_0 is its initial thickness (mm), and k is the apparent reaction rate constant. In the above equation, D_0 was set to 3 mm, and the thickness of the heater protection tube at the time of replacement was denoted by D_i. The condition for replacement was set to when the thickness is half the initial thickness (D_i/D_0=0.5). The lifespan for replacement (t) is 6 months; thus, the reaction rate (corresponding to constant k) was calculated to be 0.067578 mm per month by substituting these values.

On the other hand, in the case of the ceramic protection tube, the tube is replaced after 7 years, which is the same as the lifespan of the entire furnace. However, the reaction of the ceramic itself hardly progressed, and the details are not known. In this study, the value of D_i/D_0 was assumed to be 0.8, and the lifespan was assumed to be 200 months. The reaction rate constant was calculated in a similar way as above, and the reaction rate was determined to be 0.000912 mm per month. The calculated thickness change according to elapsed time is shown in Fig. 4.

The exergy is consumed by dissolution loss. The consumed exergy was calculated according to the following equation using the obtained reaction rate constant and the results shown in Fig. 4:

$$E_x = E_{x0} \times \exp(kt) \qquad (9)$$

Here, E_x is the consumed exergy, E_{x0} the initial exergy, and t the elapsed time. In the case of the steel protection tube, one tube (19 kg) has an exergy of 126.0 MJ (see Table I); thus, the exergy per unit weight is 6.6 MJ/kg. The consumed exergy was calculated in a similar way for the silicon nitride protection tube. In this case, the exergy per unit weight is 12.1 MJ/kg based on the data in Table II. The change in consumed exergy with time is shown in Fig. 5. A steel heater protection tube needs exchange every 6 months; on the contrary, a ceramic tube can be used continuously for 7 years.

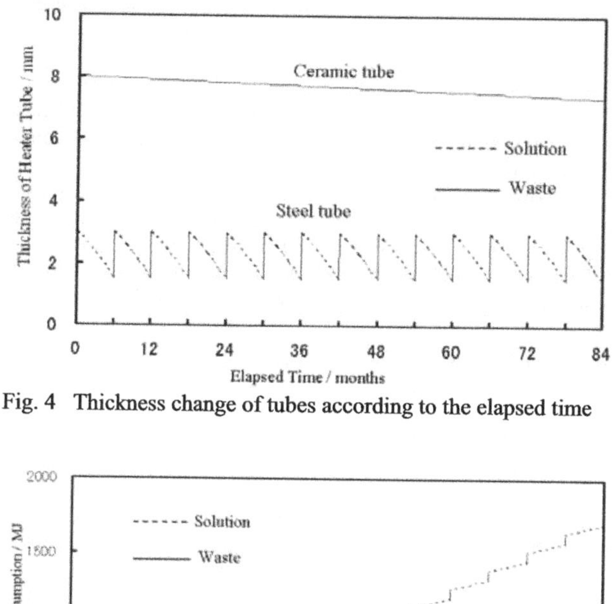

Fig. 4 Thickness change of tubes according to the elapsed time

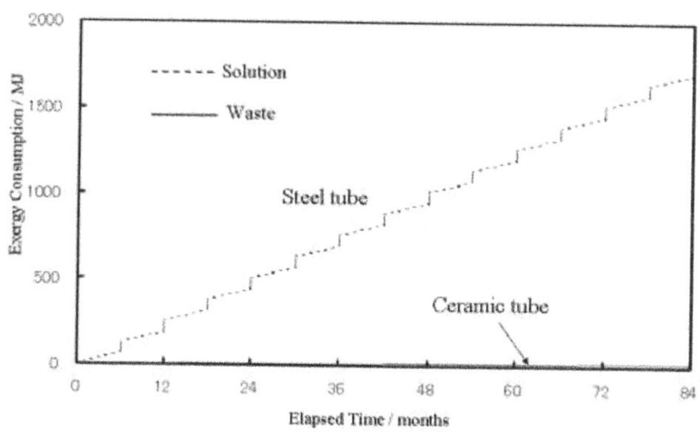

Fig. 5 Calculated lost exergy of tubes according to the elapsed time

The value of the completed ceramic heater protection tube is its high conservation stability (no diffusion). That is to say, while the steel tube consumes exergy as a result of the chemical reaction during operation, the ceramic tube hardly consumes any exergy. Because the steel tube is replaced every 6 months, 14 steel tubes are used in 7 years. On the other hand, only one ceramic tube is used in 7 years. Therefore, the respective consumed exergies are as follows:

- when a steel tube is used:
  126.0 (MJ/tube) x 14 (tubes) = 1764 MJ
- when a silicon nitride tube is used:
  229.4 (MJ/tube) x 1 (tube) = 229.4 MJ

Steel diffuses in molten aluminum and becomes an impurity, whereas the chemically stable ceramic does not diffuse in molten aluminum. As a result, the quality of the ceramic is excellent and the aluminum product is easily recyclable. An additional advantage of the ceramic is that its operation period can be shortened.

Exergy Necessary for Furnace Operation: Table III shows a comparison of the electric power consumed during operation and that consumed during down-time.[8] When a steel tube is used, it would be a vertical soaking type. In the case of a ceramic tube, a horizontal soaking type is possible (Fig. 1). For the vertical soaking type, 9.4 kW is necessary during operation and 4.0 kW is necessary during down-time. The thermal efficiency of the horizontal soaking type, in which the ceramic tube is used, can be improved. The consumed electric powers during operation and during down-time are 6.8 kW and 3.8 kW, respectively. If the operation time is assumed to be 60% of a day (*i.e.*, 40% down-time) and the equipment is assumed to be operated for 360 days per year, the total consumed electric power (exergy) in 7 years will be as follows:

- when a steel tube is used:
  (9.4 x 0.6 x 24+4.0 x 0.4 x 24) x 360 x 7 x 3.6/1000 = 1576 GJ
- when a silicon nitride tube is used:
  (6.8 x 0.6 x 24+3.8 x 0.4 x 24) x 360 x 7 x 3.6/1000 = 1219 GJ

That is, for the vertical soaking type, the consumed exergy is 1576 GJ. On the other hand, the consumed exergy is 1219 GJ for the horizontal soaking type when a silicon nitride tube is used.

Comparison of Exergy Consumptions for Production, Operation, and Disposal

As mentioned above, when a furnace is operated for 7 years, 14 steel protection tubes are needed because of dissolution loss. On the other hand, only one silicon nitride tube is required for the same period. The ceramic tube does not have to be changed for 7 years; however, it is replaced because of wearing by the furnace material.

As mentioned above, the disposal exergy in the production process is 495.0 and 3779.7 MJ for one steel tube and for one silicon nitride tube, respectively (refer to Tables I and II). Therefore, the total exergy can be calculated, as shown below, by multiplying the number of tubes used in 7 years:

- When a steel tube is used:

  495.0 (MJ/tube) x 14 (tubes) = 6930.0 MJ

Table III Electric Power Consumption for Vertical and Horizontal Soaking Type Furnace [8]

Stage	**(A) Vertical Soaking Type** (Steel Tube)	**(B) Horizontal Soaking Type** (Ceramic Tube)
Running	9.4 kW	6.8 kW
Down time	4.0 kW	3.8 kW

Table IV Comparison of Consumed Exergy for Steel and Ceramic Tubes through Life Cycles

Stage / Type	**Steel Tube**	**Ceramic Tube**
[1] Production stage	6,930 MJ =495(MJ/P)*14(P)	3,780 MJ =3780(MJ/P)*1(P)
[2] Usage stage (A)*1	1,764 MJ =126(MJ/P)*14(P)	229 MJ =229(MJ/P)*1(P)
[3] Usage stage (B)*2	1,576 GJ =(9.4*0.6*24+4.0*0.4*24)*360*7*3.6/1000	1,219 GJ =(6.8*0.6*24+3.8*0.4*24)*360*7*3.6/1000
Total([1]+[2]+[3])	1,585 GJ	1,223 GJ

*1: Lost exergy by reaction with molten aluminum plus disposal
*2: Input energy for running the furnace

- When a silicon nitride tube is used:

  3779.7 (MJ/tube) x 1 (tube) = 3779.7 MJ (10)

In the following, the disposed exergy as a result of operational wear is also shown.

- When a steel tube is used:

  126.0 (MJ/tube) x 14 (tubes) = 1764 MJ

- When a silicon nitride tube is used:

  229.4 (MJ/tube) x 1 (tube) = 229. 4 MJ (11)

As shown in "Exergy Necessary for Furnace Operation," the consumed electric powers during usage for the steel tube and the silicon nitride tube are 1576 and 1219 GJ, respectively. By adding (10) and (11) to the consumed electric power, the consumed exergies when using the steel tube and the silicon nitride tube are calculated to be 1585 and 1223 GJ, respectively. Thus, when a silicon nitride tube is used, the consumed exergy is 362 GJ less than when a steel tube is used. In other words, the silicon nitride tube consumes a larger exergy per tube during the production process; however, it is replaced less frequently and its consumption of electric power can be smaller than the steel tube because the silicon nitride tube has high conservation stability. As a result, the total exergy consumption through a life cycle of production, operation, and disposal is smaller than that of the steel tube. Silicon nitride is hard-to-recycle material. Therefore, when the silicon nitride parts reach the end of its usefulness, in many cases, they will be discarded and buried in the ground. The disposal of silicon nitride could have a very low impact on environment because they are extremely stable in the earth (Table IV).

CONCLUSIONS

Exergy analysis was conducted on ceramic and steel heater protection tube used in aluminum casting, and the amount of exergy consumed through the life cycle was calculated. In the production stage, a ceramic heater consumes much higher exergy than the steel one, 495 and 3780 MJ for one piece, respectively. Analysis through the life cycle in 7 years shows that exergy consumptions for the ceramic tube and steel tube are 1223 and 1585 GJ, respectively. These results reveal that ceramics are effective in the reduction of environmental burden.

REFERENCES

1. H. Asada and M. Shukuya, ''Exergy–Entropy Process of Electric Lighting Systems Using Fluorescent Lamps,'' J. Archit. Plann. Environ. Eng., 483 91– 100 (1996) (in Japanese).

2. M. Saito, I. Takahashi, and M. Shukuya, ''Numerical Analysis of Exergy Balance of A Leaf—Research on the Passive Cooling Effect of Trees,'' Archit. Plann. Environ. Eng., 505 51–58 (1998) (in Japanese).

3. I. Takahashi and M. Shukuya, ''An Exergy Analysis of the Circulation of Matter Obtained from Composing Garbage in an Urban Area,'' J. Archit. Plann. Environ. Eng., 510 23–29 (1998) (in Japanese).

4. J. Dewulf and H. R. Langenhove, ''Quantitative Assessment of Solid Waste Treatment Systems in the Industrial Ecology Perspective by Exergy Analysis,'' Environ. Sci. Tech., 36 1130–1135 (2002).

5. T. Morihana, I. Takahashi, and M. Shukuya, ''Exergy Consumption of Reinforced Concrete Walls in the Courses of Production and Use,'' J. Archit. Plann. Environ. Eng., 520 99–105 (1999) (in Japanese).

6. Y. Soeno, T. Akashi, H. Ino, K. Shiratori, K. Nakajima, and K. Harada, ''Exergy Analysis for the Integrated Evaluation of Environmental Impacts,'' J. Japan Inst. Metals, 66 [9] 885–888 (2002) (in Japanese).

7. H. Kita, et al. ''Exergy Analysis on the Ceramic Manufacturing Process,'' J. Ceram. Soc. Jpn., 115 12 987–992.

8. Tounetsu Corporation, ''Ceramic Heater Protection Tube,'' http://www.tounetsu.co.jp/index-j.html (in Japanese).

9. K. Karakida, ''Ekuserugi no kiso,'' Ohm-sya, 2005 (in Japanese).

10. ''General Rules for Energy Evaluation Method by Available Energy,'' JIS Z 9204 1991 (in Japanese).

11. MALT (Netsuriki-Gaku Data Base), Kagaku Gijyutsu-Sya.

12. Z. Rant, ''Exergy, A New Word for ''Technical Available Work,'' Forsch. Ing-Wes, 22 36–37 (1956).

13. J. Szargut and T. Styrylska, ''Angenäherte Bestimmung der Exergie von Brennstoffen,'' Brennst.Wärme-Kraft, 16 [12] 589–596 (1964) (in German).

14. T. Nobusawa, ''Ekuserugi no Jitsuyo Keisanhou,'' Nenryou oyobi Nensyou, 43 [11] 49–79 (1976) (in Japanese).

ANION EXCHANGE PROPERTY OF As(III), As(V), Se(IV), CrR(VI) AND B(III) WITH HYDROTALCITE-LIKE COMPOUNDS

Junji Shibata, Norihiro Murayama and Shigeno Matsumoto
Department of Chemical, Energy and Environmental Engineering, Kansai University
Suita, Osaka, 564-8680 Japan

ABSTRACT

Hydrotalcite-like compounds (HT) is one of the inorganic anion exchangers and have a layered structure of metal complex hydroxides. In this study, the anion exchange properties of various toxic anions composed of As(III), As(V), Se(IV), Cr(VI) and B(III) were investigated by using the hydrotalcite-like compounds and its calcination products.

Various toxic metal ions can be removed from aqueous solution by the anion exchange reaction in the pH region where toxic metal ions exist as the anionic species like AsO_2^-, $HAsO_4^{2-}$, CrO_4^{2-}, SeO_3^{2-} etc. Anion exchange selectivity of the hydrotalcite-like compounds is mainly based on the ionic radius and the valency of chemical species. When the hydrotalcite-like compounds containing CrO_4^{2-} after an anion exchange operation were calcined at 773K, it is found that the anionic species in hydrotalcite-like compounds can be immobilized. It is possible to use the hydrotalcite-like compounds as an excellent water treatment agent for toxic anionic species.

INTRODUCTION

The amount of aluminum consumed in Japan occupies about 15% in the world. A large amount of electrical energy is needed to produce aluminum by the molten salt electrolysis, but the energy for aluminum recycling is about 3% compared with that for molten salt electrolysis. The aluminum recycling is important from the viewpoint of resources and energy conservations. However, the various wastes such as aluminum dross, $MgCl_2$ waste solution and so on are exhausted in the recycling process of various aluminum products[1)]. The main component of aluminum dross, which floats on the surface of molten aluminum in the furnace, comprises metal aluminum, aluminum oxides and nitrides. The $MgCl_2$ waste solution is discharged in a magnesium removal process from Mg/Al alloys using Cl_2 gas. Especially, the aluminum dross is appointed as the managed wastes because of the possibility of ammonia generation and spontaneous ignition by the reaction with moisture in air. Only a small part of the aluminum dross is used as a deoxidizer for steel making and a cement mixture. It is important to develop the new technology to use aluminum dross effectively.

In this study, the synthesis of hydrotalcite-like materials (HT), which have a structure of layered double hydroxides and are one of the inorganic anion exchangers[2-11)], was carried out using reagents and the wastes discharged in an aluminum recycling process as a raw material. The removal and immobilization of various toxic anions such as As(III), As(V), Se(IV), Cr(VI), B(III) and F were studied by using various HT and its calcination products derived from reagent and the above waste.

EXPERIMENTAL

Synthesis of HT from reagent and waste

Five gram of aluminum dross and 200cm^3 of 1.0mol/dm^3 HCl or aqua regia were put into a beaker, and leaching of the aluminum component was conducted for 3h. After the leaching, the filtrate was recovered by a solid-liquid separation with a vacuum filtration equipment. Then, $MgCl_2$ waste solution was added to the filtrate containing Mg and Al. The Mg/Al molar ratio in the solution was set to 2.5. The Mg/Al mixed solution was quantitatively added to Na_2CO_3 solution at pH 10.5 under stirring. In order to maintain the reaction pH at about 10.5, 30w/v% NaOH was dropped in the Mg/Al mixed solution. After these operations, an agitation was continued for 2 days. The solid product obtained by a vacuum filtration method was washed with distilled water and then dried at 343K. In the case of synthesis from reagents, the Mg/Al mixed solution was prepared using $MgCl_2 \cdot 6H_2O$ and $AlCl_3 \cdot 6H_2O$, and the above operation was carried out according to the same way. Identification of crystal structure and analysis of the chemical composition of the reaction products were carried out with an X-ray diffraction equipment (Nihon Denshi Co. Ltd., JDX-3530) and an energy-distributed fluorescence X-ray analysis equipment (Nihon Denshi Co. Ltd., JED-2110), respectively.

Removal and immobilization of toxic ions with HT

In order to investigate the removal property for As(III), As(V), Se(IV), Cr(VI) and B(III) in aqueous solution with the HT products, 0.3g of various HTs and HT products calcined at 673K were put into 30cm^3 of various solutions containing toxic ions mentioned above. After the contact operation for 1h at room temperature, the amounts of toxic ions remained in the mother liquor were measured. Concentration of metal ions in aqueous solution was measured with an inductively-coupled plasma spectrometer (Shimazu Co. Ltd., ICPS-1000III).

Heat treatment of HT containing various toxic ions was conducted to immobilize toxic ions in HT. The HT after a removal operation of toxic ions was dried at 343K, and then heat treatment at 573-1173K was carried out in N_2 gas atmosphere for 1h. The product, 0.1g, obtained by the heat treatment was added to pure water of 30cm^3, and the mixture was shaken for 24h. After a solid-liquid separation, the amounts of toxic ions remained in the mother liquor were measured by the same analytical method. The change in crystal structure of the product before and after heat treatment was investigated by an X-ray diffraction method.

RESULTS AND DISCUSSION

The chemical composition of aluminum dross and Mg-containing waste solution is shown in Table I. The weight percent for each metal component was calculated by defining the total metal content in leachants as 100%. The contents of Al and Mg in the aluminum dross, which are also the main metal composition of HT, are 85.3% and 5.6%, respectively, in case of leaching with aqua regia. On the other hand, 83.8% of Al and 4.7% of Mg are contained in 1.0mol/dm^3 HCl used as a leachant. Impurity components such as Fe, Si, Ca, K, Na and so on exist in the aluminum dross, and these are originally contained in Al scrap. The $MgCl_2$ waste solution contains 98.2% of Mg. The ratio of Cl/Mg and Mg^{2+} concentration of the waste solution are 2.2 and 3.2mol/dm^3, respectively. The mixed solution of the leachant (b) and the $MgCl_2$ waste solution (c) are used as a raw material of HT synthesis.

The SEM photograph of aluminum dross used in this study is shown in Photo.1. The fine aluminum dross particles are coagulated, and large particles with over 20 micrometer are formed as shown in Photo.1. The aluminum dross particle size has also wide distribution.

The XRD pattern of HT synthesized from reagent and aluminum dross is shown in Fig.1. The peak pattern of hydrotalcite-like materials ($[Mg_{1-x}Al_x(OH)_2]^{x+}$ $[(A^{n-})_{x/n}\cdot mH_2O]^{x-}$, x=0.2-0.33) is identified for the products obtained from aluminum dross (Fig.1(b)). The reaction product are identified as CO_3^{2-} type HT having CO_3^{2-} anions as an exchangeable ion, and this is almost the same for HT produced from reagent (Fig.1(a)). The results indicate that it is possible to synthesize HT with almost the same ability of HT reagent by using the aluminum dross and $MgCl_2$ waste solution.

Table I. Chemical composition of aluminum dross and Mg-containing waste solution

Material	Leachant	Metal composition [wt%]							
		Al	Mg	Si	Ca	Fe	Na	K	Pb
(a) aluminum dross	aqua regia	85.3	5.6	1.8	1.9	4.4	0.5	0.4	0.1
(b) aluminum dross	1.0mol/dm^3 HCl	83.8	4.7	2.6	1.5	5.3	1.2	0.8	0.1
(c) $MgCl_2$ waste solution		—	98.2	—	1.5	—	0.2	0.1	—

Photograph 1. SEM photograph of aluminum dross

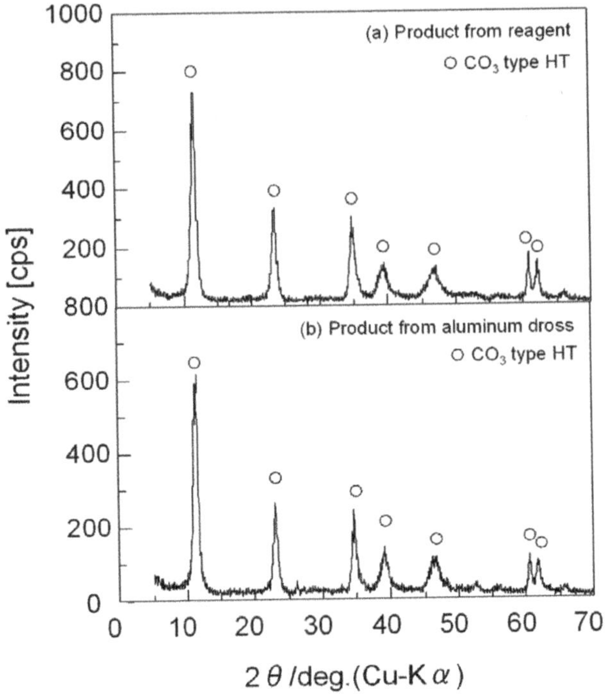

Figure 1. XRD pattern of HT synthesized from reagent and aluminum dross

Generally, CO_3^{2-} is very stable in the HT structure, and the anion exchange amounts between CO_3^{2-} in the HT and various oxoacid anions are not so large. From our previous studies[2-6] or the other reports[7-11] , the removal amounts of various anions with the HT calcination product (Mg/Al complex oxide) are larger than those with the CO_3^{2-} type HT. There are two processes for anion uptake, that is, the direct uptake reaction with rehydration and the anion exchange reaction, in using the HT calcination product for the removal of various anions. These anions are mainly taken by the HT calcination product with a rehydration reaction to neutralize the positive charges in HT.

The removal of various toxic anion species such as As(III), As(V), Se(IV), Cr(VI) and B(III) was carried out using the calcination product of CO_3^{2-} type HT. The relationship between the removal amount of anion species with the HT calcination product and the equilibrium concentration is shown in Fig.2. In this pH region (pH8-11), the toxic metal ions exist as a form of anionic species like AsO_2^-, $HAsO_4^{2-}$, CrO_4^{2-}, SeO_3^{2-} etc. The removal amounts of As(III), As(V) and Se(VI) with the HT calcination product increase with an increase in the equilibrium concentration. On the other hand, in case of Cr(VI) and B(III), the removal amounts are drastically increasing with increasing equilibrium concentration in the region below $100 mg/dm^3$, and then the removal amounts do not change any more over $100 mg/dm^3$ of equilibrium concentration. The order of anion exchange amount is as follows;

B(III) < Cr(VI) < As(III) < As(V) < Se(VI).

From the results, various toxic metal ions can be removed from aqueous solution. It is considered that the anion exchange selectivity of HT compounds is mainly based on the hydration ionic radius and the valency of electric charge.

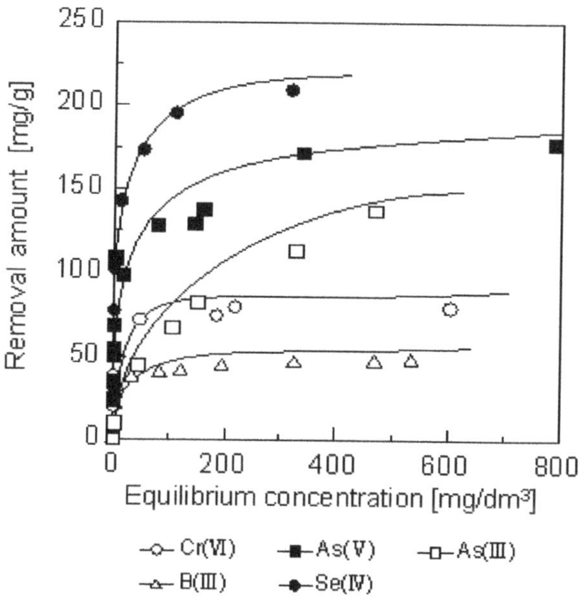

Figure 2. Relationship between removal amount of anion species and equilibrium concentration

In order to estimate the minimum required amount of HT calcination product to obtaine the outlet solution containing $0.5 mg/dm^3$ of Cr(VI) by multi-stage batch operations, McCabe-Thiele analysis is applied to the adsorption isotherm. That is, the objective of this analysis is to determine the optimized additional amounts of adsorbent for every stage so as to minimize the sum of additional amount in each stage.

The adsorption isotherm of Cr(VI) with the HT calcined at 673K for 6h is shown in Fig.3. When the concentration of Cr(VI) in the inlet solution (C_0) and in the outlet solution (C_3), the number of stage and treatment volume are respectively assumed to be $100 mg/dm^3$, $0.5 mg/dm^3$, 3

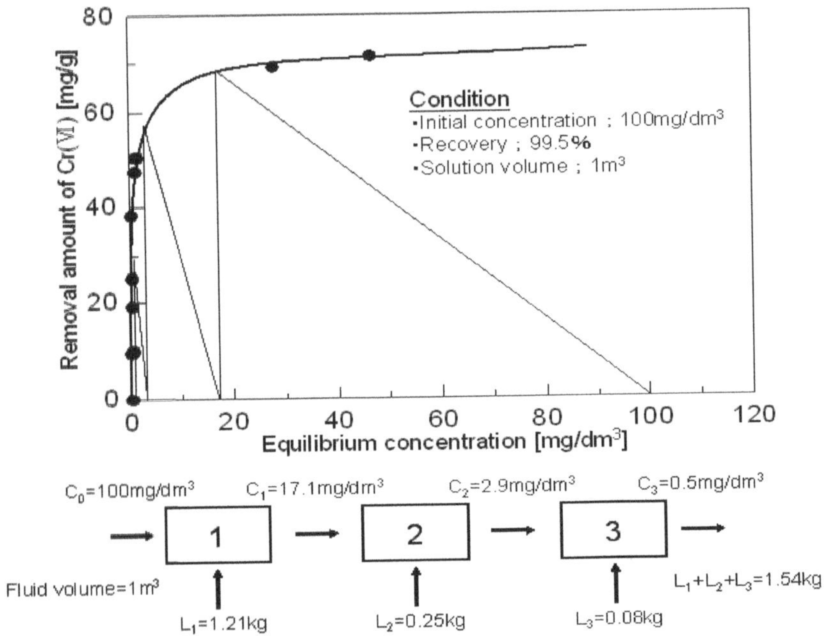

Figure 3. Adsorption isotherm of Cr(VI) with the HT calcination product

stages and 1dm^3, the required amount of HT calcination product in each stage (L_1, L_2, L_3) and the Cr(VI) concentrations of outlet solution in the 1st and 2nd stages (C_1, C_2), are theoretically determined to be as follows; L_1=1.21kg, L_2=0.25kg, L_3=0.08kg, C_1=17.1mg/dm^3 and C_2=2.9mg/dm^3. From the results, the theoretical minimum value of additional amount is estimated to be 1.54kg ($L_1+L_2+L_3$).

To immobilize the Cr(VI) in the HT component, the effect of heat treatment was investigated for the HT product containing Cr(VI). The dissolved amount and dissolved pate of Cr(VI) are shown in Fig.4. The dissolved rates of Cr(VI) for the original HT (dried at 343K after removal operation of Cr(VI)) and its calcination product at 573K are about 38% and 12%, respectively. The dissolved amount and dissolved rate of Cr(VI) are linearly decreasing with an increase in calcination temperature, and the Cr(VI) dissolution can not be found by the heat treatment over 773K. Though it is not shown as a figure, according to the results of XRD for the heat-treated product, MgO and$MgAl_2O_4$ are formed by the heat treatment at 773K. It is considered that the immobilization of Cr(VI) in the HT after heat treatment is mainly caused by the formation of solid solution composed of Cr(VI) and MgO or $MgAl_2O_4$.

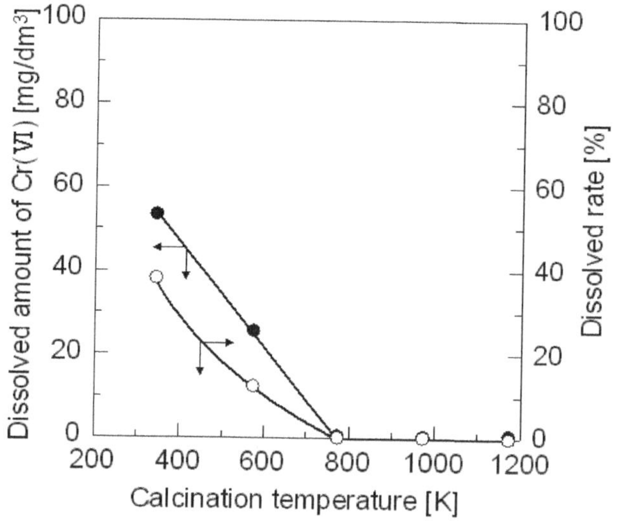

Figure 4. Effect of calcination temperature on immobilization of Cr(VI) in HT

CONCLUSION

As one of effective usages of the wastes generated in an Al regeneration process, the syntheses of HT were conducted using aluminum dross and $MgCl_2$ waste solution as a raw material. The removal and immobilization of As(III), As(V), Se(IV), Cr(VI) and B(III) were investigated with the HT calcination product.

The CO_3^{2-} type HT can be synthesized from the above wastes by the combination of acid leaching and co-precipitation. The XRD pattern of HT obtained from the wastes are almost similar to that from reagents. Toxic oxoacid anions such as As(III), As(V), Se(IV), Cr(VI) and B(III) in aqueous solution can be adsorbed by using the HT or HT calcination product. The immobilization of Cr(VI) in HT can be achieved by the heat treatment over 773K. The HT derived from the above wastes can be used as an excellent material for the removal and immobilization of various toxic heavy metal ions.

ACKNOWLEDGMENT

This work was supported by the MEXT, Grant-in-Aid for Scientific Research (A)(1), (17206091) and Grant-in-Aid for Young Scientist (B), (18760633).

REFERENCES
[1]H. Harada, T. Kashiwabara, M. Takaoka, K. Oshita, N. Takeda: *Resources Processing*, **50**, 140-148 (2003).
[2]J. Shibata, N. Murayama, M. Tanabe, H. Yamamoto, Synthesis of Hydrotalcite from Wastes Discharged in Aluminum Regeneration Process and its Physical Properties, *Kagaku Kogaku Ronbunshu*, 31, 74-79 (2005).
[3]J. Shibata, N. Murayama, M. Tanabe, H. Yamamoto, pH Buffer Action of Layered Double Hydroxide, *Kagaku Kogaku Ronbunshu*, 33, 273-277 (2007).
[4]N. Murayama, H. Yamamoto, J. Shibata, Synthesis of Hydrotalcite and its Anion Exchange Properties, *Resources Processing*, 51, 92-98 (2004).
[5]N. Murayama, M. Tanabe, R. Shibata, H. Yamamoto, J. Shibata, Removal of Toxic Heavy Metal Ions in Aqueous Solution with Mg/Al Type Hydrotalcite Derived from Wastes, *Kagaku Kogaku Ronbunshu*, 31, 285-290 (2005).
[6]N. Murayama, H. Yamamoto, J. Shibata, Synthesis of Hydrotalcite-like Materials from Various Wastes in Aluminum Regeneration Process, *Resources Processing*, 53, 92-98 (2006).
[7]G. Carja, R. Nakamura, T. Aida, H. Niiyama, Textural Properties of Layered Double Hydroxides: Effect of Magnesium Substitution by Copper or Iron, *Micropor. Mesopor. Mat.*, **47**, 275-284 (2001).
[8]L. Hickey, J. T. Kloprogge, R. L. Frost, The Effects of Various Hydrothermal Treatments on Magnesium- Aluminium Hydrotalcites, *J. Mat. Sci.*, **35**, 4347-4355 (2000).
[9]V. Rives, S. Kannan, Layered Double Hydroxides with the Hydrotalcite-type Structure Containing Cu^{2+}, Ni^{2+} and Al^{3+}, *J. Mat. Chem.*, **10**, 489-495 (2000).
[10]J. V. Sanchez, F. Figueras, M. Gravelle, P. Kumbhar, J. Lopez, Basic Properties of the Mixed Oxides Obtained by Thermal Decomposition of Hydrotalcites Containing Different Metallic Compositions, *J. Catal.*, **189**, 370-381 (2000).
[11]Y. Seida, Y. Nakano, Removal of Phosphate in Dissolution-Coagulation Process of Layered Double Hydroxide, *J. Chem. Eng. Japan*, **34**, 906-911 (2001).

PREPARATION OF METAL OXIDE PHOTOCATALYST BY SOFT SOLUTION PROCESS WITH ANION EXCHANGE RESIN

Yoshiyuki Kamo, Masayoshi Uno, and Shinsuke Yamanaka
Division of Sustainable Energy and Environmental Engineering, Graduate School of Engineering, Osaka University
2-1 Yamadaoka, Suita, Osaka 565-0871, Japan

ABSTRACT

Photocatalysts are expected for removing an environmental pollutant only with solar energy. Generally, the photocatalic powders are synthesized with high energy process such as the solid-phase synthesis or gas-phase process. On the other hand, there is a concept for making high performance material directly from solution, in soft and environmentally friendly condition, which is called soft solution process. In this study, zinc hydroxide and oxide were synthesized by ion exchange resin. By this method, we succeeded in harmless and energy saving process. Afterwards, the basic physical properties and photocatalic performance of the obtained sample were evaluated. The as-prepared photocatalyst was characterized by TG-DTA, XRD, FE-SEM, BET, and UV-vis and ZnO calcined at 573 K showed the highest photocatalic activity.

INTRODUCTION

In the past decades, the preparation of nanoscale oxide semiconductors has been made such as ZnO, TiO_2 and SnO_2 [1~3]. Among them, ZnO is of great importance due to its properties and novel applications ranging from gas sensors to ceramic varistors [4]. Zinc oxide is a direct band-gap semiconductor with applications in the field of photocatalyst [5~6]. Although there is a fault that the zinc oxide is dissolved to solution after the light irradiation, it is possible to use it by mixing with the wallpaper [7] for the purpose of removing acetaldehyde because it is a gas phase reaction. Here, high performance photocatalysts like Degussa P25 are synthesized by the methods requiring high energy [8] as solid state reaction and gas phase reaction. To develop more green process, the newly synthesis process of metal oxide particles is proposed in this study. Following reaction is applied in the liquid phase deposition (LPD) method [9].

$$MF_x^{(x-2n)-} + nH_2O \rightleftharpoons MO_n + xF^- + 2nH^+ \qquad (a)$$

However, because a harmful fluorine ion is exhausted in this method, it is not friendly to environment. Therefore, we focused on the OH type anion exchange resin as the removal materials of anion. The ion exchange reaction by OH type anion exchange resin is described as follows;

$$2R-OH + nX \rightleftharpoons R-X + OH^- \qquad (b)$$

In the equation (b), to move right side, we apply the ion exchange reaction of the anion exchange resin instead of using fluorine ion. The mechanism of the acceleration of the hydrolysis reaction of the metal complex ions by the removal of acidic anion could be explained by the solubility constant. The supersaturated solution of metal oxide or hydroxide was adjusted by the exchanging of anion to hydroxyl ion. This mechanism is same as the ion exchange isothermal supersaturation (IXISS) method [10]. In this paper, we reported a new soft solution process [11] applying the IXISS to the synthesis of zinc oxide or hydroxide particles and evaluated its photoactivity.

EXPERIMENTAL

Preparation of ZnO photocatalyst

Zinc sulfate heptahydrate ($Zn(SO_4)\cdot 7H_2O$, Kishida Chemical, 98%) and anion exchange resin (DOWEX Monosphere 550A; Dow chemical company) were selected to synthesize the nanoparticles of ZnO. To compare with commercial and high performance photocatalyst, ZnO (Aldrich, 99.99%) and P25 (Degussa) were chosen. Zinc sulfate was dissolved in distilled water to 0.01 mol/dm^3 as a treatment solution. The desired amount of the anion exchange resin was added to the treatment solution with stirring, and the reaction was carried out at 293 K for 24 h. The precipitation was dried at 353 K for 12 h. Finally, the dry precipitate calcined at 573, 673 and 873 K for 2 h, then named ZnO-573, ZnO-673 and ZnO-873 respectively.

Charactarizations

The surface structure of the sample was observed by a field emission-type scanning electron microscopy (FE-SEM) (JSM-6500F; JEOL. Ltd.). The crystal structures of the samples were characterized by X-ray diffraction (XRD) using Cu Kα radiation (2θ ranges from 20 to 80 degree) (RINT2000; Rigaku. Ltd.). The textural structures were measured by N_2 adsorption-desorption at 77 K, and specific surface areas of samples were measured using Brunauer–Emmett–Teller (BET) method. The pore size distributions (average pore diameter and mean pore volume) were measured from N_2 isotherm applying cylindrical pore model (GEMINI 2360; Shimadzu). All the samples were degassed at 373 K for 2 h before the actual measurement. Simultaneous thermal gravimetric (TG) and differential thermal analysis (DTA) measurements were performed between room temperature and 1273 K using Al_2O_3 as a reference (2020 Unit; MAC Science). The samples were heated at a rate of 10 K min^{-1} under dry air atmosphere of 30 mL min^{-1}. UV–visible spectra (UV–vis) were achieved using a UV–visible spectrometer (V-570, ISN-470; JASCO) using MgO as a reference sample.

Photoactivity test

Acetaldehyde is a typical indoor pollutant and exists widely in modern building materials [12]. Therefore, the photocatalytic activity of the nanocrystalline zinc oxides was evaluated from the rate of photodegradation of acetaldehyde by a gaschromatograph (GC-8A; Shimadzu). The detailed experimental procedure can be described as follows. 0.10 g samples were put in 500 mL dry air containing 100 ppm acetaldehyde gas, and used 365 nm UV lamp (0.1W). All photodecomposition experiments were conducted in a batch reactor, with 140 mm inner diameter, 35 mm its height and 500 mL vessel. The obtained samples were kept in the container for 30 minutes to confirm adsorption of the acetaldehyde gas, and the concentration of the gas was measured every ten minutes. Afterwards, the samples were irradiated under UV light and the change of the concentration in every ten minutes was measured.

RESULTS AND DISCUSSION

Charactoristics of ZnO nanoparticles

TG-DTA analyses were employed to follow the decomposition of precursor. Figure1 shows the TG-DTA curve of the precursor powder at the temperature range from 373 K to 1273 K. Taking into account the amounts of starting materials, the weight loss must be 23.6%. It was found that the dehydration reaction into the zinc oxide took place at around 523 K because the TG and DTA curves suddenly dropped. Therefore, we decided to sinter the samples at over 573 K.

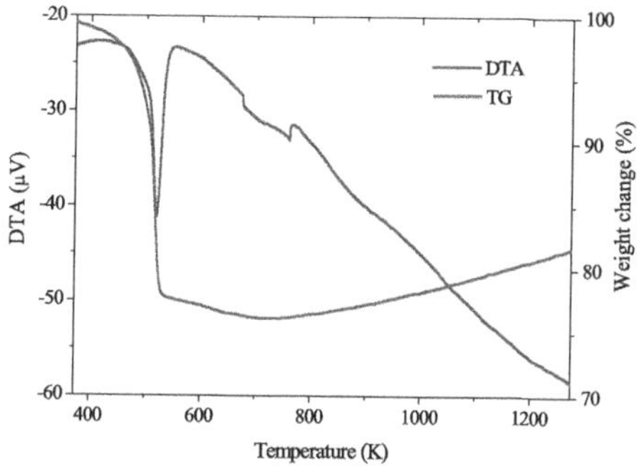

Fig.1 TG and DTA analysis of the precursor used to prepare ZnO nanoparticles

The crystal structure of the sample depends on the temperature of heat treatment. XRD patterns of the samples before and after sintering are shown in Fig. 2. The sample without sintering was zinc hydroxide, and the peaks were not able to be specified. As the calcination temperature increases, the peak width and the intensity became sharper and stronger. It indicates the formation of larger crystallite size and higher crystallinity by increasing the calcination temperature.

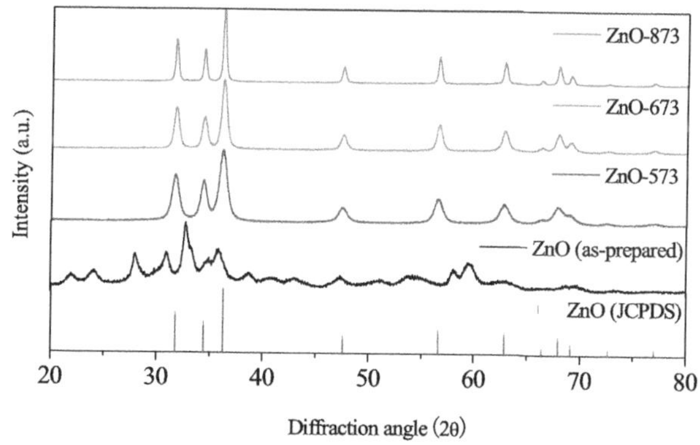

Fig.2 XRD patterns of nanocrystalline ZnO sintering at 573, 673 and 873 K

Figure 3 shows FE-SEM pictures of samples calcined at different temperatures. A typical characteristic of the samples had tendency to form aggregates. The smallest particle diameter was detected around 9 nm, and observed in the sample before calcination. The particle size slightly increased as the sintering temperature, and it became more spherical.

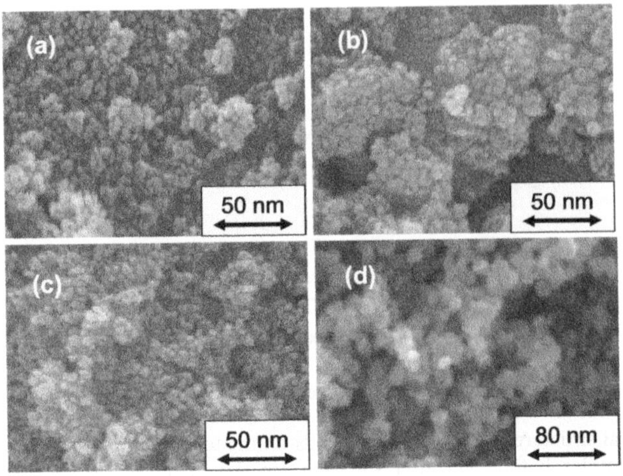

Fig.3 FE-SEM images of ZnO samples. (a)ZnO-as prepared; (b)ZnO-573; (c)ZnO-673 and (d)ZnO-873

The UV–visual spectra of the nanocrystalline ZnO samples are shown in Fig. 4. The absorbance increases a little with the rise of calcination temperature. However, the absorption wavelength is almost the same at any sintering temperature. All band gaps of the samples calculated from each wavelength were about 3.1 eV, and it is thought that it has the same photo-absorption character as the commercial zinc oxide.

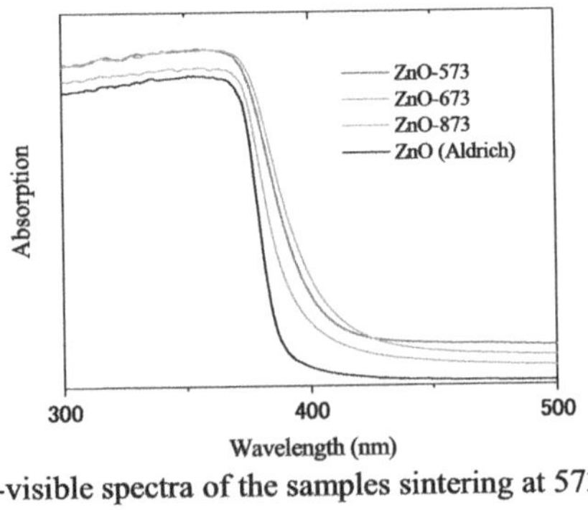

Fig.4 UV-visible spectra of the samples sintering at 573, 673 and 873 K

Moreover, the differences of physical properties by the sintering temperature were summarized in Table I.

Table I. Surface and optical properties of nanocristalline ZnO annealed at different temperatures

Sample	Calcination temperature (K)	BET surface area (m^2/g)	Pore volume (cm^3/g)	Particle diameter (nm)	Band gap (eV)
ZnO-573	573	4.7	0.00224	11	3.1
ZnO-673	673	6.8	0.00374	14	3.1
ZnO-873	873	5.1	0.00133	29	3.1
Commercial ZnO	-	7.9	0.00412	41	3.1
Degussa P25	-	52.1	0.0161	21	3.2

Photocatalitic performance of nanocrystalline ZnO

To evaluate the photocatalytic activity of the nanocrystalline ZnO, the reactions of acetaldehyde degradation were performed under UV irradiation. The behavior of acetaldehyde photodecomposition is given in Fig. 5(a). Degussa P25 and ZnO (Aldrich) were tested for comparison. Because the particle diameter increased as the annealing temperature and the specific surface area remained, the photocatalytic activity decreased. However, the highest photocatalytic performance of ZnO-573 was found to be comparable to P25 [13]. It is thought that the ZnO-573 had the highest photocatalic performance because the quantum size effect [14] works remarkably, though crystallinity is lower than that calcinated at higher temperatures.

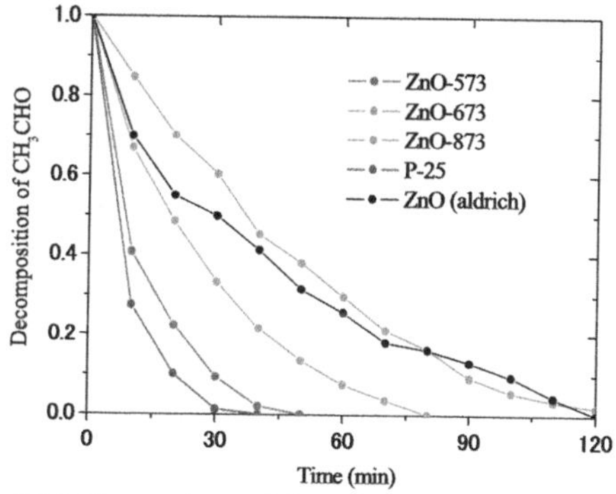

Fig.5 (a) Comparison of degradation ratio of acetaldehyde

In addition, the same degree of photoactivity as the commercial ZnO was shown when it was calcined at 873 K. It is thought that crystallization and the particle size were similar as commercial ZnO. Then, to examine photocatalitic efficiency more strictly, the photocatalyst reaction was analyzed as the first-order reaction. Differential type equation (c) can be applied by assuming this gas phase reaction to be primary order reaction of the dependence for the density of the gas.

$$r = -\frac{dC}{dt} = kC \qquad (c)$$

The photocatalyst performance was strictly analyzed by comparing the first reaction constant by using expression (d) that integrated expression (c).

$$r = \ln\left(\frac{C}{C_0}\right) = kt \qquad (d)$$

Equation (d) was applied to initial photodegration rate up to 80% decomposition [15]. Here, r is the reaction rate, k is the first reaction constant (min^{-1}), C_0 is first concentration of the acetaldehyde gas (ppm), and C is gas concentration after irradiation (ppm).

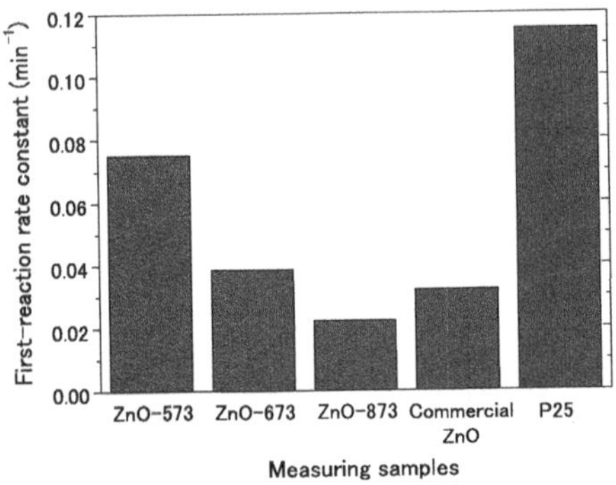

Fig.5 (b) Comparison of decomposition rate constants

The result is shown in Fig.5 (b). As the above-mentioned, ZnO-573 had very high reaction efficiency about 2/3 of P25 in concentration standard when the decomposition rate constants are compared. The reason why ZnO-573 has very high photocatalisis activity should be investigated in future.

CONCLUSION

In summary, we made the zinc hydroxide nanoparticles from the zinc solution using the anion exchange resin. From the results of the TG-DTA and XRD measurement, the zinc hydroxide nanoparticles were dehydrated at about 523 K, then, crystallized afterwards. The particle size has grown at the same time though the crystallization has advanced as the sintering temperature is raised. Then, ZnO-573 has the highest photocatalisis activity. Further work for the examination of the high photocatalitic activity of samples with anion exchange regin is being under way.

REFERENCE

[1] Yue Wu, Zhonghe Xi, Julan Zhang, and Qiang Zhang, Two types of ZnO-tubular nanostructures fabricated by stepped gas-phase reaction, *Materials Chemistry and Physics,* **110**, 445–48 (2008)

[2] Azam Anaraki Firooz, Ali Reza Mahjoub, and Abbas Ali Khodadadi, Preparation of SnO_2 nanoparticles and nanorods by using a hydrothermal method at low temperature, *Materials Letters.,* **62**, 2008, 1789–92 (2008)

[3] M.G. Manera, J. Spadavecchia, D. Buso, C. de Juli′an Fern′andez, G. Mattei,A. Martucci, P. Mulvaney, J. P′erez-Juste, R. Rella, L. Vasanelli, and P. Mazzoldi, Optical gas sensing of TiO_2 and TiO_2/Au nanocomposite thin films, *Sensors and Actuators B.,* **132**, 107–15(2008)

[4] J. Wienke, B. van der Zanden, M. Tijssen, and M. Zeman, Performance of spray-deposited ZnO:In layers as front electrodes in thin-film silicon solar cells, *Solar Energy Materials & Solar Cells,* **92**, 884– 90(2008)

[5] Narayanasamy Sobana, Manickavasakam Muruganandam, and Meenakshisundaram Swaminathan, Characterization of AC–ZnO catalyst and its photocatalytic activity on 4-acetylphenol degradation, *Catalysis Communications,* **9**, 2008, 262–68(2008)

[6] HaiqiangWang, ZhongbiaoW, Yue Liu, and Zhongyi Sheng, The characterization of ZnO–anatase–rutile three-component semiconductor and enhanced photocatalytic activity of nitrogen oxides, *J. Mol. Cat. A*; Chemical, **287**, 176–81 (2008)

[7] Yumi Iguchi, Hideaki Ichiura, Takuya Kitaoka, and Hiroo Tanaka, Preparation and characteristics of high performance paper containing titanium dioxide photocatalyst supported on inorganic fiber matrix, *Chemosphere,* **53**, 2003, 1193–99(2003)

[8] W. Thongsuwan, T. Kumpika, and P. Singjai, Photocatalytic property of colloidal TiO_2 nanoparticles prepared by sparking process, *Current Applied Physics,* **8**, 563–68(2008)

[9] Akiyoshi Nakata, Minoru Mizuhata, and Shigehito Deki, Novel fabrication of highly crystallized nanoparticles in the confined system by the liquid phase deposition (LPD) method, *Electrochimica Acta,* **53**, 2007, 179–85(2007)

[10] F. Mijangos, M. Kamel, G. Lesmes and D. N. Muraviev, Synthesis of struvite by ion exchange isothermal supersaturation technique, *Reactive & Functional Polymers,* **60**, 151–61(2004)

[11] Tsuyoshi Hamaguchi, *Doctor thesis*, Chapter 4 (2007)

[12] Michael J. Duryee, Lynell W. Klassen, Bonnie L. Jones, Monte S. Willis, Dean J. Tuma, and Geoffrey M. Thiele, Increased immunogenicity to P815 cells modified with malondialdehyde and acetaldehyde, *International Immunopharmacology,* **8**, 1112–18(2008)

[13] N. Daneshvar, S. Aber, M.S. Seyed Dorraji, A.R. Khataee, and M.H. Rasoulifard, Photocatalytic degradation of the insecticide diazinon in the presence of prepared nanocrystalline ZnO powders under irradiation of UV-C light, *Separation and Purification Technology,* **58**, 91–98(2007)

[14] Amy L. Linsebigler, Guangquan Lu, and John T. Yates, Jr, Photocatalysis on TiO_2 Surfaces: Principles, Mechanisms, and Selected Results, *Chem. Rev.*, **95**, 735-758(1995)

[15] Seiji Yamazoe, Yutaka Hitomi, Tetsuya Shishido, and Tsunehiro Tanaka, Kinetic study of photo-oxidation of NH_3 over TiO_2, *Applied Catalysis B: Environmental,* **82**, 67–76(2008)

PHOTOCATALYTICALLY EFFICIENT ZINC OXIDE MICROSTRUCTURAL ASSEMBLY

I.H.J. Arellano and L.M. Payawan, Jr.

Nanostructures and Surface Chemistry Research Laboratory, Institute of Chemistry, University of the Philippines, Diliman, Quezon City 1101 Philippines

ABSTRACT

Immobilized zinc oxide microstructural architecture composed of wires, rods, tubes and tetrapods was synthesized via carbothermal reduction at 900^{0}C in an oxidative muffle furnace which was oven quenched after 15-minute reaction time. These diverse ZnO crystal habits were characterized using field emission scanning electron microscope (FESEM), Xray diffraction (XRD) and photoluminescence (PL). A strong band gap emission was observed corresponding to 3.20 eV with suppressed visible emission indicative of good crystal quality as supported by XRD data showing a wurtzite structure with a = 3.25A and c = 5.28A. The synthesized architecture was used as photocatalyst in the degradation of methylene blue (MB) and rhodamine B (RB). The photocatalytic degradation rate constants for methylene blue are 0.0406, 0.0156 and 0.0138 min^{-1} for 10 ppm, 25 ppm and 50 ppm, respectively. The rate constants for rhodamine B are 0.0132, 0.0242 and 0.0192 min^{-1}, for 5 ppm, 10 ppm and 25 ppm, respectively. For uncatalyzed reaction, a rate constant of 0.0010 and 0.0009 min^{-1} for MB and RB were calculated, respectively. The efficiency of the degradation is dependent on the initial pH of the solution and an optimal activity was observed at pH 7.

INTRODUCTION

The current scenario of devastatingly rapid environmental degradation fuels multidisciplinary efforts to decrease pollutants and to lessen the effect of bioavailable contaminants via environmental remediation. The problem of pollutants, particularly the serious environmental impact of their residues and the low efficiency of current remediation processes, has led to several attempts to find more efficient degradation alternatives.[1] The search for an efficient catalyst is central to such endeavor.[2] Semiconductor inorganic oxides have been found to be efficient photocatalyst promoting advanced oxidative processes. The most commonly used is titania. Their efficiency relies on the ability to produce stable excitons that aids in the production of free radicals which are responsible for the degradation pathway. The use of ZnO over titania is based on the ease of processability of the former since precursors of titania are far more expensive, unstable and highly sensitive to water. In here, we report a photocatalyst architecture composed of diverse zinc oxide crystal habits that showed efficient activity towards the photodegradation of several model pollutants namely methylene blue (MB) and rhodamine B (RB).

The determination of kinetic parameters such as degradation half-life and rate constant are important in the study of the reaction kinetics. Simple degradation kinetics follows a first order rate law described by the equation 1.

$$\text{rate} = k[A] \qquad [1]$$

where k is the first order rate constant in per unit time and [A] is the concentration of the species being degraded in molar concentration. Using the integrated rate law for 1st order reaction kinetics

$$\ln[A]_n = -kt + \ln[A]_0 \qquad [2]$$

where $[A]_n$ is the normalized concentration at time t equal to $[A]_n = [A_t/A_0]$, $[A]_0$ is the concentration at t = 0, k is the rate constant and t is the reaction time, it is clear that a plot of $\ln[A]_n$ versus t will give a linear plot with a slope equal to the negative of the rate constant. The concentration of these model pollutants are monitored using UV-Vis spectroscopy using Beer's law. Thus, by measuring the absorbance at regular time intervals, one can monitor the decrease in concentration of the absorbing species which is indicative of photodegradation.

EXPERIMENTAL DETAIL

Zinc oxide microstructural assembly was synthesized using carbothermal reduction route previously reported by our group.[3] Briefly, finely ground ZnO (HiMedia) and technical grade activated charcoal (Merck) were placed in an unglazed crucible and heated in an oxidative muffle furnace for 15 minutes, afterwhich the system was oven-quenched. The substrate was placed on top of the crucible. The substrates were prepared prior to use via the standard RCA method. Surface morphology was investigated using Field Emission Scanning Electron Microscope, FESEM (Philips XL30) and the crystal structure was determined using Xray Diffractometer (Siemens Kristalloflex 760). The photoluminescence was taken using 157 nm excimer laser as the excitation source.

The photodegradation experiment was carried using the setup shown in Figure 1. The solution to be degraded was prepared according to the concentration indicated in a 1:5 30%H_2O_2/ distilled H_2O volume ratio. An aliquot was taken every 30 minutes and the UV-Vis absorbance of the solution was recorded. The absorbance of the solution prior to irradiation was also recorded as reference for the normalization.

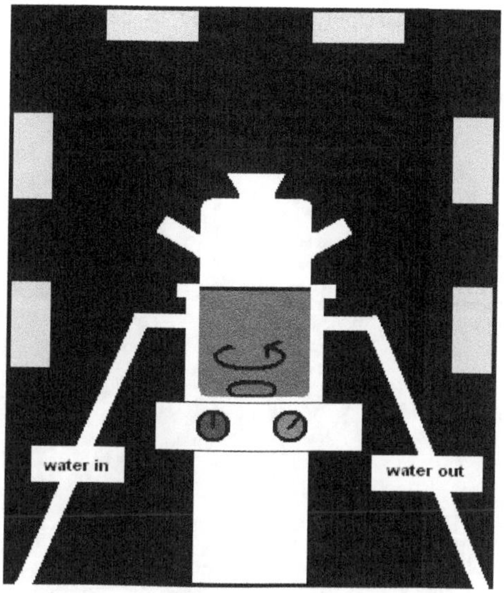

Figure 1. Photodegradation setup using 6 UV lamps in a black box.

RESULTS AND DISCUSSION

Shown in Figure 2 are field emission scanning electron microscope (FESEM) images showing the microstructural architecture consisting of various crystal habits. The deposition on the substrate forms an assembly resulting to a mesoporous layer. The diversity of the crystal habits formed is due to the inhomogeniety of the boundary diffusion layer due to a vapor current-induced during the deposition time arising from concentration gradient of the vapor of the starting materials. Nanowires grew anisotropically on the substrate's surface above the sheets. Hexagonal microrods adapted the hexagonal lattice type of ZnO. Tubes also adapt the hexagonal structure with a hollow central region that runs throughout the structures length. Tetrapods are three-dimensional structures composed of rods that share a common center. The preferential growth of ZnO structures to ZnO seed is more favorable despite the steric hindrance due to its lower surface energy relative to the silicon substrate. The difference in surface energy is due to the lattice mismatch of ZnO (wiurtzite) and Si (diamond).

Figure 3 shows the XRD pattern of the catalyst layer exhibiting an intense peak at (002) and a preferential growth along the *c*-axis. The lattice parameters of the wurtzite structure are a = 3.25A and c = 5.28A. Photoluminescence studies were conducted to determine the band gap emission of the assembly. As shown in Figure 4, an intense band gap emission at 390 nm was observed corresponding to a 3.20 eV band gap energy. A minimal visible emission at around 500 nm indicates a suppressed electron-hole recombination due to low crystal defect. As characterized, the photocatalyst assembly is expected to be efficient oxidizing agent upon excitation under UV radiation.

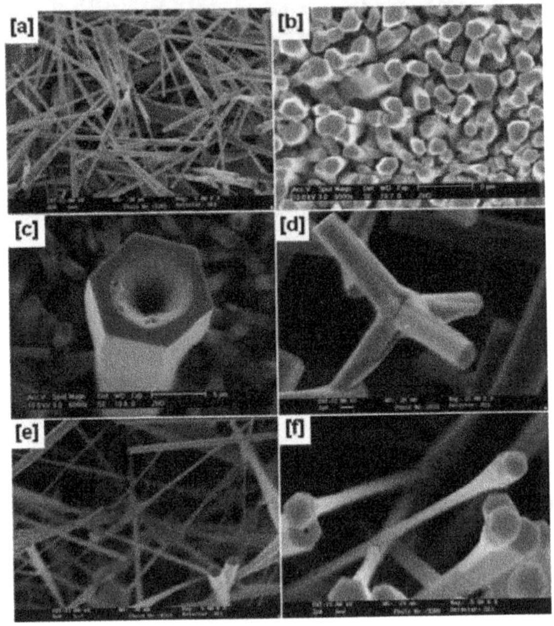

Figure 2. Diverse ZnO crystal habit namely; a) wires, b) rods, c) tubes, d) tetrapods, e) network homojunction, and f) wires with large hexagonal crown.

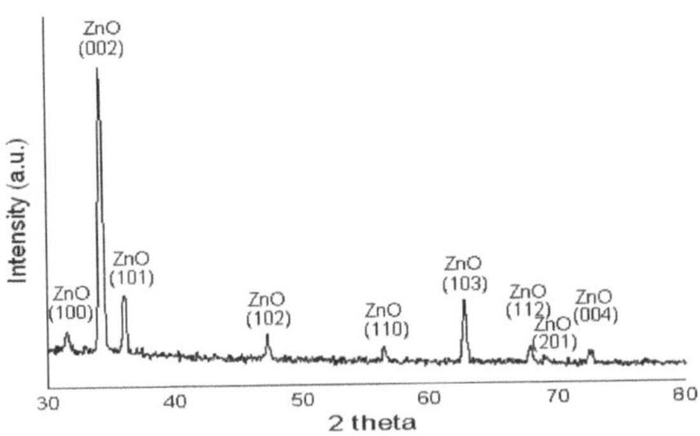

Figure 3. XRD pattern of the as-synthesized assembly showing the signature peaks for a wurtzite structure.

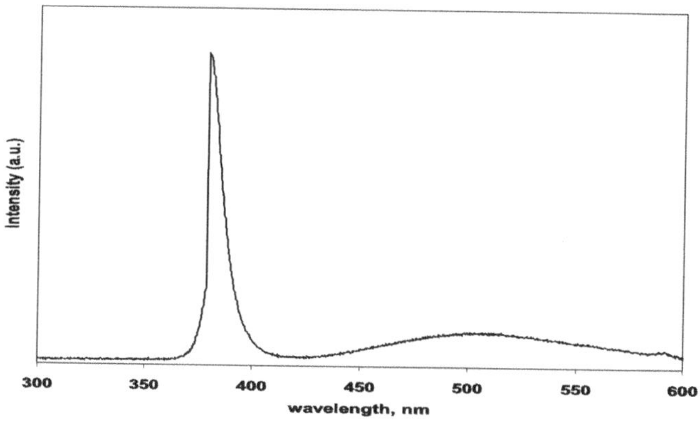

Figure 4. Room temperature photoluminescence spectrum showing a band edge emission at ca. 390 nm and a minimal visible emission at around 500 nm.

Organic pollutants can be vaguely classified as ionic and nonionic. It is a given fact that the ease of degradation is a function of the complexity of the structure of the molecules. Furthermore, the stability of the molecule and the degradation products also rely on the solvent used. The structures of the model pollutants, methylene blue and rhodamine B are shown in Figure 5.

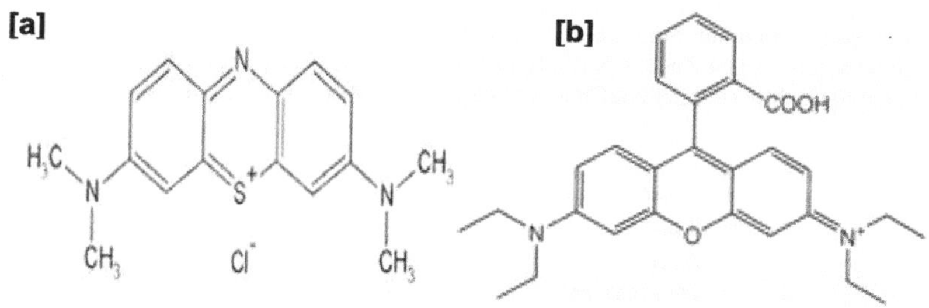

Figure 5. Structure of the model pollutants. a) methylene blue and b) rhodamine B.

The degradation of Rhodamine B with different concentration is shown in Figure 6. It is clear that at these conditions, the solution containing 10 ppm RB was optimally degraded. The trend can be explained by the diffusion limited nature of the photodegradation reaction. Catalysis occurs at the surfaces of the catalysts that were immobilized in a substrate. Thus, if the molecule is not within the vicinity of the catalyst, then no photocatalytic degradation can occur. As such, at very low concentration of the molecule, there is a significantly low probability that the molecule is within the catalyst architecture. On the other hand, at very high concentration, saturation will occur and the reaction will be limited by the rate at which the degradation products will be desorbed from the catalyst surface so that it can accommodate other undegraded molecules. It is worth to note that without the catalyst, there is an insignificant decrease in the concentration of the species which can be attributed to photobleaching.

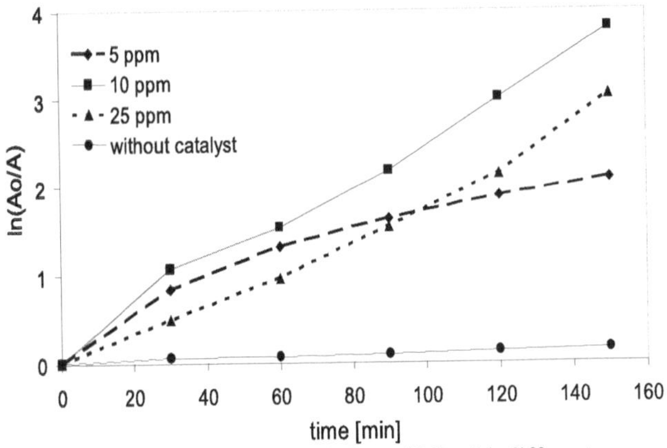

Figure 6. Kinetic plot of the degradation of RB with different concentrations.

Figure 7 shows the degradation of methylene blue. The same trend can be observed which can be explained by the same line of reasoning as stated in the case of RB. The photodegradation behavior of the two model compounds are similar because at neutral pH where the reaction was carried out, both are in their ionic form. Thus, their interaction with the catalyst can be slightly ionic since ZnO is partially positive due to oxygen vacancies present which can be attributed to the low sticking coefficient of oxygen at 900^0C.

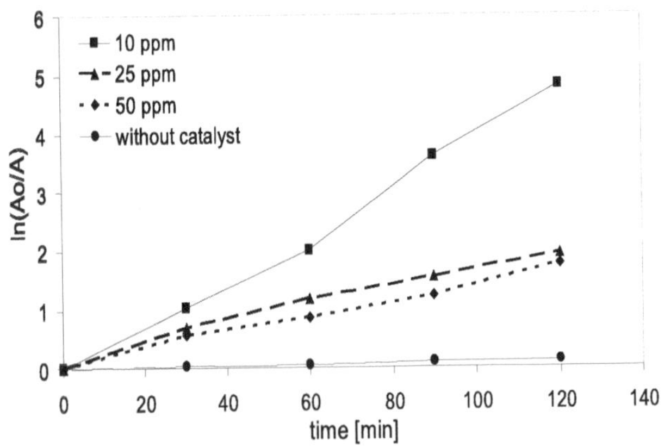

Figure 7. Kinetic plot of the degradation of MB with different concentrations.

All photocatalytic degradation reaction follows a first order kinetics. Thus, the rate constant and half-life can be calculated from the slope of the best fit curve. These kinetic parameters were tabulated in Table 1.

Table 1. Calculated rate constants and half-lives of the degradation of RB and MB.

	RB		MB	
	k (min^{-1})	$t_{1/2}$ (min)	k (min^{-1})	$t_{1/2}$ (min)
w/o catalyst	0.0009	770	0.0010	693
5 ppm	0.0132	52.5	- - -	- - -
10 ppm	0.0242	28.6	0.0406	17.07
25 ppm	0.0192	36.1	0.0156	44.41
50 ppm	- - -	- - -	0.0138	50.23

It has been established that photocatalytic degradation is pH dependent. The pH of the solution becomes more important in the system using ZnO because it is an amphiprotic oxide, a species that can act as either an acid or a base. It can be seen from Figure 8 that the maximum degradation occurs at neutral pH. This is because under acidic or basic conditions, the catalyst is decomposed according to the equations below.

$$ZnO + 2H^{+} \longrightarrow Zn^{2+} + H_2O \quad [3]$$

$$ZnO + H_2O + 2OH^{-} \longrightarrow Zn(OH)_4^{2-} \quad [4]$$

Under acidic conditions, the catalyst will be degraded to zinc ion while under basic conditions, the tetrahydroxozincate will be produced both will significantly lower the amount of catalyst present in the system and will also interact with the dyes being degraded.

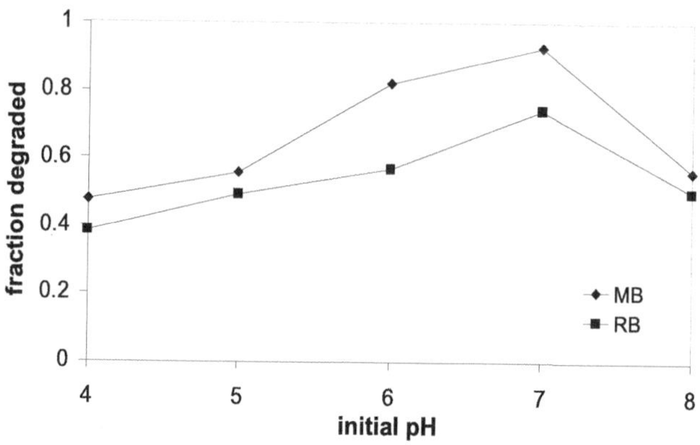

Figure 8. pH dependence of the photodegradation reaction.

The catalyst exhibited selectivity over simpler molecules. It is shown in Figure 9 that methylene blue is degraded faster than rhodamine B if the two are present in a 1:1 concentration ratio. The structure of MB is simpler than that RB. The site of possible electrostatic interaction between the catalyst and the species to be degraded is important in the efficiency of the process. It has been found in literature that the degradation reaction of the dyes is just a deamination reaction.[4] Because MB is less sterically hindered, the molecule can easily penetrate the microstructural architecture and be in contact with the catalyst. RB, however, will have a significant ionic interaction between the catalyst and the carboxyl moiety thus suppressing the rapid desorption of the degradation product which in effect slows down the subsequent decomposition of new molecules. The carboxyl group acts as an anchor to the catalyst which poisons the catalyst at a given fragment of time. The poisoning is not permanent due to the high energy supplied by the irradiation which apparently is sufficient to break the interaction produced.

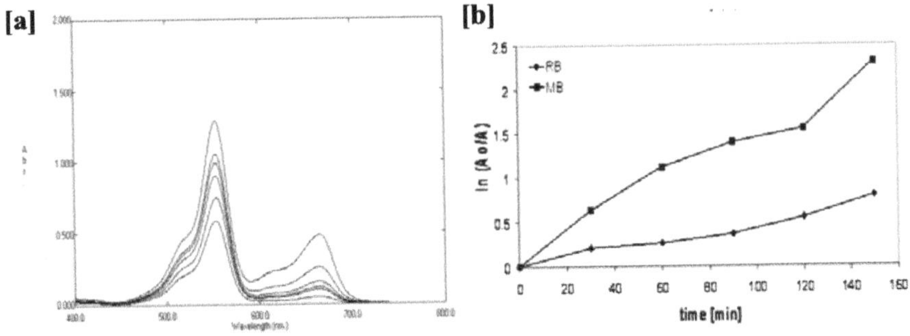

Figure 9. a.) UV Vis spectra of the degradation showing preferential decrease in absorbance of MB over RB, and b) kinetic plot of the system at λ_{max}.

The stability of the catalyst was likewise analyzed. This was done by using the catalyst several times and monitoring the percentage degraded after 150 minutes. It can be seen from Figure 10 that the catalyst is more stable under the photodegradation of MB than with RB which further justifies the effect of the complexity of the structure of the compounds being degraded and the additional strain imposed by the presence of an ionic interaction between the molecule and the catalyst.

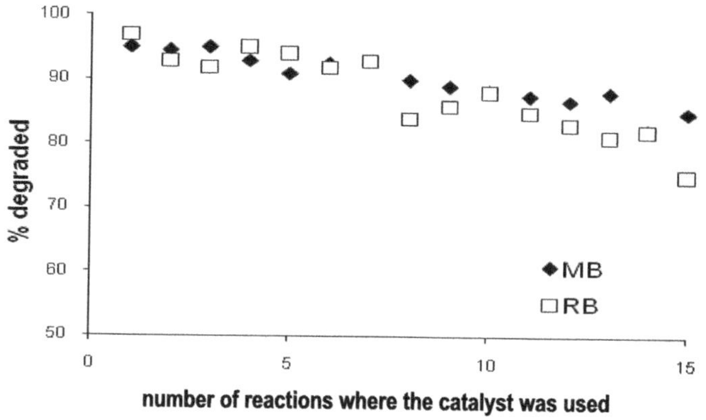

Figure 10. Catalyst stability.

CONCLUSIONS

We have reported the successful synthesis and utilization of ZnO microstructural assembly as efficient photocatalyst in the degradation of model pollutants, rhodamine B and methylene blue MB. The kinetic parameters such as rate constants and half-life of the 1st order reactions were calculated. The pH dependence of the photodegradation reaction was governed by the amphiprotic nature of the catalyst. Likewise, the synthesized catalyst assembly is robust and can be used repeatedly with only a slight decrease in its catalytic efficiency.

ACKNOWLEDGMENTS

The authors would like to acknowledge the financial support of the Office of the Vice-Chancellor for Research and Development of the University of the Philippines and NXP Semiconductors, Inc.

REFERENCES

[1] XH Ou, CH Wu and SL Lo. *React.* 2006. *Kinet. Catal. Lett.* **88**, 89-95.
[2] AD Paola, E Garcia-Lopez, S Ikeda, G Marci, B Ohtani and L Palmisano. 2002. *Catal. Today*. **78**, 87.
[3] IHJ Arellano, ER Magdaluyo, RV Sarmago and LM Payawan. 2006. *Proceedings of the 24th SPP Congress.*
[4] ER Magdaluyo, IHJ Arellano, AK Tapia, RV Sarmago and LM Payawan. 2008. *Adv Mat. Research* **31** 92-94.

Author Index

Printed and bound by CPI Group (UK) Ltd, Croydon, CR0 4YY

16/09/2026

14963461-0002